An Experimental Approach to Nonlinear Dynamics and Chaos

Studies in Nonlinearity

Series Editor: Robert L. Devaney

Ralph H. Abraham and Christopher D. Shaw, *Dynamics: The Geometry of Behavior* (1992)

Robert L. Devaney, *A First Course in Chaotic Dynamical Systems: Theory and Experiment* (1992)

Robert L. Devaney, James F.Georges, Delbert L. Johnson, *Chaotic Dynamical Systems Software* (1992)

An Experimental Approach to Nonlinear Dynamics and Chaos

Nicholas B. Tufillaro
Woods Hole Oceanographic Institution

Tyler Abbott

Jeremiah Reilly

Addison-Wesley Publishing Company
The Advanced Book Program
Redwood City, California • Menlo Park, California • Reading, Massachusetts
New York • Don Mills, Ontario • Wokingham, United Kingdom • Amsterdam
Bonn • Sydney • Singapore • Tokyo • Madrid • San Juan

Mathematics and Physics Editor: *Barbara Holland*
Production Manager: *Pam Suwinsky*
Production Assistant: *Karl Matsumoto*
Editorial Assistant: *Diana Tejo*
Copy Editor: *Andrew L. Alden*
Cover Designer: *Iva Frank*

Library of Congress Cataloging-in-Publication Data

Tufillaro, Nicholas
    An experimental approach to nonlinear dynamics and chaos/
    Nicholas Tufillaro, Tyler Abbott, Jeremiah Reilly.
        p. cm.—(Addison-Wesley studies in nonlinearity)
    Includes bibliographical references.
    1. Dynamics.   2. Nonlinear theories.   3. Chaotic behavior
    in systems.   I. Abbott, Tyler.   II. Reilly, Jeremiah.
    III. Title.   IV. Series
    QC133.T84 1991   533'.2–dc20   91-37157
    ISBN 0-201-55441-0

This book was typeset by the authors, using the TEX typesetting language.

1 2 3 4 5 6 7 8 9 10-MA-95 94 93 92

Contents

CONTENTS

Preface

An Experimental Approach to Nonlinear Dynamics and Chaos is a textbook and a reference work designed for advanced undergraduate and beginning graduate students. This book provides an elementary introduction to the basic theoretical and experimental tools necessary to begin research into the nonlinear behavior of mechanical, electrical, optical, and other systems. A focus of the text is the description of several desktop experiments, such as the nonlinear vibrations of a current-carrying wire placed between the poles of an electromagnet and the chaotic patterns of a ball bouncing on a vibrating table. Each of these experiments is ideally suited for the small-scale environment of an undergraduate science laboratory.

In addition, the book includes software that simulates several systems described in this text. The software provides the student with the opportunity to immediately explore nonlinear phenomena outside of the laboratory. The feedback of the interactive computer simulations enhances the learning process by promoting the formation and testing of experimental hypotheses. Taken together, the text and associated software provide a hands-on introduction to recent theoretical and experimental discoveries in nonlinear dynamics.

Studies of nonlinear systems are truly interdisciplinary, ranging from experimental analyses of the rhythms of the human heart and brain to attempts at weather prediction. Similarly, the tools needed to analyze nonlinear systems are also interdisciplinary and include techniques and methodologies from all the sciences. The tools presented in the text include those of:

theoretical and applied mathematics (dynamical systems theory and perturbation theory),

theoretical physics (development of models for physical phenomena, application of physical laws to explain the dynamics, and the topological characterization of chaotic motions),

experimental physics (circuit diagrams and desktop experiments),

engineering (instabilities in mechanical, electrical, and optical systems), and

computer science (numerical algorithms in C and symbolic computations with *Mathematica*).

A major goal of this project is to show how to integrate tools from these different disciplines when studying nonlinear systems.

Many sections of this book develop one specific "tool" needed in the analysis of a nonlinear system. Some of these tools are mathematical, such as the application of symbolic dynamics to nonlinear equations; some are experimental, such as the necessary circuit elements required to construct an experimental surface of section; and some are computational, such as the algorithms needed for calculating fractal dimensions from an experimental time series. We encourage students to try out these tools on a system or experiment of their own design. To help with this, Appendix I provides an overview of possible projects suitable for research by an advanced undergraduate. Some of these projects are in acoustics (oscillations in gas columns), hydrodynamics (convective loop—Lorenz equations, Hele-Shaw cell, surface waves), mechanics (oscillations of beams, stability of bicycles, forced pendulum, compass needle in oscillating B-field, impact-oscillators, chaotic art mobiles, ball in a swinging track), optics (semiconductor laser instabilities, laser rate equations), and other systems showing complex behavior in both space and time (video-feedback, ferrohydrodynamics, capillary ripples).

This book can be used as a primary or reference text for both experimental and theoretical courses. For instance, it can be used in a junior level mathematics course that covers dynamical systems or as a reference or lab manual for junior and senior level physics labs. In addition, it can serve as a reference manual for demonstrations and, perhaps more importantly, as a source book for undergraduate research projects. Finally, it could also be the basis for a new interdisciplinary course in nonlinear dynamics. This new course would contain an equal mixture of mathematics, physics, computing, and laboratory work. The primary goal of this new course is to give students the desire, skills, and confidence needed to begin their own research into nonlinear systems.

Regardless of her field of study, a student pursuing the material in this book should have a firm grounding in Newtonian physics and a course in differential equations that introduces the qualitative theory of ordinary differential equations. For the latter chapters, a good dose of mathematical maturity is also helpful.

To assist with this new course we are currently designing labs and software, including complementary descriptions of the theory, for the bouncing ball system, the double scroll LRC circuit, and a nonlinear string vibrations apparatus. The bouncing ball package has been completed and consists of a mechanical apparatus (a loudspeaker driven by a function generator and a ball bearing), the Bouncing Ball simulation system for the Macintosh computer, and a lab manual. This package has been used in the Bryn Mawr College Physics Laboratory since 1986.

This text is the first step in our attempt to integrate nonlinear theory with easily accessible experiments and software. It makes use of numerical algorithms, symbolic packages, and simple experiments in showing how to attack and unravel nonlinear problems. Because nonlinear effects are commonly observed in everyday phenomena (avalanches in sandpiles, a dripping faucet, frost on a window pane), they easily capture the imagination and, more importantly, fall within the research capabilities of a young scientist. Many experiments in nonlinear dynamics are individual or small group projects in which it is common for a student to follow an experiment from conception to completion in an academic year.

In our opinion nonlinear dynamics research illustrates the finest aspects of small science. It can be the effort of a few individuals, requiring modest funding, and often deals with "homemade" experiments which are intriguing and accessible to students at all levels. We hope that this book helps its readers in making the transition from studying science to doing science.

We thank Neal Abraham, Al Albano, and Paul Melvin for providing detailed comments on an early version of this manuscript. We also thank the Department of Physics at Bryn Mawr College for encouraging and supporting our efforts in this direction over several years. We would also like to thank the text and software reviewers who gave us detailed comments and suggestions. Their corrections and questions guided our revisions and the text and software are better for their scrutiny.

One of the best parts about writing this book is getting the chance to thank all our co-workers in nonlinear dynamics. We thank Neal Abraham, Kit Adams, Al Albano, Greg Alman, Ditza Auerbach, Remo Badii, Richard Bagley, Paul Blanchard, Reggie Brown, Paul Bryant, Gregory Buck, Josefina Casasayas, Lee Casperson, R. Crandall, Predrag Cvitanović, Josh Degani, Bob Devaney, Andy Dougherty, Bonnie Duncan, Brian Fenny, Neil Gershenfeld, Bob Gilmore, Bob Gioggia, Jerry Gollub, David Griffiths, G. Gunaratne, Dick Hall, Kath Hartnett, Doug Hayden, Gina Luca and Lois Hoffer-Lippi, Phil Holmes, Reto Holzner, Xin-Jun Hou, Tony Hughes, Bob Jantzen, Raymond Kapral, Kelly and Jimmy Kenison-Falkner, Tim Kerwin, Greg King, Eric Kostelich, Pat Langhorne, D. Lathrop, Wentian Li, Barbara Litt, Mark Levi, Pat Locke, Amy Lorentz, Takashi Matsumoto, Bruce McNamara, Tina Mello, Paul Melvin, Gabriel Mindlin, Tim Molteno, Ana Nunes, Oliver O'Reilly, Norman Packard, R. Ramshankar, Peter Rosenthal, Graham Ross, Miguel Rubio, Melora and Roger Samelson, Wes Sandle, Peter Saunders, Frank Selker, John Selker, Bill Sharpf, Tom Shieber, Francesco Simonelli, Lenny Smith, Hernán Solari, Tom Solomon, Vernon Squire, John Stehle, Rich Superfine, M. Tarroja, Mark Taylor, J. R. Tredicce, Hans Troger, Jim Valerio, Don Warrington, Kurt Wiesenfeld, Stephen Wolfram, and Kornelija Zgonc.

We would like to express a special word of thanks to Bob Gilmore, Gabriel Mindlin, Hernán Solari, and the Drexel nonlinear dynamics group for freely sharing and explaining their ideas about the topological characterization of strange sets. We also thank Tina Mello for experimental expertise with the bouncing ball system, Tim Molteno for the design and construction of the string apparatus, and Amy Lorentz for programing expertise with *Mathematica*.

Nick would like to acknowledge agencies supporting his research in nonlinear dynamics, which have included Sigma Xi, the Fulbright Foundation, Bryn Mawr College, Otago University Research Council, the Beverly Fund, and the National Science Foundation.

Nick would like to offer a special thanks to Mom and Dad for lots of home cooked meals during the writing of this book, and to Mary Ellen and Tracy for encouraging him with his chaotic endeavors from an early age.

Tyler would like to thank his own family and the Duncan and Dill families for their support during the writing of the book.

Jeremy and Tyler would like to thank the exciting teachers in our lives. Jeremy and Tyler dedicate this book to all inspiring teachers.

An Experimental Approach to Nonlinear Dynamics and Chaos

Introduction

What is nonlinear dynamics?

A dynamical system consists of two ingredients: a rule or "dynamic," which specifies how a system evolves, and an initial condition or "state" from which the system starts. The most successful class of rules for describing natural phenomena are differential equations. All the major theories of physics are stated in terms of differential equations. This observation led the mathematician V. I. Arnold to comment, "consequently, differential equations lie at the basis of scientific mathematical philosophy," our scientific world view. This scientific philosophy began with the discovery of the calculus by Newton and Leibniz and continues to the present day.

Dynamical systems theory and nonlinear dynamics grew out of the qualitative study of differential equations, which in turn began as an attempt to understand and predict the motions that surround us: the orbits of the planets, the vibrations of a string, the ripples on the surface of a pond, the forever evolving patterns of the weather. The first two hundred years of this scientific philosophy, from Newton and Euler through to Hamilton and Maxwell, produced many stunning successes in formulating the "rules of the world," but only limited results in finding their solutions. Some of the motions around us—such as the swinging of a clock pendulum—are regular and easily explained, while others—such as the shifting patterns of a waterfall—are irregular and initially appear to defy any rule.

The mathematician Henri Poincaré (1892) was the first to appreciate the true source of the problem: the difficulty lay not in the rules, but rather in specifying the initial conditions. At the beginning of this

1

century, in his essay *Science and Method*, Poincaré wrote:

> A very small cause which escapes our notice determines
> a considerable effect that we cannot fail to see, and then
> we say that that effect is due to chance. If we knew ex-
> actly the laws of nature and the situation of the universe at
> the initial moment, we could predict exactly the situation
> of that same universe at a succeeding moment. But even
> if it were the case that the natural laws had no longer any
> secret for us, we could still only know the initial situation
> approximately. If that enabled us to predict the succeed-
> ing situation with the same approximation, that is all we
> require, and we should say that the phenomenon had been
> predicted, that it is governed by laws. But it is not always
> so; it may happen that small differences in the initial con-
> ditions produce very great ones in the final phenomena. A
> small error in the former will produce an enormous error in
> the latter. Prediction becomes impossible, and we have the
> fortuitous phenomenon.

Poincaré's discovery of *sensitive dependence on initial conditions* in
what are now termed *chaotic* dynamical systems has only been fully
appreciated by the larger scientific community during the past three
decades. Mathematicians, physicists, chemists, biologists, engineers,
meteorologists—indeed, individuals from all fields have, with the help
of computer simulations and new experiments, discovered for themselves the cornucopia of chaotic phenomena existing in the simplest
nonlinear systems.

Before we proceed, we should distinguish nonlinear dynamics from
dynamical systems theory.[1] The latter is a well-defined branch of
mathematics, while nonlinear dynamics is an interdisciplinary field that
draws on all the sciences, especially mathematics and the physical sciences.

---

[1]For an outline of the mathematical theory of dynamical systems see D. V.
Anosov, I. U. Bronshtein, S.Kh. Aranson, and V. Z. Grines, Smooth dynamical
systems, in *Encyclopaedia of Mathematical Sciences*, Vol. 1, edited by D. V. Anosov
and V. I. Arnold (Springer-Verlag: New York, 1988).

Scientists in all fields are united by their need to solve nonlinear equations, and each different discipline has made valuable contributions to the analysis of nonlinear systems. A meteorologist discovered the first strange attractor in an attempt to understand the unpredictability of the weather.[2] A biologist promoted the study of the quadratic map in an attempt to understand population dynamics.[3] And engineers, computer scientists, and applied mathematicians gave us a wealth of problems along with the computers and programs needed to bring nonlinear systems alive on our computer screens. Nonlinear dynamics is interdisciplinary, and nonlinear dynamicists rely on their colleagues throughout all the sciences.

To define a nonlinear dynamical system we first look at an example of a linear dynamical system. A linear dynamical system is one in which the dynamic rule is linearly proportional to the system variables. Linear systems can be analyzed by breaking the problem into pieces and then adding these pieces together to build a complete solution. For example, consider the second-order linear differential equation

$$\frac{d^2 x}{dt^2} = -x.$$

The dynamical system defined by this differential equation is linear because all the terms are linear functions of x. The second derivative of x (the acceleration) is proportional to $-x$. To solve this linear differential equation we must find some function $x(t)$ with the following property: the second derivative of x (with respect to the independent variable t) is equal to $-x$. Two possible solutions immediately come to mind,

$$x_1(t) = \sin(t) \quad \text{and} \quad x_2(t) = \cos(t),$$

since

$$\frac{d^2}{dt^2} x_1(t) = -\sin(t) = -x_1(t)$$

and

$$\frac{d^2}{dt^2} x_2(t) = -\cos(t) = -x_2(t),$$

[2] E. N. Lorenz, Deterministic nonperiodic flow, J. Atmos. Sci. **20**, 130–141 (1963).

[3] R. M. May, Simple mathematical models with very complicated dynamics, Nature **261**, 459–467 (1976).

that is, both x_1 and x_2 satisfy the linear differential equation. Because the differential equation is linear, the sum of these two solutions defined by

$$x(t) = x_1(t) + x_2(t)$$

is also a solution.[4] This can be verified by calculating

$$
\begin{aligned}
\frac{d^2}{dt^2} x(t) &= \frac{d^2}{dt^2} x_1(t) + \frac{d^2}{dt^2} x_2(t) \\
&= -[x_1(t) + x_2(t)] \\
&= -x(t).
\end{aligned}
$$

Any number of solutions can be added together in this way to form a new solution; this property of linear differential equations is called the *principle of superposition*. It is the cornerstone from which all linear theory is built.

Now let's see what happens when we apply the same method to a nonlinear system. For example, consider the second-order nonlinear differential equation

$$\frac{d^2 x}{dt^2} = -x^2.$$

Let's assume we can find two different solutions to this nonlinear differential equation, which we will again call $x_1(t)$ and $x_2(t)$. A quick calculation,

$$
\begin{aligned}
\frac{d^2 x}{dt^2} &= \frac{d^2 x_1}{dt^2} + \frac{d^2 x_2}{dt^2} \\
&= -(x_1^2 + x_2^2) \\
&\neq -(x_1^2 + x_2^2 + 2x_1 x_2) \\
&= -(x_1 + x_2)^2 \\
&= -x^2,
\end{aligned}
$$

shows that the solutions of a nonlinear equation cannot usually be added together to build a larger solution because of the "cross-terms" $(2x_1 x_2)$. The principle of superposition fails to hold for nonlinear systems.

---

[4]The full definition of a linear system also requires that the sum of scalar products $x(t) = a_1 x_1(t) + a_2 x_2(t)$, where a_1 and a_2 are constants, is also a solution.

Traditionally, a differential equation is "solved" by finding a function that satisfies the differential equation. A trajectory is then determined by starting the solution with a particular initial condition. For example, if we want to predict the position of a comet ten years from now we need to measure its current position and velocity, write down the differential equation for its motion, and then integrate the differential equation starting from the measured initial condition. The traditional view of a solution thus centers on finding an individual orbit or trajectory. That is, given the initial condition and the rule, we are asked to predict the future position of the comet. Before Poincaré's work it was thought that a nonlinear system would always have a solution; we just needed to be clever enough to find it.

Poincaré's discovery of chaotic behavior in the three-body problem showed that such a view is wrong. No matter how clever we are we won't be able to write down the equations that solve many nonlinear systems. This is not wholly unexpected. After all, in a (bounded) closed form solution we might expect that any small change in initial conditions should produce a proportional change in the predicted trajectories. But a chaotic system can produce large differences in the long-term trajectories even when two initial conditions are close. Poincaré realized the full implications of this simple discovery, and he immediately redefined the notion of a "solution" to a differential equation.

Poincaré was less interested in an individual orbit than in all possible orbits. He shifted the emphasis from a local solution—knowing the exact motion of an individual trajectory—to a global solution—knowing the qualitative behavior of all possible trajectories for a given class of systems. In our comet example, a qualitative solution for the differential equation governing the comet's trajectory might appear easier to achieve since it would not require us to integrate the equations of motion to find the exact future position of the comet. The qualitative solution is often difficult to completely specify, though, because it requires a global view of the dynamics, that is, the possible examination of a large number of related systems.

Finding individual solutions is the traditional approach to solving a differential equation. In contrast, recurrence is a key theme in Poincaré's quest for the qualitative solution of a differential equation. To understand the recurrence properties of a dynamical system, we need to know what regions of space are visited and how often the orbit returns to those regions. We can seek to statistically characterize how often a region of space is visited; this leads to the so-called ergodic[5] theory of dynamical systems. Additionally, we can try to understand the geometric transformations undergone by a group of trajectories; this leads to the so-called topological theory of dynamical systems emphasized in this book.

There are many different levels of recurrence. For instance, the comet could crash into a planet. After that nothing much happens (unless you're a dinosaur). Another possibility is that the comet could go into an orbit about a star and from then on follow a periodic motion. In this case the comet will always return to the same points along the orbit. The recurrence is strictly periodic and easily predicted. But there are other possibilities. In particular, the comet could follow a chaotic path exhibiting a complex recurrence pattern, visiting and revisiting different regions of space in an erratic manner.

To summarize, Poincaré advocated the qualitative study of differential equations. We may lose sight of some specific details about any individual trajectory, but we want to sketch out the patterns formed by a large collection of different trajectories from related systems. This global view is motivated by the fact that it is nonsensical to study the orbit of a single trajectory in a chaotic dynamical system. To understand the motions that surround us, which are largely governed by nonlinear laws and interactions, requires the development of new qualitative techniques for analyzing the motions in nonlinear dynamical systems.

What is in this book?

This book introduces qualitative (bifurcation theory, symbolic dynamics, etc.) and quantitative (perturbation theory, numerical methods, etc.) methods that can be used in analyzing a nonlinear dynamical system. Further, it provides a basic set of experimental techniques required to set up and observe nonlinear phenomena in the laboratory.

---

[5]V. I. Arnold and A. Avez, *Ergodic problems of classical mechanics* (W. A. Benjamin: New York, 1968).

Some of these methods go back to Poincaré's original work in the last century, while many others, such as computer simulations, are more recent. A wide assortment of seemingly disparate techniques is used in the analysis of the humblest nonlinear dynamical system. Whereas linear theory resembles an edifice built upon the principle of superposition, nonlinear theory more closely resembles a toolbox, in which many of the essential tools have been borrowed from the laboratories of many different friends.

To paraphrase Tolstoy, all linear systems resemble one another, but each nonlinear system is nonlinear in its own way. Therefore, on our first encounter with a new nonlinear system we need to search our toolbox for the proper diagnostic tools (power spectra, fractal dimensions, periodic orbit extraction, etc.) so that we can identify and characterize the nonlinear and chaotic structures. And next, we need to analyze and unfold these structures with the help of additional tools and methods (computer simulations, simplified geometric models, universality theory, etc.) to find those properties that are common to a large class of nonlinear systems.

The tools in our toolbox are collected from scientists in a wide range of disciplines: mathematics, physics, computing, engineering, economics, and so on. Each discipline has developed a different dialect, and sometimes even a new language, in which to discuss nonlinear problems. And so one challenge facing a new researcher in nonlinear dynamics is to develop some fluency in these different dialects.

It is typical in many fields, from cabinet making to mathematics, to introduce the tyro first to the tedious elements and next, when these basic elements are mastered, to introduce her to the joys of the celestial whole. The cabinet maker first learns to sweep and sand and measure and hold. Likewise, the aspiring mathematician learns how to express limits, take derivatives, calculate integrals, and make substitutions.

All too often the consequence of an introduction through tedium is the destruction of the inquisitive, eager spirit of inquiry. We hope to diminish this tedium by tying the eagerness of the student to projects and experiments that illustrate nonlinear concepts. In a few words: we want the student to get her hands dirty. Then, with maturity and insight born from firsthand experience, she will be ready to fill in the big picture with rigorous definitions and more comprehensive study.

The study of nonlinear dynamics is eclectic, selecting what appears to be most useful among various and diverse theories and methods. This poses an additional challenge since the skills required for research in nonlinear dynamics can range from a knowledge of some sophisticated mathematics (hyperbolicity theory) to a detailed understanding of the nuts and bolts of computer hardware (binary arithmetic, digitizers). Nonlinear dynamics is not a tidy subject, but it is vital.

The common thread throughout all nonlinear dynamics is the need and desire to solve nonlinear problems, by hook or by crook. Indeed, many tools in the nonlinear dynamicist's toolbox originally were crafted as the solution to a specific experimental problem or application. Only after solving many individual nonlinear problems did the common threads and structures slowly emerge.

The first half of this book, Chapters 1, 2, and 3, uses a similar experimental approach to nonlinear dynamics and is suitable for an advanced undergraduate course. Our approach seeks to develop and motivate the study of nonlinear dynamics through the detailed analysis of a few specific systems that can be realized by desktop experiments: the period doubling route to chaos in a bouncing ball, the symbolic analysis of the quadratic map, and the quasiperiodic and chaotic vibrations of a string. The detailed analysis of these examples develops intuition for— and motivates the study of—nonlinear systems. In addition, analysis and simulation of these elementary examples provide ample practice with the tools in our nonlinear dynamics toolbox. The second half of the book, Chapters 4 and 5, provides a more formal treatment of the theory illustrated in the desktop experiments, thereby setting the stage for an advanced or graduate level course in nonlinear dynamics. In addition, Chapters 4 and 5 provide the more advanced student or researcher with a concise introduction to the mathematical foundations of nonlinear dynamics, as well as introducing the topological approach toward the analysis of chaos.

The pedagogical approach also differs between the first and second half of the book. The first half tends to introduce new concepts and vocabulary through usage, example, and repeated exposure. We believe this method is pedagogically sound for a first course and is reminiscent of teaching methods found in an intensive foreign language course. In the second half of the book examples tend to follow formal definitions, as is more common in a traditional mathematics course.

Although a linear reading of the first three chapters of this text is the most useful, it is also possible to pick and choose material from different sections to suit specific course needs. A mixture of mathematical, theoretical, experimental, and computational methods are employed in the first three chapters. The following table provides a rough road map to the type of material found in each section:

Mathematical	Theoretical	Experimental	Computational
1.5	1.2, 1.3, 1.4	1.1	1.6
2.2, 2.3, 2.4,	2.7, 2.8,	2.1	2.6
2.5, 2.9, 2.11	2.10, 2.12		*Mathematica* usage
	3.3, 3.4, 3.6, 3.7	3.2, 3.5, 3.8	3.8.3, 3.8.4, 3.8.5
	Appendix B		Appendices
			A, C, D, E, F, G

The mathematical sections present the core material for a mathematical dynamical systems course. The experimental sections present the core experimental techniques used for laboratory work with a low-dimensional chaotic system. For instance, those interested in experimental techniques could turn directly to the experimental sections for the information they seek.

Some Terminology: Maps, Flows, and Fractals

In this section we heuristically introduce some of the basic terminology used in nonlinear dynamics. This material should be read quickly, as background to the rest of the book. It might also be helpful to read Appendix H, Historical Comments, before delving into the more technical material. For precise definitions of the mathematical notions introduced in this section we highly recommend V. I. Arnold's masterful introduction to the theory of ordinary differential equations.[6] The goal in this section is to begin using the vocabulary of nonlinear dynamics even before this vocabulary is precisely defined.

---

[6]V. I. Arnold, *Ordinary differential equations* (MIT Press: Cambridge, MA, 1973). Also see D. K. Arrowsmith and C. M. Place, *Ordinary differential equations* (Chapman and Hall: New York, 1982).

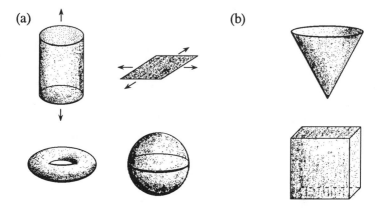

Figure 0.1: (a) The surfaces of the three-dimensional objects are two-dimensional manifolds. (b) Examples of objects that are not manifolds.

Flows and Maps

A geometric formulation of the theory of differential equations says that a *differential equation is a vector field on a manifold*. To understand this definition we present an informal description of a manifold and a vector field.

A *manifold* is any smooth geometric space (line, surface, solid). The smoothness condition ensures that the manifold cannot have any sharp edges. An example of a one-dimensional manifold is an infinite straight line. A different one-dimensional manifold is a circle. Examples of two-dimensional manifolds are the surface of an infinite cylinder, the surface of a sphere, the surface of a torus, and the unbounded real plane (Fig. 0.1). Three-dimensional manifolds are harder to visualize. The simplest example of a three-dimensional manifold is unbounded three-space, $\mathbf{R}^3$. The surface of a cone is an example of a two-dimensional surface that is not a manifold. At the apex of the cone is a sharp point, which violates the smoothness condition for a manifold. Manifolds are useful geometric objects because the smoothness condition ensures that a local coordinate system can be erected at each and every point on the manifold.

A *vector field* is a rule that smoothly assigns a vector (a directed line segment) to each point of a manifold. This rule is often written as a system of first-order differential equations. To see how this works, consider again the linear differential equation

$$\frac{d^2x}{dt^2} = -x.$$

Let us rewrite this second-order differential equation as a system of two first-order differential equations by introducing the new variable v, velocity, defined by $dx/dt = v$, so that

$$\frac{dx}{dt} = v,$$
$$\frac{dv}{dt} = -x.$$

The manifold in this example is the real plane, $\mathbf{R}^2$, which consists of the ordered pair of variables (x, v). Each point in this plane represents an individual *state*, or possible initial condition, of the system. And the collection of all possible states is called the *phase space* of the system. A process is said to be *deterministic* if both its future and past states are uniquely determined by its present state. A process is called *semideterministic* when only the future state, but not the past, is uniquely determined by the present state. Not all physical systems are deterministic, as the bouncing ball system (which is only semideterministic) of Chapter 1 demonstrates. Nevertheless, full determinism is commonly assumed in the classical scientific world view.

A system of first-order differential equations assigns to each point of the manifold a vector, thereby forming a vector field on the manifold (Fig. 0.2). In our example each point of the phase plane (x, v) gets assigned a vector $(v, -x)$, which forms rings of arrows about the origin (Fig. 0.3). A solution to a differential equation is called a *trajectory* or an *integral curve*, since it results from "integrating" the differential equations of motion. An individual vector in the vector field determines how the solution behaves locally. It tells the trajectory to "go thataway." The collection of all solutions, or integral curves, is called the *flow* (Fig. 0.3).

When analyzing a system of differential equations it is important to present both the equations and the manifold on which the equations

Figure 0.2: Examples of vector fields on different manifolds.

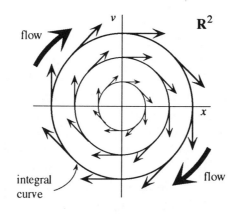

Figure 0.3: Vector field and flow for a linear differential equation.

are specified. It is often possible to simplify our analysis by transferring the vector field to a different manifold, thereby changing the topology of the phase space (see section 3.4.3). *Topology* is a kind of geometry which studies those properties of a space that are unchanged under a reversible continuous transformation. It is sometimes called rubber sheet geometry. A basketball and a football are identical to a topologist. They are both "topological" spheres. However, a torus and a sphere are different topological spaces as you cannot push or pull a sphere into a torus without first cutting up the sphere. Topology is also defined as the study of closeness within neighborhoods. Topological spaces can be analyzed by studying which points are "close to" or "in the neighborhood of" other points. Consider the line segment between 0

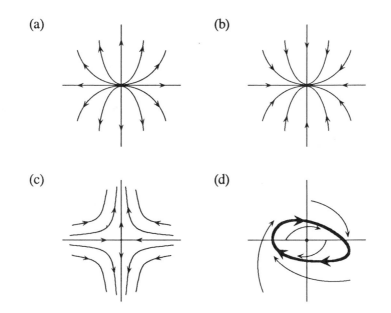

Figure 0.4: Typical motions in a planar vector field: (a) source, (b) sink, (c) saddle, and (d) limit cycle.

and 1. The endpoints 0 and 1 are far away; they aren't neighbors. But if we glue the ends together to form a circle, then the endpoints become identical, and the points around 0 and 1 have a new set of neighbors.

In its grandest form, Poincaré's program to study the qualitative behavior of ordinary differential equations would require us to analyze the generic dynamics of all vector fields on all manifolds. We are nowhere near achieving this goal yet. Poincaré was inspired to carry out this program by his success with the Swedish mathematician Ivar Bendixson in analyzing all typical behavior for differential equations in the plane. As illustrated in Figure 0.4, the *Poincaré-Bendixson Theorem* says that typically no more than four kinds of motion are found in a planar vector field, those of a *source*, *sink*, *saddle*, and *limit cycle*. In particular, *no chaotic motion is possible in time-independent planar vector fields*. To get chaotic motion in a system of differential equations one needs three dimensions, that is, a vector field on a three-dimensional manifold.

The asymptotic motions ($t \rightarrow \infty$ limit sets) of a flow are characterized by four general types of behavior. In order of increasing complexity these are *equilibrium points, periodic solutions, quasiperiodic solutions*, and *chaos*.

An *equilibrium point* of a flow is a constant, time-independent solution. The equilibrium solutions are located where the vector field vanishes. The source in Figure 0.4(a) is an example of an unstable equilibrium solution. Trajectories near to the source move away from the source as time goes by. The sink in Figure 0.4(b) is an example of a stable equilibrium solution. Trajectories near the sink tend toward it as time goes by.

A *periodic solution* of a flow is a time-dependent trajectory that precisely returns to itself in a time T, called the *period*. A periodic trajectory is a closed curve. Like an equilibrium point, a periodic trajectory can be stable or unstable, depending on whether nearby trajectories tend toward or away from the periodic cycle. One illustration of a stable periodic trajectory is the limit cycle shown in Figure 0.4(d). A *quasiperiodic* solution is one formed from the sum of periodic solutions with incommensurate periods. Two periods are incommensurate if their ratio is irrational. The ability to create and control periodic and quasiperiodic cycles is essential to modern society: clocks, electronic oscillators, pacemakers, and so on.

An asymptotic motion that is not an equilibrium point, periodic, or quasiperiodic is often called *chaotic*. This catchall use of the term chaos is not very specific, but it is practical. Additionally, we require that a chaotic motion is a bounded asymptotic solution that possesses sensitive dependence on initial conditions: two trajectories that begin arbitrarily close to one another on the chaotic limit set start to diverge so quickly that they become, for all practical purposes, uncorrelated. Simply put, a chaotic system is a deterministic system that exhibits random (uncorrelated) behavior. This apparent random behavior in a deterministic system is illustrated in the bouncing ball system (see section 1.4.5). A more rigorous definition of chaos is presented in section 4.10.

All of the stable asymptotic motions (or limit sets) just described (e.g., sinks, stable limit cycles), are examples of *attractors*. The unstable limit sets (e.g., sources) are examples of *repellers*. The term *strange attractor* (strange repeller) is used to describe attracting (repelling) limit sets that are chaotic. We will get our first look at a strange attractor in a physical system when we study the bouncing ball system in Chapter 1.

Maps are the discrete time analogs of flows. While flows are specified by differential equations, maps are specified by difference equations. A point on a trajectory of a flow is indicated by a real parameter t, which we think of as the time. Similarly, a point in the orbit of a map is indexed by an integer subscript n, which we think of as the discrete analog of time. Maps and flows will be the two primary types of dynamical systems studied in this book.

Maps (difference equations) are easier to solve numerically than flows (differential equations). Therefore, many of the earliest numerical studies of chaos began by studying maps. A famous map exhibiting chaos studied by the French astronomer Michel Hénon (1976), now known as the *Hénon map*, is

$$\begin{aligned} x_{n+1} &= \alpha - x_n^2 + \beta y_n, \\ y_{n+1} &= x_n, \end{aligned}$$

where n is an integer index for this pair of nonlinear coupled difference equations, with $\alpha = 1.4$ and $\beta = 0.3$ being the parameter values most commonly studied. The Hénon map carries a point in the plane, (x_0, y_0), to some new point, (x_1, y_1). An *orbit* of a map is the sequence of points generated by some initial condition of a map. For instance, if we start the Hénon map at the point $(x_0, y_0) = (0.0, 0.5)$, we find that the orbit for this pair of initial conditions is

$$\begin{aligned} x_1 &= 1.4 - (0.0 * 0.0) + 0.3 * 0.5 = 1.55, \\ y_1 &= 0.0, \end{aligned}$$

$$\begin{aligned} x_2 &= 1.4 - (1.55 * 1.55) + 0.3 * 0.0 = -1.0025, \\ y_2 &= 1.55, \end{aligned}$$

and so on to generate (x_3, y_3), (x_4, y_4), etc. Unlike planar differential equations, this two-dimensional difference equation can generate chaotic orbits. In fact, in Chapter 2 we will study a one-dimensional difference equation called the quadratic map, which can also generate chaotic orbits.

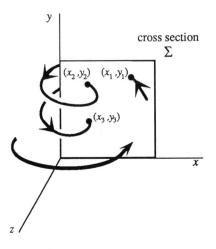

Figure 0.5: A Poincaré map for a three-dimensional flow with a two-dimensional cross section.

The Hénon map is an example of a *diffeomorphism* of a manifold (in this case the manifold is the plane $\mathbf{R}^2$). A map is a *homeomorphism* if it is bijective (one-to-one and onto), continuous, and has a continuous inverse. A *diffeomorphism* is a differentiable homeomorphism. A map with an inverse is called *invertible*. A map without an inverse is called *noninvertible*.

Maps exhibit similar types of asymptotic behavior as flows: equilibrium points, periodic orbits, quasiperiodic orbits, and chaotic orbits. There are many similarities and a few important differences between the theory and language describing the dynamics of maps and flows. For a detailed comparison of these two theories see Arrowsmith and Place, *An introduction to dynamical systems.*

The dynamics of flows and maps are closely related. The study of a flow can often be replaced by the study of a map. One prescription for doing this is the so-called Poincaré map of a flow. As illustrated in Figure 0.5, a *cross section* of the flow is obtained by choosing some surface transverse to the flow. A cross section for a three-dimensional flow is shown in the illustration and is obtained by choosing the x–y plane, $(x, y, z = 0)$. The flow defines a map of this cross section to itself, and this map is an example of a *Poincaré map* (also called a first return map). A trajectory of the flow carries a point $(x(t_1), y(t_1))$ into a new point $(x(t_2), y(t_2))$. And this in turn goes to the point $(x(t_3), y(t_3))$. In this way the flow generates a map of a portion of the plane, and an orbit of this map consists of the sequence of points

$$
\begin{aligned}
(x_1, y_1) &= (x(t_1), y(t_1), z = 0, \dot{z} < 0), \\
(x_2, y_2) &= (x(t_2), y(t_2), z = 0, \dot{z} < 0), \\
(x_3, y_3) &= (x(t_3), y(t_3), z = 0, \dot{z} < 0),
\end{aligned}
$$

and so on.

There is another reason for studying maps. To quote Steve Smale on the "diffeomorphism problem,"[7]

> [T]here is a second and more important reason for study-
> ing the diffeomorphism problem (besides its great natural
> beauty). That is, the same phenomena and problems of
> the qualitative theory of ordinary differential equations are
> present in their simplest form in the diffeomorphism prob-
> lem. Having first found theorems in the diffeomorphism
> case, it is usually a secondary task to translate the results
> back into the differential equations framework.

The first dynamical system we will study, the bouncing ball system, illustrates more fully the close connection between maps and flows.

Binary Arithmetic

Before turning to nonlinear dynamics proper, we need some familiarity with the binary number system. Consider the problem of converting a fraction between 0 and 1 ($x_0 \in [0, 1]$) written in decimal (base 10) to a binary number (base 2). The formal expansion for a binary fraction in powers of 2 is

$$
x_0 = \frac{\beta_1}{2} + \frac{\beta_2}{2^2} + \frac{\beta_3}{2^3} + \frac{\beta_4}{2^4} + \frac{\beta_5}{2^5} + \cdots \quad \text{(decimal)}
$$

---

[7]S. Smale, Differential dynamical systems, Bull. Am. Math. Soc. **73**, 747–817 (1967).

$$= \frac{\beta_1}{2} + \frac{\beta_2}{4} + \frac{\beta_3}{8} + \frac{\beta_4}{16} + \frac{\beta_5}{32} + \cdots$$
$$= 0.\beta_1\beta_2\beta_3\beta_4\beta_5\ldots \qquad \text{(binary)}$$

where $\beta_i \in \{0,1\}$. The goal is to find the β_i's for a given decimal fraction. For example, if $x_0 = 3/4$ then

$$x_0 = \frac{3}{4}$$
$$= \frac{1}{2} + \frac{1}{4} + \frac{0}{8} + \frac{0}{16} + \frac{0}{32} + \cdots$$
$$= 0.11 \quad \text{(binary)}.$$

The general procedure for converting a decimal fraction less than one to binary is based on repeated doublings in which the ones or "carry" digit is used for the β_i's. This is illustrated in the following calculation for $x_0 = 0.314$:

$$2 \times 0.314 = 0.628 \longrightarrow \beta_1 = 0$$
$$2 \times 0.628 = 1.256 \longrightarrow \beta_2 = 1$$
$$2 \times 0.256 = 0.512 \longrightarrow \beta_3 = 0$$
$$2 \times 0.512 = 1.024 \longrightarrow \beta_4 = 1$$
$$2 \times 0.024 = 0.048 \longrightarrow \beta_5 = 0$$
$$2 \times 0.048 = 0.096 \longrightarrow \beta_6 = 0$$

so

$$x_0 = 0.010100\ldots \quad \text{(binary)}.$$

Fractals

Nature abounds with intricate fragmented shapes and structures, including coastlines, clouds, lightning bolts, and snowflakes. In 1975 Benoit Mandelbrot coined the term *fractal* to describe such irregular shapes. The essential feature of a fractal is the existence of a similar structure at all length scales. That is, a fractal object has the property that a small part resembles a larger part, which in turn resembles the whole object. Technically, this property is called *self-similarity* and is theoretically described in terms of a *scaling* relation.

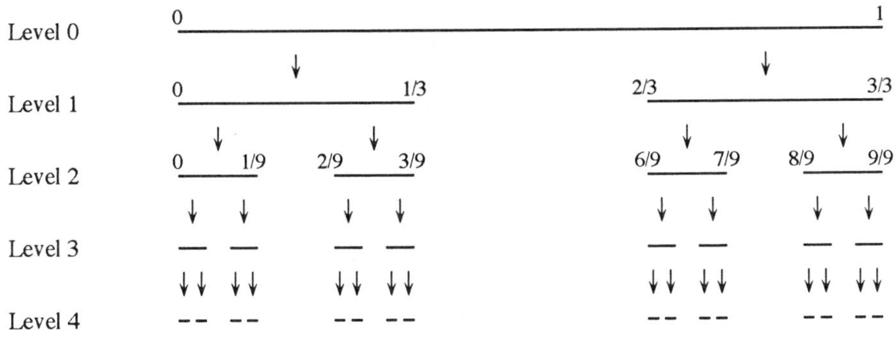

Figure 0.6: Construction of Cantor's middle thirds set.

Chaotic dynamical systems almost inevitably give rise to fractals. And fractal analysis is often useful in describing the geometric structure of a chaotic dynamical system. In particular, fractal objects can be assigned one or more fractal dimensions, which are often *fractional*; that is, they are not integer dimensions.

To see how this works, consider a *Cantor set*, which is defined recursively as follows (Fig. 0.6). At the zeroth level the construction of the Cantor set begins with the unit interval, that is, all points on the line between 0 and 1. The first level is obtained from the zeroth level by deleting all points that lie in the "middle third," that is, all points between 1/3 and 2/3. The second level is obtained from the first level by deleting the middle third of each interval at the first level, that is, all points from 1/9 to 2/9, and 7/9 to 8/9. In general, the next level is obtained from the previous level by deleting the middle third of all intervals at the previous level. This process continues forever, and the result is a collection of points that are tenuously cut out from the unit interval. At the nth level the set consists of 2^n segments, each of which has length $l_n = (1/3)^n$, so that the length of the Cantor set is

$$\lim_{n \to \infty} 2^n \left(\frac{1}{3}\right)^n = 0.$$

In the 1920s the mathematician Hausdorff developed another way to "measure" the size of a set. He suggested that we should examine the number of small intervals, $N(\epsilon)$, needed to "cover" the set at a scale ϵ. The measure of the set is calculated from

$$\lim_{\epsilon \to 0} N(\epsilon) = \left(\frac{1}{\epsilon}\right)^{d_f}.$$

An example of a fractal dimension is obtained by inverting this equation,

$$d_f = \lim_{\epsilon \to 0} \left(\frac{\ln N(\epsilon)}{\ln\left(\frac{1}{\epsilon}\right)}\right).$$

Returning to the Cantor set, we see that at the nth level the length of the covering intervals are $\epsilon = \left(\frac{1}{3}\right)^n$, and the number of intervals needed to cover all segments at the nth level is $N(\epsilon) = 2^n$. Taking the limits $n \to \infty$ ($\epsilon \to 0$), we find

$$d_f = \lim_{\epsilon \to 0} \left(\frac{\ln N(\epsilon)}{\ln\left(\frac{1}{\epsilon}\right)}\right) = \lim_{n \to \infty} \frac{\ln 2^n}{\ln 3^n} = \frac{\ln 2}{\ln 3} \approx 0.6309.$$

The middle-thirds Cantor set has a simple scaling relation, because the factor $1/3$ is all that goes into determining the successive levels. A further elementary discussion of the middle-thirds Cantor set is found in Devaney's *Chaos, fractals, and dynamics*. In general, fractals arising in a chaotic dynamical system have a far more complex scaling relation, usually involving a range of scales that can depend on their location within the set. Such fractals are called *multifractals*.

References and Notes

Some popular accounts of the history and folklore of nonlinear dynamics and chaos include:

A. Fisher, Chaos: The ultimate asymmetry, MOSAIC **16** (1), pp. 24–33 (January/February 1985).

J. P. Crutchfield, J. D. Farmer, N. H. Packard, and R. S. Shaw, Chaos, Sci. Am. **255** (6), pp. 46–57 (1986).

J. Gleick, *Chaos: Making a new science* (Viking: New York, 1987).

I. Stewart, *Does god play dice? The mathematics of chaos* (Basil Blackwell: Cambridge, MA, 1989).

The following books are helpful references for some of the material covered in this book:

R. Abraham and C. Shaw, *Dynamics—The geometry of behavior*, Vol. 1–4 (Aerial Press: Santa Cruz, CA, 1988).

D. K. Arrowsmith and C. M. Place, *An introduction to dynamical systems* (Cambridge University Press: New York, 1990).

G. L. Baker and J. P. Gollub, *Chaotic dynamics* (Cambridge University Press: New York, 1990).

P. Bergé, Y. Pomeau, and C. Vidal, *Order within chaos* (John Wiley: New York, 1984).

R. L. Devaney, *An introduction to chaotic dynamical systems*, second ed. (Addison-Wesley: New York, 1989).

E. A. Jackson, *Perspectives of nonlinear dynamics*, Vol. 1–2 (Cambridge University Press: New York, 1990).

F. Moon, *Chaotic vibrations* (John Wiley: New York, 1987).

J. Thompson and H. Stewart, *Nonlinear dynamics and chaos* (John Wiley: New York, 1986).

S. Rasband, *Chaotic dynamics of nonlinear systems* (John Wiley: New York, 1990).

S. Wiggins, *Introduction to applied nonlinear dynamical systems and chaos* (Springer-Verlag: New York, 1990).

These review articles provide a quick introduction to the current research problems and methods of nonlinear dynamics:

J.-P. Eckmann, Roads to turbulence in dissipative dynamical systems, Rev. Mod. Phys. **53** (4), pp. 643–654 (1981).

J.-P. Eckmann and D. Ruelle, Ergodic theory of chaos and strange attractors, Rev. Mod. Phys. **57** (3), pp. 617–656 (1985).

C. Grebogi, E. Ott, and J. Yorke, Chaos, strange attractors, and fractal basin boundaries in nonlinear dynamics, Science **238**, pp. 632–638 (1987).

E. Ott, Strange attractors and chaotic motions of dynamical systems, Rev. Mod. Phys. **53** (4), pp. 655–671 (1981).

T. Parker and L. Chua, Chaos: A tutorial for engineers, Proc. IEEE **75** (8), pp. 982–1008 (1987).

R. Shaw, Strange attractors, chaotic behavior, and information flow, Z. Naturforsch. **36a**, pp. 80–112 (1981).

Advanced theoretical results are described in the following books:

V. I. Arnold, *Geometrical methods in the theory of ordinary differential equations*, second ed. (Springer-Verlag: New York, 1988).

J. Guckenheimer and P. Holmes, *Nonlinear oscillations, dynamical systems, and bifurcations of vector fields*, second printing (Springer-Verlag: New York, 1986).

D. Ruelle, *Elements of differentiable dynamics and bifurcation theory* (Academic Press: New York, 1989).

S. Wiggins, *Global bifurcations and chaos* (Springer-Verlag: New York, 1988).

Chapter 1

Bouncing Ball

1.1   Introduction

Consider the motion of a ball bouncing on a periodically vibrating table. The *bouncing ball system* is illustrated in Figure 1.1 and arises quite naturally as a model problem in several engineering applications. Examples include the generation and control of noise in machinery such as jackhammers, the transportation and separation of granular solids such as rice, and the transportation of components in automatic assembly devices, which commonly employ oscillating tracks. These vibrating tracks are used to transport parts much like a conveyor belt [1].

Assume that the ball's motion is confined to the vertical direction and that, between impacts, the ball's height is determined by Newton's laws for the motion of a particle in a constant gravitational field. A nonlinear force is applied to the ball when it hits the table. At impact, the ball's velocity suddenly reverses from the downward to the upward direction (Fig. 1.1).

The bouncing ball system is easy to study experimentally [2]. One experimental realization of the system consists of little more than a ball bearing and a periodically driven loudspeaker with a concave optical lens attached to its surface. The ball bearing will rattle on top of this lens when the speaker's vibration amplitude is large enough. The curvature of the lens is chosen so as to help focus the ball's motion in the vertical direction.

Impacts between the ball and lens can be detected by listening to

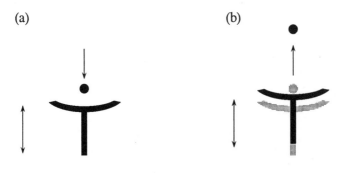

Figure 1.1: Ball bouncing on an oscillating table.

the rhythmic clicking patterns produced when the ball hits the lens. A piezoelectric film, which generates a small current every time a stress is applied, is fastened to the lens and acts as an impact detector. The piezoelectric film generates a voltage spike at each impact. This spike is monitored on an oscilloscope, thus providing a visual representation of the ball's motion. A schematic of the bouncing ball machine is shown in Figure 1.2. More details about its construction are provided in reference [3].

The ball's motion can be described in several equivalent ways. The simplest representation is to plot the ball's height and the table's height, measured from the ground, as a function of time. Between impacts, the graph of the ball's vertical displacement follows a parabolic trajectory as illustrated in Figure 1.3(a). The table's vertical displacement varies sinusoidally. If the ball's height is recorded at discrete time steps,

$$\{x(t_0), x(t_1), \ldots, x(t_i), \ldots, x(t_n)\}, \tag{1.1}$$

then we have a *time series* of the ball's height where $x(t_i)$ is the height of the ball at time t_i.

Another view of the ball's motion is obtained by plotting the ball's height on the vertical axis, and the ball's velocity on the horizontal axis. The plot shown in Figure 1.3(b) is essentially a *phase space* representation of the ball's motion. Since the ball's height is bounded, so is the ball's velocity. Thus the phase space picture gives us a description of the ball's motion that is more compact than that given by a plot of

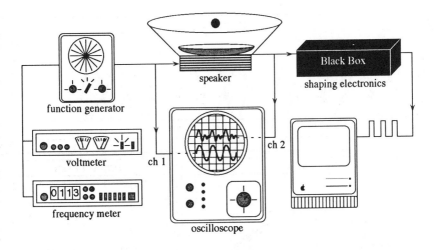

Figure 1.2: Schematic for a bouncing ball machine.

the time series. Additionally, the sudden reversal in the ball's velocity at impact (from positive to negative) is easy to see at the bottom of Figure 1.3(b). Between impacts, the graph again follows a parabolic trajectory.

Yet another representation of the ball's motion is a plot of the ball's velocity and the table's forcing phase at each impact. This is the so-called *impact map* and is shown in Figure 1.3(c). The impact map goes to a single point for the simple periodic trajectory shown in Figure 1.3. The vertical coordinate of this point is the ball's velocity at impact and the horizontal coordinate is the table's forcing phase. This phase, θ, is defined as the product of the table's angular frequency, ω, and the time, t:

$$\theta = \omega t, \quad \omega = 2\pi/T, \tag{1.2}$$

where T is the forcing period. Since the table's motion is 2π-periodic in the phase variable θ, we usually consider the phase mod 2π, which means we divide θ by 2π and take the remainder:

$$\theta \bmod 2\pi = \mathrm{remainder}(\theta/2\pi). \tag{1.3}$$

A time series, phase space, and impact map plot are presented together in Figure 1.4 for a complex motion in the bouncing ball system.

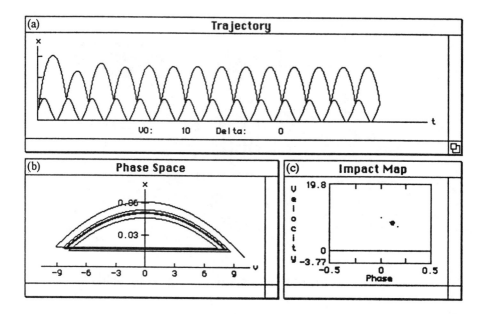

Figure 1.3: Simple periodic orbit of a bouncing ball: (a) height vs. time, (b) phase space (height vs. velocity), (c) impact map (velocity and forcing phase at impact). (Generated by the Bouncing Ball program.)

This particular motion is an example of a nonperiodic orbit known as a *strange attractor*. The impact map, Figure 1.4(c), is a compact and abstract representation of the motion. In this particular example we see that the ball never settles down to a periodic motion, in which it would impact at only a few points, but rather explores a wide range of phases and velocities. We will say much more about these strange trajectories throughout this book, but right now we turn to the details of modeling the dynamics of a bouncing ball.

1.2 Model

To model the bouncing ball system we assume that the table's mass is much greater than the ball's mass and the impact between the ball and the table is instantaneous. These assumptions are realistic for the

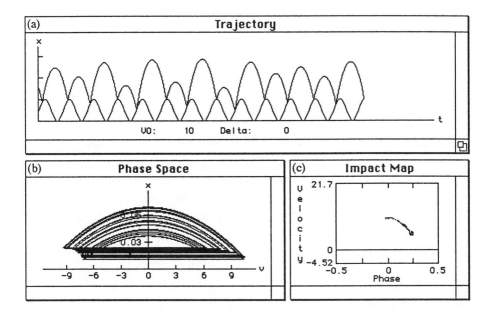

Figure 1.4: "Strange" orbit of a bouncing ball: (a) height vs. time, (b) phase space, (c) impact map. (Generated by the Bouncing Ball program.)

experimental system described in the previous section and simply mean that the table's motion is not affected by the collisions. The collisions are usually *inelastic*; that is, a little energy is lost at each impact. If no energy is lost then the collisions are called *elastic*. We will examine both cases in this book: the case in which energy is dissipated (*dissipative*) and the case in which energy is conserved (*conservative*).

1.2.1 Stationary Table

First, though, we must figure out how the ball's velocity changes at each impact. Consider two different reference frames: the ball's motion as seen from the ground (the ground's reference frame) and the ball's motion as seen from the table (the table's reference frame). Begin by considering the simple case where the table is stationary and the two reference frames are identical. As we will show shortly, understanding the stationary case will solve the nonstationary case.

Let v_k' be the ball's velocity right before the kth impact, and let v_k be the ball's velocity right after the kth impact. The prime notation indicates a velocity immediately before an impact. If the table is stationary and the collisions are elastic, then $v_k = -v_k'$: the ball reverses direction but does not change speed since there is no energy loss. If the collisions are inelastic and the table is stationary, then the ball's speed will be reduced after the collision because energy is lost: $v_k = -\alpha v_k'$ ($0 \le \alpha < 1$), where α is the *coefficient of restitution*. The constant α is a measure of the energy loss at each impact. If $\alpha = 1$, the system is conservative and the collisions are elastic. The coefficient of restitution is strictly less than one for inelastic collisions.[1]

1.2.2   Impact Relation for the Oscillating Table

When the table is in motion, the ball's velocity immediately after an impact will have an additional term due to the kick from the table. To calculate the change in the ball's velocity, imagine the motion of the ball from the table's perspective. The key observation is that in the table's reference frame the table is always stationary. The ball, however, appears to have an additional velocity which is equal to the opposite of the table's velocity in the ground's reference frame. Therefore, to calculate the ball's change in velocity we can calculate the change in velocity in the table's reference frame and then add the table's velocity to get the ball's velocity in the ground's reference frame. In Figure 1.5 we show the motion of the ball and the table in both the ground's and the table's reference frames.

Let u_k be the table's velocity in the ground's reference frame. Further, let $\bar{v}_k'$ and $\bar{v}_k$ be the velocity in the table's reference frame immediately before and after the kth impact, respectively. The bar denotes measurements in the table's reference frame; the unbarred coordinates are measurements in the ground's reference frame. Then, in the table's reference frame,

$$\bar{v}_k = -\alpha \bar{v}_k', \tag{1.4}$$

---

[1]The coefficient of restitution α is called the *damping coefficient* in the Bouncing Ball program.

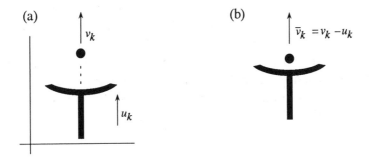

Figure 1.5: Motion of the ball in the reference frame of the ground (a) and the table (b).

since the table is always stationary. To find the ball's velocity in the ground's reference frame we must add the table's velocity to the ball's apparent velocity,

$$v_k = \bar{v}_k + u_k, \quad v_k' = \bar{v}_k' + u_k,$$

or equivalently,

$$\bar{v}_k = v_k - u_k, \quad \bar{v}_k' = v_k' - u_k. \tag{1.5}$$

Therefore, in the ground's reference frame, equation (1.4) becomes

$$v_k - u_k = -\alpha[v_k' - u_k], \tag{1.6}$$

when it is rewritten using equation (1.5). Rewriting equation (1.6) gives the velocity v_k after the kth impact as

$$v_k = [1 + \alpha]u_k - \alpha v_k'. \tag{1.7}$$

This last equation is known as the *impact relation*. It says the kick from the table contributes $[1 + \alpha]u_k$ to the ball's velocity.

1.2.3 The Equations of Motion: Phase and Velocity Maps

To determine the motion of the ball we must calculate the times, hence phases (from eq. (1.2)), when the ball and the table collide. An impact occurs when the difference between the ball position and the table position is zero. Between impacts, the ball goes up and down according to Newton's law for the motion of a projectile in a constant gravitational field of strength g. Since the motion between impacts is simple, we will present the motion of the ball in terms of an impact map, that is, some rule that takes as input the current values of the impact phase and impact velocity and then generates the next impact phase and impact velocity.

Let[2]

$$x(t) = x_k + v_k(t - t_k) - \frac{1}{2}g(t - t_k)^2 \qquad (1.8)$$

be the ball's position at time t after the kth impact, where x_k is the position at the kth impact and t_k is the time of the kth impact, and let

$$s(t) = A[\sin(\omega t + \theta_0) + 1] \qquad (1.9)$$

be the table's position with an amplitude A, angular frequency ω, and phase θ_0 at $t = 0$. We add one to the sine function to ensure that the table's amplitude is always positive. The difference in position between the ball and table is

$$d(t) = x(t) - s(t), \qquad (1.10)$$

which should always be a non-negative function since the ball is never below the table. The first value at which $d(t) = 0$, $t > t_k$, implicitly defines the time of the next impact. Substituting equations (1.8) and (1.9) into equation (1.10) and setting $d(t)$ to zero yields

$$0 = x_k + v_k(t_{k+1} - t_k) - \frac{1}{2}g(t_{k+1} - t_k)^2 - A[\sin(\omega t_{k+1} + \theta_0) + 1]. \quad (1.11)$$

Equation (1.11) can be rewritten in terms of the phase when the identification $\theta = \omega t + \theta_0$ is made between the phase variable and the time variable. This leads to the implicit *phase map* of the form,

$$0 = A[\sin(\theta_k) + 1] + v_k\left[\frac{1}{\omega}(\theta_{k+1} - \theta_k)\right]$$
$$-\frac{1}{2}g\left[\frac{1}{\omega}(\theta_{k+1} - \theta_k)\right]^2 - A[\sin(\theta_{k+1}) + 1], \qquad (1.12)$$

---

[2]For a discussion of the motion of a particle in a constant gravitational field see any introductory physics text such as R. Weidner and R. Sells, *Elementary Physics*, Vol. 1 (Allyn and Bacon: Boston, 1965), pp. 19–22.

where θ_{k+1} is the *next* θ for which $d(\theta) = 0$. In deriving equation (1.12) we used the fact that the table position and the ball position are identical at an impact; that is, $x_k = A[\sin(\theta_k) + 1]$.

An explicit *velocity map* is derived directly from the impact relation, equation (1.7), as

$$v_{k+1} = (1 + \alpha)\omega A \cos(\omega t_{k+1} + \theta_0) - \alpha[v_k - g(t_{k+1} - t_k)], \qquad (1.13)$$

or, in the phase variable,

$$v_{k+1} = (1 + \alpha)\omega A \cos(\theta_{k+1}) - \alpha\left\{v_k - g\left[\frac{1}{\omega}(\theta_{k+1} - \theta_k)\right]\right\}, \qquad (1.14)$$

noting that the table's velocity is just the time derivative of the table's position, $u(t) = \dot{s}(t) \equiv ds/dt = A\omega \cos(\omega t + \theta_0)$, and that, between impacts, the ball is subject to the acceleration of gravity, so its velocity is given by $v_k - g(t - t_k)$. The overdot is Newton's original notation denoting differentiation with respect to time.

The implicit phase map (eq. (1.12)) and the explicit velocity map (eq. (1.14)) constitute the exact model for the bouncing ball system. The dynamics of the bouncing ball are easy to simulate on a computer using these two equations. Unfortunately, the phase map is an implicit algebraic equation for the variable θ_{k+1}; that is, θ_{k+1} cannot be isolated from the other variables. To solve the phase function for θ_{k+1} a numerical algorithm is needed to locate the zeros of the phase function (see Appendix A). Still, this presents little problem for numerical simulations, or even, as we shall see, for a good deal of analytical work.

1.2.4 Parameters

The parameters for the bouncing ball system should be determined before we continue our analysis. The relevant parameters, with typical experimental values, are listed in Table 1.1. In an experimental system, the table's frequency or the table's amplitude of oscillation is easy to adjust with the function generator. The coefficient of restitution can also be varied by using balls composed of different materials. Steel balls, for instance, are relatively hard and have a high coefficient of restitution. Balls made from brass, glass, plastic, or wood are softer and tend to dissipate more energy at impact.

Parameter	Symbol	Experimental values
Coefficient of restitution	α	0.1–0.9
Table's amplitude	A	0.01–0.1 cm
Table's period	T	0.1–0.01 s
Gravitational acceleration	g	981 cm/s^2
Frequency	f	$f = 1/T$
Angular frequency	ω	$\omega = 2\pi f$
Normalized acceleration	β	$\beta = 2\omega^2(1 + \alpha)A/g$

Table 1.1: Reference values for the Bouncing Ball System.

As we will show in the next section, the physical parameters listed in Table 1.1 are related. By rescaling the variables, it is possible to show that there are only two fundamental parameters in this model. For our purposes we will take these to be α, the coefficient of restitution, and a new parameter β, which is essentially proportional to $A\omega^2$. The parameter β is, in essence, a normalized acceleration and it measures the violence with which the table oscillates up and down.

1.3   High Bounce Approximation

In the high bounce approximation we imagine that the table's displacement amplitude is always small compared to the ball's maximum height. This approximation is depicted in Figure 1.6 where the ball's trajectory is perfectly symmetric about its midpoint, and therefore

$$v'_{k+1} = -v_k. \tag{1.15}$$

The velocity of the ball between the kth and $k + 1$st impacts is given by

$$v(t) = v_k - g(t - t_k). \tag{1.16}$$

At the $k + 1$st impact, the velocity is v'_{k+1} and the time is t_{k+1}, so

$$v'_{k+1} = v_k - g(t_{k+1} - t_k). \tag{1.17}$$

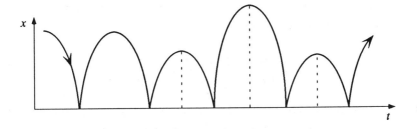

Figure 1.6: Symmetric orbit in the high bounce approximation.

Using equation (1.15) and simplifying, we get

$$t_{k+1} = t_k + \frac{2}{g}v_k, \tag{1.18}$$

which is the *time map* in the high bounce approximation.

To find the velocity map in this approximation we begin with the impact relation (eq. (1.7)),

$$
\begin{aligned}
v_{k+1} &= (1+\alpha)u_{k+1} - \alpha v'_{k+1} \\
&= (1+\alpha)u_{k+1} + \alpha v_k, \tag{1.19}
\end{aligned}
$$

where the last equality follows from the high bounce approximation, equation (1.15). The table's velocity at the $k+1$st impact can be written as

$$
\begin{aligned}
u_{k+1} &= \omega A \cos(\omega t_{k+1} + \theta_0) \\
&= \omega A \cos[\omega(t_k + 2v_k/g) + \theta_0], \tag{1.20}
\end{aligned}
$$

when the time map, equation (1.18), is used. Equations (1.19) and (1.20) give the velocity map in the high bounce approximation,

$$v_{k+1} = \alpha v_k + \omega(1+\alpha)A \cos[\omega(t_k + 2v_k/g) + \theta_0]. \tag{1.21}$$

The impact equations can be simplified somewhat by changing to the dimensionless quantities

$$
\begin{aligned}
\theta &= \omega t + \theta_0, \tag{1.22} \\
\nu &= 2\omega v/g, \text{ and} \tag{1.23} \\
\beta &= 2\omega^2(1+\alpha)A/g, \tag{1.24}
\end{aligned}
$$

which recasts the time map (eq. (1.18)) and the velocity map (eq. (1.21)) into the explicit mapping form

$$f = f_{\alpha,\beta} \begin{cases} \theta_{k+1} & = \theta_k + \nu_k, \\ \nu_{k+1} & = \alpha\nu_k + \beta\cos(\theta_k + \nu_k). \end{cases} \qquad (1.25)$$

In the special case where $\alpha = 1$, this system of equations is known as the *standard map* [4]. The subscripts of $f_{\alpha,\beta}$ explicitly show the dependence of the map on the parameters α and β. The mapping equation (1.25) is easy to solve on a computer. Given an initial condition (θ_0, ν_0), the map explicitly generates the next impact phase and impact velocity as $f^1(\theta_0, \nu_0) = (\theta_1, \nu_1)$, and this in turn generates $f^2(\theta_0, \nu_0) = f^1(\theta_1, \nu_1) = f \circ f(\theta_0, \nu_0) = (\theta_2, \nu_2)$, etc., where, in this notation, the superscript n in f^n indicates functional composition (see section 2.2). Unlike the exact model, both the phase map and the velocity map are explicit equations in the high bounce approximation.

The high bounce approximation shares many of the same qualitative properties of the exact model for the bouncing ball system, and it will serve as the starting point for several analytic calculations. However, for comparisons with experimental data, it is worthwhile to put the extra effort into numerically solving the exact equations because the high bounce model fails in at least two major ways to model the actual physical system [5]. First, the high bounce model can generate solutions that cannot possibly occur in the real system. These unphysical solutions occur for very small bounces at negative table velocities, where it is possible for the ball to be projected downward beneath the table. That is, the ball can pass through the table in this approximation. Second, this approximation cannot reproduce a large class of real solutions, called "sticking solutions," which are discussed in section 1.4.3. Fundamentally, this is because the map in the high bounce approximation is invertible, whereas the exact model is not invertible. In the exact model there exist some solutions—in particular the sticking solutions—for which two or more orbits are mapped to the same identical point. Thus the map at this point does not have a unique inverse.

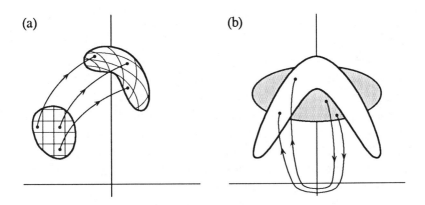

Figure 1.7: (a) Evolution of a region in phase space. (b) Recurrent regions in the phase space of a nonlinear system.

1.4 Qualitative Description of Motions

In specifying an individual solution to the bouncing ball system, we need to know both the initial condition, that is, the initial impact phase and impact velocity of the ball (θ_0, v_0), and the relevant system parameters, α, A, and T. Then, to find an individual trajectory, all we need to do is iterate the mapping for the appropriate model. However, finding the solution for a single trajectory gives little insight into the global dynamics of the system. As stressed in the Introduction, we are not interested so much in solving an individual orbit, but rather in understanding the behavior of a large collection of orbits and, when possible, the system as a whole.

An individual solution can be represented by a curve in phase space. In considering a collection of solutions, we will need to understand the behavior not of a single curve in phase space, but rather of a bundle of adjacent curves, a region in phase space. Similarly, in the impact map we want to consider a collection of initial conditions, a region in the impact map. In general, the future of an orbit is well defined by a flow or mapping. The fate of a region in the phase space or the impact map is defined by the collective futures of each of the individual curves or points, respectively, in the region as is illustrated in Figure 1.7.

A number of questions can, and will, be asked about the evolution

of a region in phase space (or in the impact map). Do regions in the
phase space expand or contract as the system evolves? In the bouncing ball system, a bundle of initial conditions will generally contract in
area whenever the system is dissipative—a little energy is lost at each
impact, and this results in a shrinkage of our initial region, or patch, in
phase space (see section 4.4.4 for details). Since this region is shrinking, this raises many questions that will be addressed throughout this
book, such as where do all the orbits go, how much of the initial area
remains, and what do the orbits do on this remaining set once they get
to wherever they're going? This turns out to be a subtle collection of
questions. For instance, even the question of what we mean by "area"
gets tricky because there is more than one useful notion of the area,
or measure, of a set. Another related question is, do these regions intersect with themselves as they evolve (see Figure 1.7)? The answer
is generally yes, they do intersect, and this observation will lead us to
study the rich collection of recurrence structures of a nonlinear system.

A simple question we can answer is: does there exist a closed,
simply-connected subset, or region, of the whole phase space (or impact
map) such that all the orbits outside this subset eventually enter into it,
and, once inside, they never get out again? If such a subset exists, it is
called a *trapping region*. Establishing the existence of a trapping region
can simplify our general problem somewhat, because instead of considering all possible initial conditions, we need only consider those initial
conditions inside the trapping region, since all other initial conditions
will eventually end up there.

1.4.1 Trapping Region

To find a trapping region for the bouncing ball system we will first find
an upper bound for the next outgoing velocity, v_{k+1}, by looking at the
previous value, v_k. We will then find a lower bound for v_{k+1}. These
bounds give us the boundaries for a trapping region in the bouncing
ball's impact map (θ_i, v_i), which imply a trapping region in phase space.

To bound the outgoing velocity, we begin with equation (1.13) in
the form

$$v_{k+1} - \alpha v_k = (1 + \alpha)\omega A \cos(\omega t_{k+1} + \theta_0) + \alpha g(t_{k+1} - t_k). \quad (1.26)$$

The first term on the right-hand side is easy to bound. To bound the second term, we first look at the average ball velocity between impacts, which is given by

$$\bar{v}_k = v_k - \frac{1}{2}g(t_{k+1} - t_k).$$

Rearranging this expression gives

$$t_{k+1} - t_k = \frac{2}{g}(v_k - \bar{v}_k).$$

Equation (1.26) now becomes

$$v_{k+1} + \alpha v_k = (1 + \alpha)A\omega \cos(\omega t_{k+1} + \theta_0) - 2\alpha\bar{v}_k + 2\alpha v_k. \qquad (1.27)$$

Noting that the average table velocity between impacts is the same as the average ball velocity between impacts (see Prob. 1.14), we find that

$$v_{k+1} - \alpha v_k \leq (1 + 3\alpha)A\omega. \qquad (1.28)$$

If we define

$$v_{max} = \frac{1 + 3\alpha}{1 - \alpha}A\omega, \qquad (1.29)$$

and let $v_k > v_{max}$, then $v_{k+1} - \alpha v_k < (1 - \alpha)v_k$, or

$$v_{k+1} < v_k.$$

In this case it is essential that the system be dissipative ($\alpha < 1$) for a trapping region to exist. In the conservative limit no trapping region exists—it is possible for the ball to reach infinite heights and velocities when no energy is lost at impact.

To find a lower bound for v_{k+1}, we simply realize that the velocity after impact must always be at least that of the table,

$$v_{k+1} \geq -A\omega = v_{min}. \qquad (1.30)$$

For the bouncing ball system the compact trapping region, D, given by

$$D = \{(\theta, v) \mid v_{min} \leq v \leq v_{max}\} \qquad (1.31)$$

is simply a strip bounded by v_{min} and v_{max}. To prove that D is a trapping region, we also need to show that v cannot approach v_{max} asymptotically, and that once inside D, the orbit cannot leave D (these calculations are left to the reader—see Prob. 1.15). The previous calculations show that all orbits of the dissipative bouncing ball system will eventually enter the region D and be "trapped" there.

1.4.2   Equilibrium Solutions

Once the orbits enter the trapping region, where do they go next? To answer this question we first solve for the motion of a ball bouncing on a stationary table. Then we will imagine slowly turning up the table amplitude.

If the table is stationary, then the high bounce approximation is no longer approximate, but exact. Setting $A = 0$ in the velocity map, equation (1.21), immediately gives

$$v_{k+1} = \alpha v_k. \tag{1.32}$$

Using the time map, $t_{k+1} - t_k = (2/g)v_k$, the coefficient of restitution is easy to measure [6] by recording three consecutive impact times, since

$$\alpha = \frac{t_{k+2} - t_{k+1}}{t_{k+1} - t_k}. \tag{1.33}$$

To find how long it takes the ball to stop bouncing, consider the sum of the differences of consecutive impact times,

$$\Gamma = \sum_{n=0}^{\infty} \tau_n = \tau_0 + \tau_1 + \tau_2 + \cdots, \quad \tau_k \equiv t_{k+1} - t_k. \tag{1.34}$$

Since $\tau_{k+1} = \alpha \tau_k$, Γ is the summation of a *geometric series*,

$$\Gamma = \tau_0 + \tau_1 + \tau_2 + \cdots = \tau_0 + \alpha \tau_0 + \alpha^2 \tau_0 + \cdots = \sum_{n=0}^{\infty} \tau_0 \alpha^n = \frac{\tau_0}{1 - \alpha}, \tag{1.35}$$

which can be summed for $\alpha < 1$. After an infinite number of bounces, the ball will come to a halt in a finite time.

For these equilibrium solutions, all the orbits in the trapping region come to rest on the table. When the table's acceleration is small, the picture does not change much. The ball comes to rest on the oscillating table and then moves in unison with the table from then on.

1.4.3   Sticking Solutions

Now that the ball is moving with the table, what happens as we slowly turn up the table's amplitude while keeping the forcing frequency fixed?

Initially, the ball will remain *stuck* to the table until the table's maximum acceleration is greater than the earth's gravitational acceleration, g. The table's acceleration is given by

$$\ddot{s} = -A\omega^2 \sin(\omega t + \theta_0). \tag{1.36}$$

The maximum acceleration is thus $A\omega^2$. When $A\omega^2$ is greater than g, the ball becomes *unstuck* and will fly free from the table until its next impact. The phase at which the ball becomes initially unstuck occurs when

$$-g = -A\omega^2 \sin(\theta_{unstuck}) \Longrightarrow \theta_{unstuck} = \arcsin\left(\frac{g}{A\omega^2}\right). \tag{1.37}$$

Even in a system in which the table's maximum acceleration is much greater than g, the ball can become stuck. An infinite number of impacts can occur in a finite stopping time, Γ. The sum of the times between impacts converges in a finite time much less than the table's period, T. The ball gets stuck again at the end of this sequence of impacts and moves with the table until it reaches the phase $\theta_{unstuck}$. This type of sticking solution is an eventually periodic orbit. After its first time of getting stuck, it will exactly repeat this pattern of getting stuck, and then released, forever.

However, these sticking solutions are a bit exotic in several respects. Sticking solutions are not invertible; that is, an infinite number of initial conditions can eventually arrive at the same identical sticking solution. It is impossible to run a sticking solution backward in time to find the exact initial condition from which the orbit came. This is because of the geometric convergence of sticking solutions in finite time.

Also, there are an infinite number of different sticking solutions. Three such solutions are illustrated in Figure 1.8. To see how some of these solutions are formed, let's turn the table amplitude up a little so that the stopping time, Γ, is lengthened. Now, it happens that the ball does not get stuck in the first table period, T, but keeps bouncing on into the second or third period. However, as it enters each new period, the bounces get progressively lower so that the ball does eventually get stuck after several periods. Once stuck, it again gets released when the table's acceleration is greater than g, and this new pattern repeats itself forever.

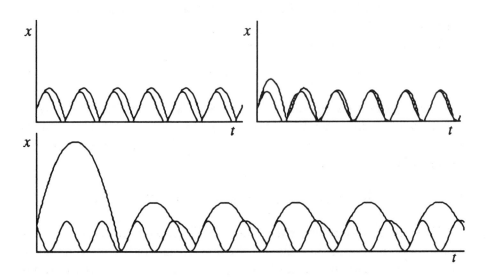

Figure 1.8: Sticking solutions in the bouncing ball system.

1.4.4   Period One Orbits and Period Doubling

As we increase the table's amplitude we often see that the orbit jumps from a sticking solution to a simple periodic motion. Figure 1.9 shows the convergence of a trajectory of the bouncing ball system to a period one orbit. The ball's motion converges toward a periodic orbit with a period exactly equal to that of the table, hence the term *period one orbit* (see Prob. 1.1).

What happens to the period one solution as the forcing amplitude

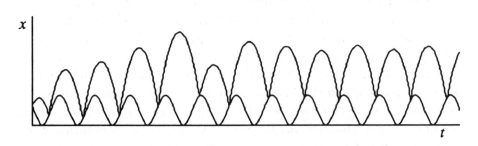

Figure 1.9: Convergence to a period one orbit.

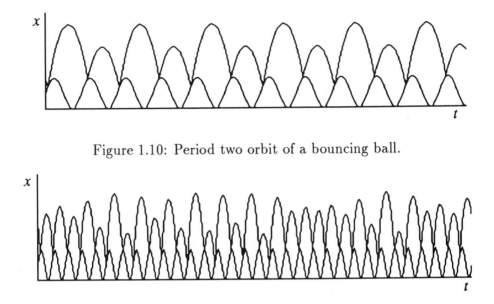

Figure 1.10: Period two orbit of a bouncing ball.

Figure 1.11: Chaotic orbit of a bouncing ball.

of the table increases further? We discover that the period one orbit bifurcates (literally, splits in two) to the period two orbit illustrated in Figure 1.10. Now the ball's motion is still periodic, but it bounces high, then low, then high again, requiring twice the table's period to complete a full cycle. If we gradually increase the table's amplitude still further we next discover a period four orbit, and then a period eight orbit, and so on. In this *period doubling cascade* we only see orbits of period

$$P = 2^n = 1, 2, 4, 8, 16 \ldots, \qquad (1.38)$$

and not, for instance, period three, five, or six.

The amplitude ranges for which each of these period 2^n orbits is observable, however, gets smaller and smaller. Eventually it converges to a critical table amplitude, beyond which the bouncing ball system exhibits the nonperiodic behavior illustrated in Figure 1.11. This last type of motion found at the end of the period doubling cascade never settles down to a periodic orbit and is, in fact, our first physical example of a chaotic trajectory known as a *strange attractor*. This motion is an *attractor* because it is the asymptotic solution arising from many

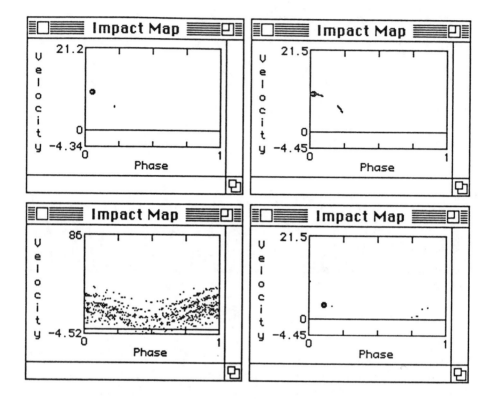

Figure 1.12: A zoo of periodic and chaotic motions seen in the bouncing ball system. (Generated by the Bouncing Ball program.)

different initial conditions: different motions of the system are attracted to this particular motion. At this point, the term *strange* is used to distinguish this motion from other motions such as periodic orbits or equilibrium points. A more precise definition of the term strange is given in section 3.8.

At still higher table amplitudes many other types of strange and periodic motions are possible, a few of which are illustrated in Figure 1.12. The type of motion depends on the system parameters and the specific initial conditions. It is common in a nonlinear system for many solutions to coexist. That is, it is possible to see several different periodic and chaotic motions for the same parameter values. These

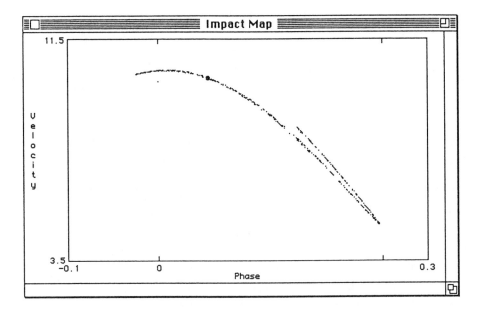

Figure 1.13: Strange attractor in the bouncing ball system arising at the end of a period doubling cascade. (Generated by the Bouncing Ball program.)

coexisting orbits are only distinguished by their initial conditions.

The "period doubling route to chaos" we saw above is common to a wide variety of nonlinear systems and will be discussed in depth in section 2.8.

1.4.5 Chaotic Motions

Figure 1.13 shows the impact map of the strange attractor discovered at the end of the period doubling route to chaos. This strange attractor looks almost like a simple curve (segment of an upside-down parabola) with gaps. Parts of this curve look chopped out or eaten away. However, on magnification, this curve appears not so simple after all. Rather, it seems to resemble an intricate web of points spread out on a narrow curved strip. Since this chaotic solution is not periodic (and hence, never exactly repeats itself) it must consist of an infinite collection of discrete points in the impact (velocity vs. phase) space.

This strange set is generated by an orbit of the bouncing ball system, and it is chaotic in that orbits in this set exhibit sensitive dependence on initial conditions. This sensitive dependence on initial conditions is easy to see in the bouncing ball system when we solve for the impact phases and velocities for the exact model with the numerical procedure described in Appendix A. First consider two slightly different trajectories that converge to the same period one orbit. As shown in Table 1.2, these orbits initially differ in phase by 0.00001. This phase difference increases a little over the next few impacts, but by the eleventh impact the orbits are indistinguishable from each other, and by the eighteenth impact they are indistinguishable from the period one orbit. Thus the difference between the two orbits decreases as the system evolves.

An attracting periodic orbit has both long-term and short-term predictability. As the last example indicates, we can predict, from an initial condition of limited resolution, where the ball will be after a few bounces (short-term) and after many bounces (long-term).

The situation is dramatically different for motion on a strange attractor. Chaotic motions may still possess short-term predictability, but they lack long-term predictability. In Table 1.3 we show two different trajectories that again differ in phase by 0.00001 at the zeroth impact. However, in the chaotic case the difference increases at a remarkable rate with the evolution of the system. That is, given a small difference in the initial conditions, the orbits diverge rapidly. By the twelfth impact the error is greater than 0.001, by the twentieth impact 0.01, and by the twenty-fourth impact the orbits show no resemblance. Chaotic motion thus exhibits sensitive dependence on initial conditions. Even if we increase our precision, we still cannot predict the orbit's future position exactly.

As a practical matter we have no exact long-term predictive power for chaotic motions. It does not really help to double the resolution of our initial measurement as this will just postpone the problem. The bouncing ball system is both deterministic and unpredictable. Chaotic motions of the bouncing ball system are unpredictable in the practical sense that initial measurements are always of limited accuracy, and any initial measurement error grows rapidly with the evolution of the system.

Hit	Phase	Phase
0	0.12001	0.12002
1	0.119553	0.119563
2	0.123625	0.123613
3	0.119647	0.119657
4	0.122627	0.122620
5	0.120645	0.120650
6	0.121893	0.121890
7	0.121140	0.121142
8	0.121584	0.121583
9	0.121327	0.121328
10	0.121474	0.121473
11	0.121391	0.121391
12	0.121437	0.121437
13	0.121412	0.121412
14	0.121426	0.121426
15	0.121418	0.121418
16	0.121422	0.121422
17	0.121420	0.121420
18	*0.121421*	*0.121421*
19	*0.121421*	*0.121421*
20	*0.121421*	*0.121421*

Table 1.2: Convergence of two different initial conditions to a period one orbit. The digits in bold are where the orbits differ. At the zeroth hit the orbits differ in phase by 0.00001. Note that the difference between the orbits decreases so that after 18 impacts both orbits are indistinguishable from the period one orbit. The operating parameters are: $A = 0.01$ cm, frequency = 60 Hz, $\alpha = 0.5$, and the initial ball velocity is 8.17001 cm/s. The impact phase is presented as $\delta = (\theta \bmod 2\pi)/(2\pi)$ so that it is normalized to be between zero and one.

Hit	Phase	Phase
0	0.12001	0.12002
1	0.119575	0.119585
2	0.203686	0.203667
3	0.044295	0.044330
4	0.245370	0.245382
5	0.979140	0.979114
6	0.163451	0.163401
7	0.151935	0.152045
8	0.133956	0.133762
9	0.170026	0.170343
10	0.106407	0.105836
11	0.210176	0.210911
12	0.034337	0.033041
13	0.240314	0.239475
14	0.989893	0.991636
15	0.183346	0.186362
16	0.108543	0.102037
17	0.202784	0.211096
18	0.048083	0.033369
19	0.245904	0.239552
20	0.977588	0.991442
21	0.160466	0.186034
22	0.158498	0.102743
23	0.122340	0.210230
24	0.188441	0.034893
25	0.073121	0.240520

Table 1.3: Divergence of initial conditions on a strange attractor illustrating sensitive dependence on initial conditions. The parameter values are the same as in Table 1.2 except for $A = 0.012$ cm. The bold digits show where the impact phases differ. The orbits differ in phase by 0.00001 at the zeroth hit, but by the twenty-third impact they differ at every digit.

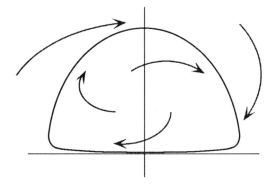

Figure 1.14: A periodic attractor and its transient.

Strange attractors are common to a wide variety of nonlinear systems. We will develop a way to name and dissect these critters in Chapters 4 and 5.

1.5 Attractors

An *attracting set* A in a trapping region D is defined as a nonempty closed set formed from some open neighborhood,

$$A = \bigcap_{n \geq 0} f^n(D). \qquad (1.39)$$

We mentioned before that for a dissipative bouncing ball system the trapping region is contracting, so the open neighborhood typically consists of a collection of smaller and smaller regions as it approaches the attracting set.

An attractor is an attempt to define the asymptotic solution of a dynamical system. It is that part of the solution that is left after the "transient" part is thrown out. Consider Figure 1.14, which shows the approach of several phase space trajectories of the bouncing ball system toward a period one cycle. The orbits appear to consist of two parts: the transient—the initial part of the orbit that is spiraling toward a closed curve—and the attractor—the closed periodic orbit itself.

In the previous section we saw examples of several different types of attractors. For small table amplitudes, the ball comes to rest on the table. For these equilibrium solutions the attractor consists of a single point in the phase space of the table's reference frame. At higher table amplitudes periodic orbits can exist, in which case the attractor is a closed curve in phase space. In a dissipative system this closed curve representing a periodic motion is also known as a limit cycle. At still higher table amplitudes, a more complicated set called a strange attractor can appear. The phase space plot of a strange attractor is a complicated curve that never quite closes. After a long time, this curve appears to sketch out a surface. Each type of attractor—a point, closed curve, or strange attractor (something between a curve and a surface)—represents a different type of motion for the system—equilibrium, periodic, or chaotic.

Except for the equilibrium solutions, each of the attractors just described in the phase space has its corresponding representation in the impact map. In general, the representation of the attractor in the impact map is a geometric object of one dimension less than in phase space. For instance, a periodic orbit is a closed curve in phase space, and this same period n orbit consists of a collection of n points in the impact map. The impact map for a chaotic orbit consists of an *infinite* collection of points.

For a nonlinear system, many attractors can coexist. This naturally raises the question as to which orbits and collections of initial conditions go to which attractors. For a given attractor, the domain of attraction, or *basin of attraction*, is the collection of all those initial conditions whose orbits approach and always remain near that attractor. That is, it is the collection of all orbits that are "captured" by an attractor. Like the attractors themselves, the basins of attraction can be simple or complex [7].

Figure 1.15 shows a diagram of basins of attraction in the bouncing ball system. The phase space is dominated by the black regions, which indicate initial conditions that eventually become sticking solutions. The white sinusoidal regions at the bottom of Figure 1.15 show unphysical initial conditions—phases and velocities that the ball cannot obtain. The gray regions represent initial conditions that approach a period one orbit. (See Plate 1 for a color diagram of basins of attraction

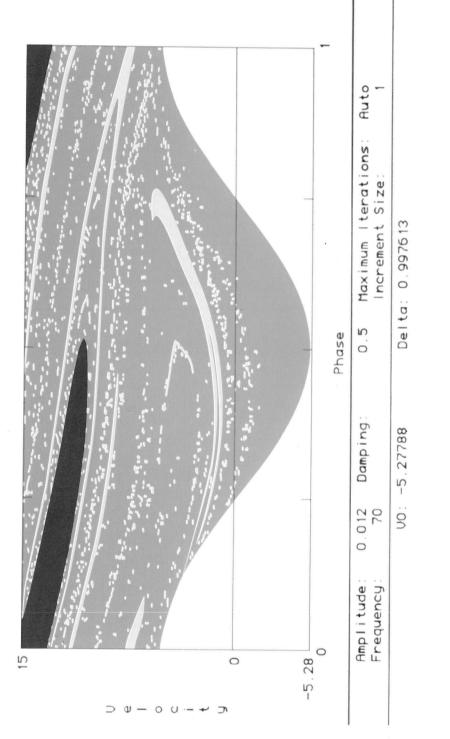

Plate 1. Basins of attraction in the bouncing ball system
(Generated by the Bouncing Ball program)

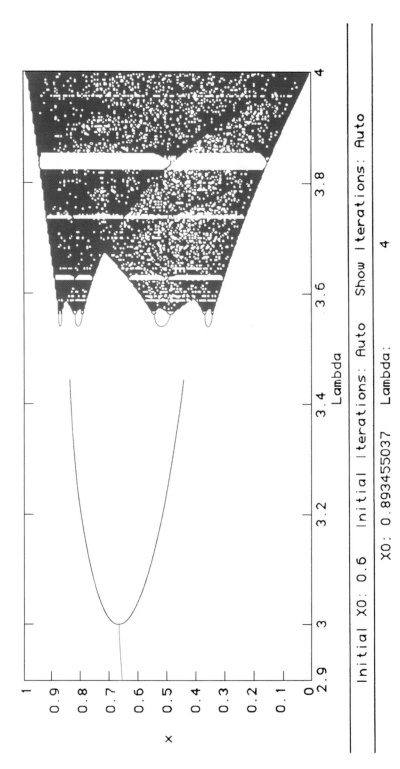

Initial X0: 0.6   Initial Iterations: Auto   Show Iterations: Auto

X0: 0.893455037   Lambda: 4

Plate 2. Bifurcation diagram for the quadratic map
(Generated by the Quadratic Map program)

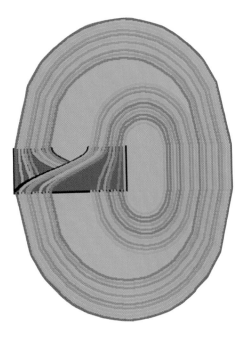

Plate 3. Lorenz Template

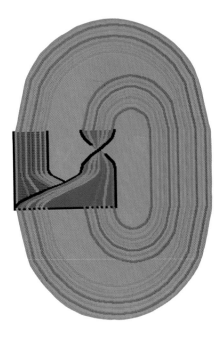

Plate 4. Horseshoe Template

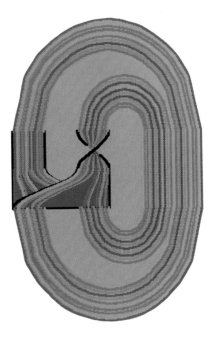

Plate 5. Horseshoe Template

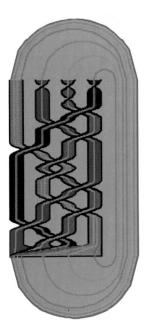

Plate 6. Four-Branch Template

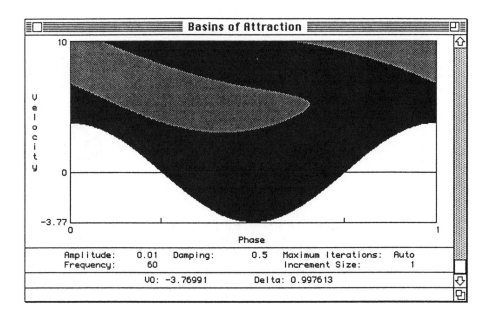

Figure 1.15: Basins of attraction in the bouncing ball system. (Generated by the Bouncing Ball program.)

in the bouncing ball system.)

1.6 Bifurcation Diagrams

A *bifurcation diagram* provides a nice summary for the transition between different types of motion that can occur as one parameter of the system is varied. A bifurcation diagram plots a system parameter on the horizontal axis and a representation of an attractor on the vertical axis. For instance, for the bouncing ball system, a bifurcation diagram can show the table's forcing amplitude on the horizontal axis and the asymptotic value of the ball's impact phase on the vertical axis, as illustrated in Figure 1.16. At a bifurcation point, the attracting orbit undergoes a qualitative change. For instance, the attractor literally splits in two (in the bifurcation diagram) when the attractor changes from a period one orbit to a period two orbit.

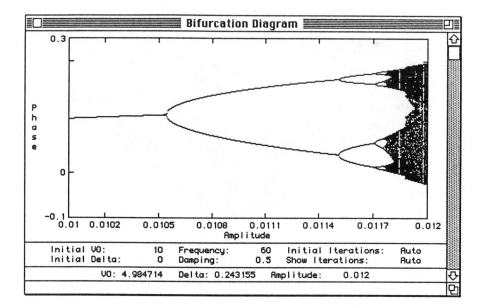

Figure 1.16: Bouncing ball bifurcation diagram. (Generated by the Bouncing Ball program.)

This bouncing ball bifurcation diagram (Fig. 1.16) shows the classic period doubling route to chaos. For table amplitudes between 0.01 cm and 0.0106 cm a stable period one orbit exists; the ball impacts with the table at a single phase. For amplitudes between 0.0106 cm and 0.0115 cm, a period two orbit exists. The ball hits the table at two distinct phases. At higher table amplitudes, the ball impacts at more and more phases. The ball hits at an infinity of distinct points (phases) when the motion is chaotic.

References and Notes

[1] For some recent examples of engineering applications in which the bouncing ball problem arises, see M.-O. Hongler, P. Cartier, and P. Flury, Numerical study of a model of vibro-transporter, Phys. Lett. A **135** (2), 106–112 (1989); M.-O. Hongler and J. Figour, Periodic versus chaotic dynamics in vibratory feeders, Helv. Phys. Acta **62**, 68–81 (1989); T. O. Dalrymple, Numerical solutions to vibroimpact via an initial value problem formulation, J. Sound Vib. **132** (1), 19–32 (1989).

[2] The bouncing ball problem has proved to be a useful system for experimentally exploring several new nonlinear effects. Examples of some of these experiments include S. Celaschi and R. L. Zimmerman, Evolution of a two-parameter chaotic dynamics from universal attractors, Phys. Lett. A **120** (9), 447–451 (1987); N. B. Tufillaro, T. M. Mello, Y. M. Choi, and A. M. Albano, Period doubling boundaries of a bouncing ball, J. Phys. (Paris) **47**, 1477–1482 (1986); K. Wiesenfeld and N. B. Tufillaro, Suppression of period doubling in the dynamics of the bouncing ball, Physica D **26**, 321–335 (1987); P. Pierański, Jumping particle model. Period doubling cascade in an experimental system, J. Phys. (Paris) **44**, 573–578 (1983); P. Pierański, Z. Kowalik, and M. Franaszek, Jumping particle model. A study of the phase of a nonlinear dynamical system below its transition to chaos, J. Phys. (Paris) **46**, 681–686 (1985); M. Franaszek and P. Pierański, Jumping particle model. Critical slowing down near the bifurcation points, Can. J. Phys. **63**, 488–493 (1985); P. Pierański and R. Bartolino, Jumping particle model. Modulation modes and resonant response to a periodic perturbation, J. Phys. (Paris) **46**, 687–690 (1985); M. Franaszek and Z. J. Kowalik, Measurements of the dimension of the strange attractor for the Fermi-Ulam problem, Phys. Rev. A **33** (5), 3508–3510 (1986); P. Pierański and J. Małecki, Noisy precursors and resonant properties of the period doubling modes in a nonlinear dynamical system, Phys. Rev. A **34** (1), 582–590 (1986); P. Pierański and J. Małecki, Noise-sensitive hysteresis loops and period doubling bifurcation points, Nuovo Cimento D **9** (7), 757–780 (1987); P. Pierański, Direct evidence for the suppression of period doubling in the bouncing ball model, Phys. Rev. A **37** (5), 1782–1785 (1988); Z. J. Kowalik, M. Franaszek, and P. Pierański, Self-reanimating chaos in the bouncing ball system, Phys. Rev. A **37** (10), 4016–4022 (1988).

[3] A description of some simple experimental bouncing ball systems, suitable for undergraduate labs, can be found in A. B. Pippard, *The physics of vibration*, Vol. 1 (New York: Cambridge University Press, 1978), p. 253; N. B. Tufillaro and A. M. Albano, Chaotic dynamics of a bouncing ball, Am. J. Phys. **54** (10), 939–944 (1986); R. L. Zimmerman and S. Celaschi, Comment on "Chaotic dynamics of a bouncing ball," Am. J. Phys. **56** (12), 1147–1148 (1988); T. M. Mello and N. B. Tufillaro, Strange attractors of a bouncing ball, Am. J. Phys. **55** (4), 316–320 (1987); R. Minnix and D. Carpenter, Piezoelectric film

reveals f versus t of ball bounce, The Physics Teacher, 280–281 (March 1985). The KYNAR piezoelectric film used for the impact detector is available from Pennwalt Corporation, 900 First Avenue, P.O. Box C, King of Prussia, PA, 19046-0018. The experimenter's kit containing an assortment of piezoelectric films costs about $50.

[4]   Some approximate models of the bouncing ball system are presented by G. M. Zaslavsky, The simplest case of a strange attractor, Phys. Lett. A **69** (3), 145–147 (1978); P. J. Holmes, The dynamics of repeated impacts with a sinusoidally vibrating table, J. Sound Vib. **84** (2), 173–189 (1982); M. Franaszek, Effect of random noise on the deterministic chaos in a dissipative system, Phys. Lett. A **105** (8), 383–386 (1984); R. M. Everson, Chaotic dynamics of a bouncing ball, Physica D **19**, 355–383 (1986); A. Mehta and J. Luck, Novel temporal behavior of a nonlinear dynamical system: The completely inelastic bouncing ball, Phys. Rev. Lett. **65** (4), 393–396 (1990).

[5]   A detailed comparison between the exact model and the high bounce model is given by C. N. Bapat, S. Sankar, and N. Popplewell, Repeated impacts on a sinusoidally vibrating table reappraised, J. Sound Vib. **108** (1), 99–115 (1986).

[6]   A. D. Bernstein, Listening to the coefficient of restitution, Am. J. Phys. **45** (1), 41–44 (1977).

[7]   H. M. Isomäki, Fractal properties of the bouncing ball dynamics, in *Nonlinear dynamics in engineering systems*, edited by W. Schiehlen (Springer-Verlag: New York, 1990), pp. 125–131.

Problems

Problems for section 1.2.

1.1.   For a period one orbit in the exact model show that

(a) $v_k = -v_k'$.

(b) $v_k = gT/2$.

(c) the impact phase is exactly given by

$$\cos(\theta_{P1}) = \frac{gT^2}{4\pi A} \left(\frac{1-\alpha}{1+\alpha} \right). \tag{1.40}$$

1.2. Assuming only that $v_k = -v'_k$,

(a) Show that the solution in Problem 1.1 can be generalized to an nth-order symmetric periodic ("equispaced") orbit satisfying $v_k = ngT/2$, for $n = 1, 2, 3 \ldots$.

(b) Show that the impact phase is exactly given by

$$\cos(\theta_{Pn}) = \frac{ngT^2}{4\pi A}\left(\frac{1-\alpha}{1+\alpha}\right). \tag{1.41}$$

(c) Draw pictures of a few of these orbits. Why are they called "equispaced"?

(d) Find parameter values at which a period one and period two $(n = 1, 2)$ equispaced orbit can coexist.

1.3. If $T = 0.01$ s and $\alpha = 0.8$, what is the smallest table amplitude for which a period one orbit can exist (use Prob. 1.1(c))? For these parameters, estimate the maximum height the ball bounces and express your answer in units of the average thickness of a human hair. Describe how you arrived at this thickness.

1.4. Describe a numerical method for solving the exact bouncing ball map. How do you determine when the ball gets stuck? How do you propose to find the zeros of the phase map, equation (1.12)?

1.5. Derive equation (1.14) from equation (1.13).

Section 1.3.

1.6. Calculate (θ_n, ν_n) in equation (1.25) for $n = 1, 2, 3, 4, 5$ when $\alpha = 0.8$, $\beta = 1$, and $(\theta_0, \nu_0) = (0.1, 1)$.

1.7. Confirm that the variables θ, ν, and β given by equations (1.22–1.24) are dimensionless.

1.8. Verify the derivation of equation (1.25), the standard map.

1.9. Calculate the inverse of the standard map (eq. (1.25)).

1.10. Write a computer program to iterate the model of the bouncing ball system given by equation (1.25), the high bounce approximation.

Section 1.4.

1.11. (a) Calculate the stopping time (eq. (1.35)) for a first impact time $\tau_0 = 1$, and a damping coefficient $\alpha = 0.5$. Also calculate τ_1, τ_2, and τ_3.

(b) Calculate τ_0 and the stopping time for an initial velocity of 10 m/s (1000 cm/s) and a damping coefficient of 0.5.

1.12. For the high bounce approximation (eq. (1.25)) show that when $\alpha < 1$,

(a) $|\nu_{j+1}| \leq \alpha |\nu_j| + \beta$.

(b) A trapping region is given by a strip bounded by $\pm v_{max}$, where $v_{max} = \beta/(1 - \alpha)$. (Note: The reader may assume, as the book does at the end of section 1.4.1, that the v_i cannot approach v_{max} asymptotically, and that once inside the strip, the orbit cannot leave.)

1.13. Calculate $\theta_{unstuck}$ (eq. (1.37)) for $A = 0.1$ cm and $T = 0.01$ s. What is the speed and acceleration of the table at this phase? Is the table on its way up or down? Are there table parameters for which the ball can become unstuck when the table is moving up? Are there table parameters for which the ball can become unstuck when the table is moving down?

1.14. Show that the average table velocity between impacts equals the average ball velocity between impacts.

1.15. These problems relate to the trapping region discussion in section 1.4.1.

(a) Prove that v_i cannot approach the v_{max} given by equation (1.29) asymptotically. (Hint: It is acceptable to increase v_{max} by some small $\epsilon > 0$.)

(b) Prove that, for the trapping region D given by equation (1.31), once the orbit enters D, it can never escape D.

(c) Use the trapping region D in the impact map to find a trapping region in phase space. Hint: Use the maximum outgoing velocity (v_{max}) to calculate a minimum incoming velocity and a maximum height.

(d) The trapping region found in the text is not unique; in fact, it is fairly "loose." Try to obtain a smaller, tighter trapping region.

1.16. Derive equation (1.33).

Section 1.5.

1.17. How many period one orbits can exist according to Problem 1.1(c), and how many of these period one orbits are attractors?

Section 1.6.

1.18. Write a computer program to generate a bifurcation diagram for the bouncing ball system.

Chapter 2

Quadratic Map

2.1   Introduction

A ball bouncing on an oscillating table gives rise to complicated phenomena which appear to defy our comprehension and analysis. The motions in the bouncing ball system are truly complex. However, part of the problem is that we do not, as yet, have the right language with which to discuss nonlinear phenomena. We thus need to develop a vocabulary for nonlinear dynamics.

A good first step in developing any scientific vocabulary is the detailed analysis of some simple examples. In this chapter we will begin by exploring the quadratic map. In linear dynamics, the corresponding example used for building a scientific vocabulary is the simple harmonic oscillator (see Figure 2.1). As its name implies, the harmonic oscillator is a simple model which illustrates many key notions useful in the study of linear systems. The image of a mass on a spring is usually not far from one's mind even when dealing with the most abstract problems in linear physics.

The similarities among linear systems are easy to identify because of the extensive development of linear theory over the past century. Casual inspection of nonlinear systems suggests little similarity. Careful inspection, though, reveals many common features. Our original intuition is misleading because it is steeped in linear theory. Nonlinear systems possess as many similarities as differences. However, the vocabulary of linear dynamics is inadequate to name these common

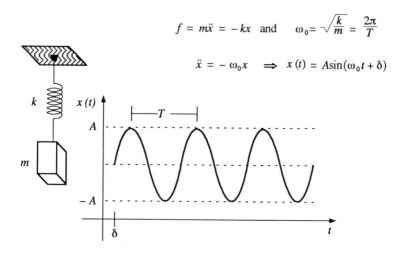

$$f = m\ddot{x} = -kx \quad \text{and} \quad \omega_0 = \sqrt{\frac{k}{m}} = \frac{2\pi}{T}$$

$$\ddot{x} = -\omega_0 x \quad \Rightarrow \quad x(t) = A\sin(\omega_0 t + \delta)$$

Figure 2.1: Simple harmonic oscillator.

structures. Thus, our task is to discover the common elements of non-linear systems and to analyze their structure.

A simple model of a nonlinear system is given by the difference equation known as a *quadratic map*,

$$\begin{aligned} x_{n+1} &= \lambda x_n - \lambda x_n^2, \\ &= \lambda x_n(1 - x_n). \end{aligned} \tag{2.1}$$

For instance, if we set the value $\lambda = 2$ and initial condition $x_0 = 1/4$ in the quadratic map we find that

$$\begin{aligned} x_0 &= 1/4, \\ x_1 &= 3/8, \\ x_2 &= 15/32, \\ &\text{etc.,} \end{aligned}$$

and in this case the value x_n appears to be approaching $1/2$.

Phenomena illustrated in the quadratic map arise in a wide variety of nonlinear systems. The quadratic map is also known as the *logistic map*, and it was studied as early as 1845 by P. F. Verhulst as a model for population growth. Verhulst was led to this difference equation by the following reasoning. Suppose in any given year, indexed by the subscript n, that the (normalized) population is x_n. Then to find the population in the next year (x_{n+1}) it seems reasonable to assume that the number of new births will be proportional to the current population, x_n, and the remaining inhabitable space, $1 - x_n$. The product of these two factors and λ gives the quadratic map, where λ is some parameter that depends on the fertility rate, the initial living area, the average disease rate, and so on.

Given the quadratic map as our model for population dynamics, it would now seem like an easy problem to predict the future population. Will it grow, decline, or vary in a cyclic pattern? As we will see, the answer to this question is easy to discover for some values of λ, but not for others. The dynamics are difficult to predict because, in addition to exhibiting cyclic behavior, it is also possible for the population to vary in a chaotic manner.

In the context of physical systems, the study of the quadratic map was first advocated by E. N. Lorenz in 1964 [1]. At the time, Lorenz was looking at the convection of air in certain models of weather prediction. Lorenz was led to the quadratic map by the following reasoning, which also applies to the bouncing ball system as well as to Lorenz's original model (or, for that matter, to any highly dissipative system). Consider a time series that comes from the measurement of a variable in some physical system,

$$\{x_0, x_1, x_2, x_3, \ldots, x_i, \ldots, x_{n-1}, x_n \ldots\}. \tag{2.2}$$

For instance, in the bouncing ball system this time series could consist of the sequence of impact phases, so that $x_0 = \theta_0, x_1 = \theta_1, x_2 = \theta_2$, and so on. We require that this time series arise from motion on an attractor. To meet this requirement, we throw out any measurements that are part of the initial transient motion. In addition, we assume no foreknowledge of how to model the process giving rise to this time series. Given our ignorance, it then seems natural to try to predict the $n+1$st element of the time series from the previous nth value. Formally, we are seeking a function, f, such that

$$x_{n+1} = f(x_n). \tag{2.3}$$

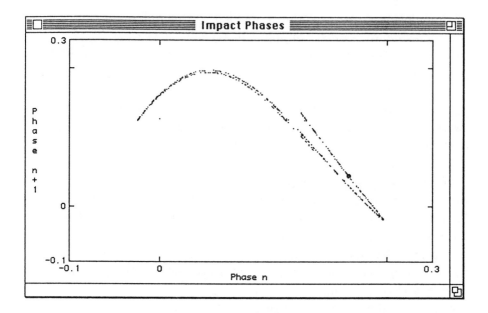

Figure 2.2: Embedded time series of chaotic motion in the bouncing ball system. (Generated by the Bouncing Ball program.)

In the bouncing ball example this idea suggests that the next impact phase would be a function of the previous impact phase; that is, $\theta_{n+1} = f(\theta_n)$.

If such a simple relation exists, then it should be easy to see by plotting $y_n = x_{n+1}$ on the vertical axis and x_n on the horizontal axis. Formally, we are taking our original time series, equation (2.2), and creating an *embedded time series* consisting of the ordered pairs, $(x, y) = (x_n, x_{n+1})$,

$$\{(x_0, x_1), (x_1, x_2), (x_2, x_3), \ldots, (x_{n-1}, x_n), (x_n, x_{n+1}), \ldots\}. \qquad (2.4)$$

The idea of embedding a time series will be central to the experimental study of nonlinear systems discussed in section 3.8.2. In Figure 2.2 we show an embedded time series of the impact phases for chaotic motions in the bouncing ball system. The points for this embedded time series appear to lie close to a region that resembles an upside-down parabola.

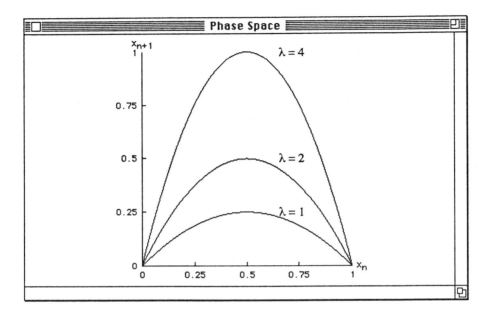

Figure 2.3: The quadratic function. (Generated by the Quadratic Map program.)

The exact details of the curve depend, of course, on the specific parameter values, but as a first approximation the quadratic map provides a reasonable fit to this curve (see Figure 2.3). Note that the curve's maximum amplitude (located at the point $x = 1/2$) rises as the parameter λ increases. We think of the parameter λ in the quadratic map as representing some parameter in our process; λ could be analogous to the table's forcing amplitude in the bouncing ball system. Such single-humped maps often arise when studying highly dissipative nonlinear systems. Of course, more complicated many-humped maps can and do occur; however, the single-humped map is the simplest, and is therefore a good place to start in developing our new vocabulary.

2.2   Iteration and Differentiation

In the previous section we introduced the equation

$$f(x) = \lambda x(1 - x), \qquad (2.5)$$

known as the quadratic map. We write $f_\lambda(x)$ when we want to make
the dependence of f on the parameter λ explicit.

In this section we review two mathematical tools we will need for
the rest of the chapter: iteration and differentiation. We think of a
map $f : x_n \longrightarrow x_{n+1}$ as generating a sequence of points. With the seed
x_0, define $x_n = f^n(x_0)$ and consider the sequence $x_0, x_1, x_2, x_3, \ldots$, as
an orbit of the map. That is, the *orbit* is the sequence of points

$$x_0, \ \ x_1 = f(x_0), \ \ x_2 = f^2(x_0), \ \ x_3 = f^3(x_0), \ \ \ldots \ ,$$

where the nth iterate of x_0 is found by functional composition n times,

$$
\begin{aligned}
f^2 &= f \circ f, \\
f^3 &= f \circ f \circ f, \\
f^n &= \overbrace{f \circ f \circ \cdots f \circ f}^{n}.
\end{aligned}
$$

When determining the stability of an orbit we will need to calculate the derivative of these composite functions (see section 2.5). The
derivative of a composite function evaluated at a point $x = x_0$ is written
as

$$(f^n)'(x_0) = \left(\frac{d}{dx} f^n(x) \right) \Big|_{x=x_0} . \qquad (2.6)$$

The left-hand side of equation (2.6) is a shorthand form for the right-
hand side that tells us to do the following when calculating the derivative. First, construct the nth composite function of f, call it f^n. Second, compute the derivative of f^n. And third, as the bar notation
($|_{x=x_0}$) tells us, evaluate this derivative at $x = x_0$. For instance, if
$f(x) = x^2$, $n = 2$, and $x_0 = 3$, then

$$
\begin{aligned}
(f^2)'(3) &= \left(\frac{d}{dx} f^2(x) \right) \Big|_{x=3}, \\
&= \frac{d}{dx}(f \circ f)(x) = \frac{d}{dx}(f(x^2)) = \frac{d}{dx}(x^4), \\
&= 4x^3 \big|_{x=3} = 108.
\end{aligned}
$$

Notice that we suppressed the bar notation during the intermediate steps. This is common practice when the meaning is clear from context. You may sometimes see the even shorter notation for evaluating the derivative at a point x_0 as

$$f^{n'}(x_0) = \left(\frac{d}{dx} f^n(x) \right) |_{x=x_0}, \tag{2.7}$$

which is sufficiently terse to be legitimately confusing.

An examination of the dynamics of the quadratic map provides an excellent introduction to the rich behavior that can exist in nonlinear systems. To find the itinerary of an individual orbit all we need is a pocket calculator or a computer program something like the following C program.[1]

```
/* quadratic.c:  calculate an orbit for the quadratic map
     input:       1   x0
     output:      1   x1
                  2   x2
                  3   x3
                  etc.
*/
#include <stdio.h>

main()
{
     int n;
     float lambda, x_zero, x_n;
     printf("Enter:  lambda x_zero\n"); scanf("%f %f", &lambda, &x_zero);
     x_n = x_zero;
     for(n = 1; n <= 100; ++n) {
          x_n = lambda * x_n * (1 - x_n); /* the quadratic map */
          printf("%d %f\n", n, x_n);
     }
}
```

---

[1]A nice, brief introduction to the C programming language sufficient for most of the programs in this book is Chapter 1: A tutorial introduction, of B. W. Kernighan and D. M. Ritchie, *The C Programming Language* (Prentice-Hall: Englewood Cliffs, NJ, 1978), pp. 5–31.

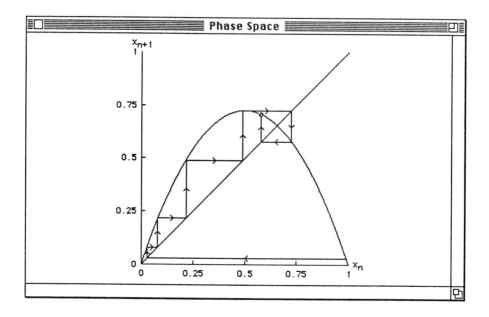

Figure 2.4: Graphical method for iterating the quadratic map. (Generated by the Quadratic Map program.)

2.3   Graphical Method

In addition to doing a calculation, there is a graphical procedure for finding the itinerary of an orbit. This graphical method is illustrated in Figure 2.4 for the same parameter value and initial condition used in the previous example and is based on the following observation. To find x_{n+1} from x_n we note that $x_{n+1} = f(x_n)$; graphically, to get $f(x_n)$ we start at x_n on the horizontal axis and move vertically until we hit the graph $y = f(x)$. Now this current value of y must be transferred from the vertical axis back to the horizontal axis so that it can be used as the next seed for the quadratic map. The simplest way to transfer the y axis to the x axis is by folding the x–y plane through the diagonal line $y = x$ since points on the vertical axis are identical to points on the horizontal axis on this line. This insight suggests the following graphical recipe for finding the orbit for some initial condition x_0:

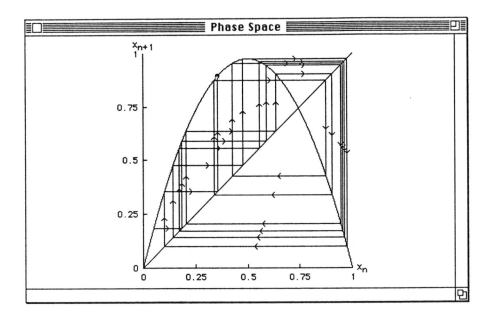

Figure 2.5: Graphical iteration of the quadratic map for a chaotic orbit. (Generated by the Quadratic Map program.)

1. Start at x_0 on the horizontal axis.
2. Move vertically up until you hit the graph $f(x)$.
3. Move *horizontally* until you hit the diagonal line $y = x$.
4. Move *vertically*—up or down—until you hit the graph $f(x)$.
5. Repeat steps 3 and 4 to generate new points.

For the example in Figure 2.4 it is clear that the orbit is converging to the point $1/2$. This same graphical technique is also illustrated in Figure 2.5 for the more complicated orbit that arises when $\lambda = 3.8$.

We can also ask again about the fate of a whole collection of initial conditions, instead of just a single orbit. In particular we can consider the transformation of all initial conditions on the *unit interval*,

$$I = \{x | x \in [0, 1]\} = [0, 1], \tag{2.8}$$

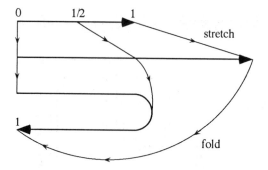

Figure 2.6: Stretching and folding in the quadratic map ($\lambda = 4$).

subject to the quadratic map, equation (2.5). As shown in Figure 2.6, the quadratic map for $\lambda = 4$ can be viewed as transforming the unit interval in two steps. The first step is a *stretch*, which takes the interval to twice its length, $f_{stretch} : [0, 1] \longrightarrow [0, 2]$. The second step is a *fold*, which takes the lower half of the interval to the whole unit interval, and the upper half of the interval also to the whole unit interval with its direction reversed, $f_{fold} : [0, 1] \longrightarrow [0, 1]$ and $[1, 2] \longrightarrow [1, 0]$. These two operations of stretching and folding are the key geometric constructions leading to the complex behavior found in nonlinear systems. The stretching operation tends to quickly separate nearby points, while the folding operation ensures that all points will remain bounded in some region of phase space.

Another way to visualize this stretching and folding process is presented in Figure 2.7. Imagine taking a rubber sheet and dividing it into two sections by slicing it down the middle. The sheet separates into two branches at the gap (the upper part of the sheet where the slice begins); the left branch is flat, while the right branch has a half-twist in it. These two branches are rejoined, or glued together again, at the branch line seen at the bottom of the diagram. Notice that the left branch passes behind the right branch. Next, imagine that there is a simple rule, indicated by the wide arrows in the diagram, that smoothly carries points at the top of the sheet to points at the bottom. In particular, the unit interval at the top of the sheet gets stretched and folded so that it ends up as the bent line segment indicated at the bottom

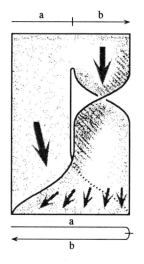

Figure 2.7: Rubber sheet model of stretching and folding in the quadratic map.

of the sheet. At this point Figure 2.7 is just a visual aid illustrating how stretching and folding can occur in a dynamical system. However, observe that the resulting folded line segment resembles a horseshoe. These horseshoes were first identified and analyzed as recurring elements in nonlinear systems by the mathematician Steve Smale. We will say much more about these horseshoes, and make more extensive use of such diagrams, in section 4.8 and Chapter 5 when we try to unravel the topological organization of strange sets.

2.4 Fixed Points

A simple linear map $f : \mathbf{R} \longrightarrow \mathbf{R}$ of the real line $\mathbf{R}$ to itself is given by $f(x) = mx$. Unlike the quadratic map, this linear map can have stretching, but no folding. The graphical analysis shown in Figure 2.8 quickly convinces us that for $x_0 > 0$ only three possible asymptotic states exist, namely:

$$\lim_{n \to \infty} x_n = +\infty, \quad \text{if } m > 1,$$
$$\lim_{n \to \infty} x_n = 0, \quad \quad \text{if } m < 1, \text{ and}$$
$$x_{n+1} = x_n, \quad \quad \quad \text{for } m = 1.$$

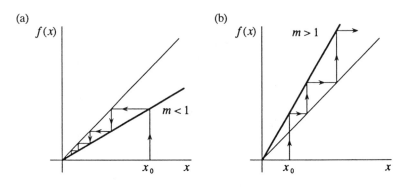

Figure 2.8: Graphical iteration of the linear map.

The period one points of a map (points that map to themselves after one iteration) are also called *fixed points*. If $m < 1$, then the origin is an *attracting fixed point* or *sink* since nearby points tend to 0 (see Figure 2.8(a)). If $m > 1$, then the origin is still a fixed point. However, because points near the origin always tend away from it, the origin is called a *repelling fixed point* or *source* (see Figure 2.8(b)). Lastly, if $m = 1$, then all initial conditions lead immediately to a period one orbit defined by $y = x$. All the periodic orbits that lie on this line have *neutral* stability.

The story for the more complicated function $f(x) = x^2$ is not much different. For this parabolic map a simple graphical analysis shows that as $n \to \infty$,

$$f^n(x) \to \infty, \quad \text{if } |x| > 1,$$
$$f^n(x) \to 0, \quad \text{if } |x| < 1,$$
$$f^n(1) = 1, \quad \text{for all } n,$$
$$f^n(-1) = 1, \quad \text{if } n \geq 1.$$

In this case all initial conditions tend to either ∞ or 0, except for the point $x = 1$, which is a repelling fixed point since all nearby orbits move away from 1. The special initial condition $x_0 = -1$ is said to be *eventually fixed* because, although it is not a fixed point itself, it goes exactly to a fixed point in a finite number of iterations. The sticking solutions of the bouncing ball system are examples of orbits that could be called eventually periodic since they arrive at a periodic orbit in a finite time.

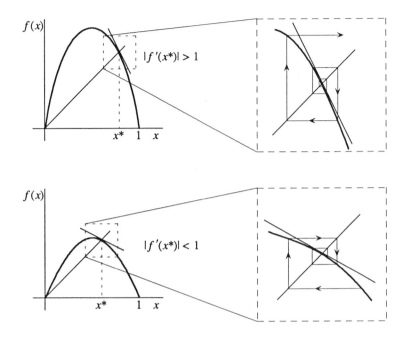

Figure 2.9: The local stability of a fixed point is determined by the slope of f at x^*.

Graphical analysis also allows us to see why certain fixed points are locally attracting and others repelling. As Figure 2.9 illustrates, the local stability of a fixed point is determined by the slope of the curve passing through the fixed point. If the absolute value of the slope is less than one—or equivalently, if the absolute value of the derivative at the fixed point is less than one—then the fixed point is locally attracting. Alternatively, if the absolute value of the derivative at the fixed point is greater then one, then the fixed point is repelling.

An orbit of a map is periodic if it repeats itself after a finite number of iterations. For instance, a point on a period two orbit has the property that $f^2(x_0) = x_0$, and a period three point satisfies $f^3(x_0) = x_0$, that is, it repeats itself after three iterations. In general a period n point repeats itself after n iterations and is a solution to the equation

$$f^n(x_0) = x_0. \tag{2.9}$$

In other words, *a period n point is a fixed point of the nth composite function of f.* Accordingly, the stability of this fixed point and of the corresponding period n orbit is determined by the derivative of $f^n(x_0)$.

Our discussion about the fixed points of a map is summarized in the following two definitions concerning fixed points, periodic points, and their stability [2]. A more rigorous account of periodic orbits and their stability in presented in section 4.5.

Definition. Let $f : \mathbf{R} \longrightarrow \mathbf{R}$. The point x_0 is a *fixed point* for f if $f(x_0) = x_0$. The point x_0 is a *periodic point* of period n for f if $f^n(x_0) = x_0$ but $f^i(x_0) \neq x_0$ for $0 < i < n$. The point x_0 is *eventually periodic* if $f^m(x_0) = f^{m+n}(x_0)$, but x_0 is not itself periodic.

Definition. A periodic point x_0 of period n is *attracting* if $|(f^n)'(x_0)| < 1$. The prime denotes differentiation with respect to x. The periodic point x_0 is *repelling* if $|(f^n)'(x_0)| > 1$. The point x_0 is *neutral* if $|(f^n)'(x_0)| = 1$.

We have just shown that the dynamics of the linear map and the parabolic map are easy to understand. By combining these two maps we arrive at the quadratic map, which exhibits complex dynamics. The quadratic map is then, in a way, the simplest map exhibiting nontrivial nonlinear behavior.

2.5   Periodic Orbits

From our definition of a period n point, namely,

$$f^n(x_0) = x_0,$$

we see that finding a period n orbit for the quadratic map requires finding the zeros for a polynomial of order 2^n. For instance, the period one orbits are given by the roots of

$$f(x) = \lambda x(1 - x) = x, \tag{2.10}$$

which is a polynomial of order 2. The period two orbits are found by evaluating

$$f^2(x) = f(f(x))$$

$$
\begin{aligned}
&= f[\lambda x(1-x)] \\
&= \lambda[\lambda x(1-x)]\left(1-[\lambda x(1-x)]\right) = x, \quad\quad (2.11)
\end{aligned}
$$

which is a polynomial of order 4. Similarly, the period three orbits are given by solving a polynomial of order 8, and so on. Unfortunately, except for small n, solving such high-order polynomials is beyond the means of both mortals and machines.

Furthermore, our definition for the stability of an orbit says that once we find a point of a period n orbit, call it x^*, we next need to evaluate the derivative of our polynomial at that point. For instance, the stability of a period one orbit is determined by evaluating

$$
f'(x^*) = \frac{d}{dx}\lambda x(1-x)\Big|_{x=x^*} = \lambda(1-2x^*). \quad\quad (2.12)
$$

Similarly, the stability of a period two orbit is determined from the equation

$$
\begin{aligned}
(f^2)'(x^*) &= \frac{d}{dx}\lambda^2 x(1-x)(1-\lambda x + \lambda x^2)\Big|_{x=x^*} \\
&= \lambda^2(1-2x^*)(1-2\lambda x^* + 2\lambda x^{*2}). \quad\quad (2.13)
\end{aligned}
$$

Again, these stability polynomials quickly become too cumbersome to analyze as n increases.

Any periodic orbit of period n will have n points in its orbit. We will generally label this collection of points by the subscript $i = 0, 1, 2, \ldots$, $n-1$, so that

$$
\mathbf{x}^* = \{x_0^*, x_1^*, x_2^*, \ldots, x_i^*, \ldots, x_{n-2}^*, x_{n-1}^*\}, \quad\quad (2.14)
$$

where i labels an individual point of the orbit. The boldface notation indicates that $\mathbf{x}^*$ is an n-tuple of real numbers. Another complication will arise: in some cases it is useful to write our indexing subscript in some base other than ten. For instance, it is useful to work in base two when studying one-humped maps. In general, it is convenient to work in base $n+1$ where n is the number of critical points of the map. It will be advantageous to label the orbits in the quadratic map according to some binary scheme.

Lastly, the question arises: which element of the periodic orbit do we use in evaluating the stability of an orbit? In Problem 2.13 we show that *all periodic points in a periodic orbit give the same value for the stability function*, $(f^n)'$ [3]. So we can use any point in the periodic sequence. This fact is good to keep in mind when evaluating the stability of an orbit.

2.5.1   Graphical Method

Although the algebra is hopeless, the geometric interpretation for the location of periodic orbits is straightforward. As we see in Figure 2.10(a), the location of the period one orbits is given by the intersection of the graphs $y = f(x)$ and $y = x$. The latter equation is simply a straight line passing through the origin with slope +1. In the case of the quadratic map, f_λ is an inverted parabola also passing through the origin. These two graphs can intersect at two distinct points, giving rise to two distinct period one orbits. One of these orbits is always at the origin and the other's exact location depends on the height of the quadratic map, that is, the specific value of λ in the quadratic map.

To find the location of the period two orbit we need to plot $y = x$ and $f^2(x)$. The graph shown in Figure 2.10(b) shows three points of intersection in addition to the origin. The middle point (the open circle) is the period one orbit found above. The two remaining intersection points are the two points belonging to a single period two orbit. A dashed line indicates where these period two points sit on the original quadratic map (the two dark circles), and the simple graphical construction of section 2.3 should convince the reader that this is, in fact, a period two orbit.

The story for higher-order orbits is the same (see Fig. 2.11(a) and (b)). The graph of the third iterate, $y = f^3(x)$, shows eight points of intersection with the straight line. Not all eight intersection points are elements of a period three orbit. Two of these points are just the pair of period one orbits. The remaining six points consist of a pair of period three orbits. The graph for the period four orbits shows sixteen points of intersection. Again, not all the intersection points are part of a period four orbit. Two intersection points are from the pair of period one orbits, and two are from the period two orbit. That leaves twelve remaining points of intersection, each of which is part of some period four orbit. Since there are twelve remaining points, there must be three

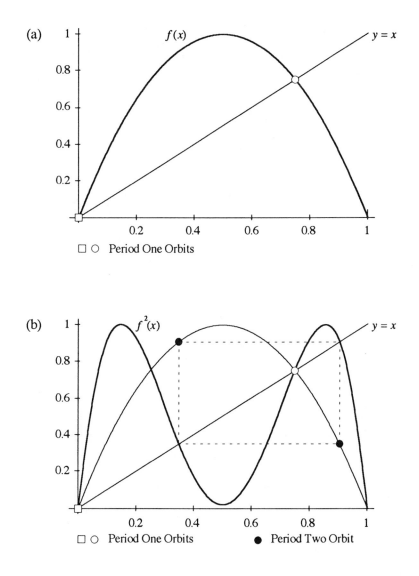

Figure 2.10: First and second iterates of the quadratic map ($\lambda = 3.98$).

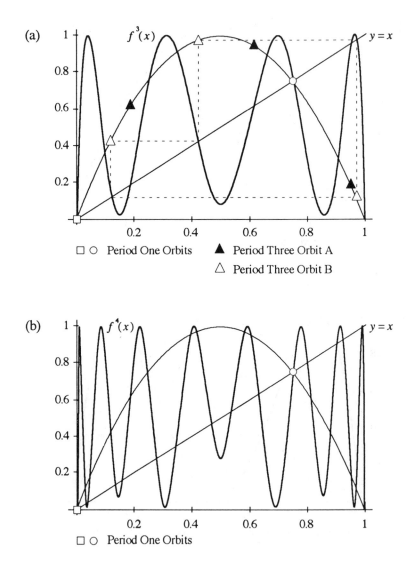

Figure 2.11: Third and fourth iterates of the quadratic map ($\lambda = 3.98$).

(12 points / 4 points per orbit) distinct period four orbits.

The number of intersection points of f_λ^n depends on λ. If $1 < \lambda < 3$ and $n \geq 2$, there are only two intersection points: the two distinct period one orbits. In dramatic contrast, if $\lambda > 4$, then it is easy to show that there will be 2^n intersection points, and counting arguments like those just illustrated allow us to determine how many of these intersection points are new periodic points of period n [4]. One fundamental question is: how can a system as simple as the quadratic map change from having only two to having an infinite number of periodic orbits? Like many aspects of the quadratic map, the answers are surprising. Before we tackle this problem, let's resume our analysis of the period one and period two orbits.

2.5.2 Period One Orbits

Solving equation (2.10) for x we find two period one solutions,[2]

$$x_0^* = 0 \tag{2.15}$$

and

$$x_1^* = 1 - \frac{1}{\lambda}. \tag{2.16}$$

The first period one orbit, labeled x_0^*, always remains at the origin, while the location of the second period one orbit, x_1^*, depends on λ. From equation (2.12), the stability of each of these orbits is determined from

$$f'(x_0^*) = \lambda \tag{2.17}$$

and

$$f'(x_1^*) = \lambda[1 - 2(1 - 1/\lambda)] = 2 - \lambda. \tag{2.18}$$

---

[2]The subscript n to x_n^* is labeling *two distinct periodic orbits*. This is potentially confusing notation since we previously reserved this subscript to label different points in the same periodic orbit. In practice this notation will not be ambiguous since this label will be a binary index, the length of which determines the period of the orbit. Different cyclic permutations of this binary index will correspond to different points on the same orbit. A noncyclic permutation must then be a point on a distinct period n orbit. The rules for this binary labeling scheme are spelled out in section 2.12.

Clearly, if $0 < \lambda < 1$ then $|f'(x_0^*)| < 1$ and $|f'(x_1^*)| > 1$, so the period one orbit x_0^* is stable and x_1^* is unstable. At $\lambda = 1$ these two orbits collide and exchange stability so that for $1 < \lambda < 3$, x_0^* is unstable and x_1^* is stable. For $\lambda > 3$, both orbits are unstable.

2.5.3   Period Two Orbit

The location of the period two orbit is found from equation (2.11),

$$x_{10}^* = \frac{1}{2\lambda}\left(1 + \lambda + \sqrt{\lambda^2 - 2\lambda - 3}\right) \tag{2.19}$$

and

$$x_{01}^* = \frac{1}{2\lambda}\left(1 + \lambda - \sqrt{\lambda^2 - 2\lambda - 3}\right). \tag{2.20}$$

These two points belong to the period two orbit. We label the left point x_{01}^* and the right point x_{10}^*. Note that the location of the period two orbit produces complex numbers for $\lambda < 3$. This indicates that the period two orbit exists only for $\lambda \geq 3$, which is obvious geometrically since $y = f_\lambda^2(x)$ begins a new intersection with the straight line $y = x$ at $\lambda = 3$.

The stability of this period two orbit is determined by rewriting equation (2.13) as

$$(f^2)'(x^*) = \lambda^2(1 - 2x_{10}^*)(1 - 2x_{01}^*), \tag{2.21}$$

where we used equations (2.19) and (2.20) for x_{10}^* and x_{01}^*. A plot of the stability for the period two orbit is presented in Figure 2.12. A close examination of this figure shows that, for $3 < \lambda < 3.45$, the absolute value of the stability function is less than one; that is, the period two orbit is stable. For $\lambda > 3.45$, the period two orbit is unstable.

The range in λ for which the period two orbit is stable can actually be obtained analytically. The period two orbit is stable as long as

$$-1 < (f^2)'(x^*) < +1. \tag{2.22}$$

The period two orbit first becomes stable when $(f^2)'(x^*) = +1$ which occurs at $\lambda = 3$, and it loses stability at $(f^2)'(x^*) = -1$ which the reader can verify takes place at $\lambda = 1 + \sqrt{6} \approx 3.449$ (see Prob. 2.17).

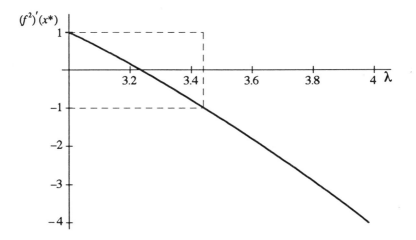

Figure 2.12: Stability of period two orbit.

2.5.4   Stability Diagram

The location and stability of the two period one orbits and the single period two orbit are summarized in the orbit stability diagram shown in Figure 2.13. The vertical axis shows the location of the periodic orbit x_n^* as a function of the parameter λ. Stable orbits are denoted by solid lines, unstable orbits by dashed lines. Two "bifurcation points" are evident in the diagram. The first occurs when the two period one orbits collide and exchange stability at $\lambda = 1$. The second occurs with the birth of a stable period two orbit from a stable period one orbit at $\lambda = 3$.

2.6   Bifurcation Diagram

To explore the dynamics of the quadratic map further, we can choose an initial condition x_0 and a parameter value λ, and then iterate the map using the program in section 2.2 to see where the orbit goes. We would notice a few general results if we play this game long enough.

First, if $\lambda \geq 1$ and $x_0 \notin [0, 1]$, then the graphical analysis of section 2.5.1 shows us that all points not in the unit interval will run off to infinity. Further, if $0 < \lambda < 1$, then the same type of graphical analysis

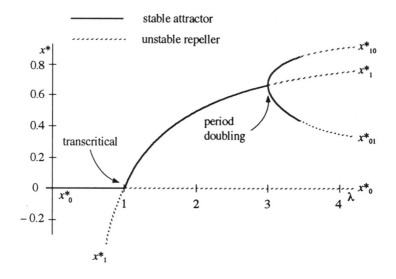

Figure 2.13: Orbit stability diagram.

shows that the dynamics of the quadratic map are simple: there is only
one attracting fixed point and one repelling fixed point. These fixed
points are the period one orbits calculated in the previous section.

Second, the initial condition we pick is usually not important in
determining the attractor, although the value of λ is very important.
We seem to end up with the same attractor no matter what $x_0 \in (0, 1)$
we pick.[3] The quadratic map usually has one, and only one, attractor,
whereas most nonlinear systems can have more than one attractor [5].
The bouncing ball system, for example, can have two or more coexisting
attractors.

Third, as we will show in section 2.11, almost all initial conditions
run off to infinity for all $\lambda > 4$. There are no attractors in this case.

Therefore, when studying the quadratic map, it will usually suffice
to pick a single initial condition from the unit interval. If $f^n(x_0)$ ever

---

[3]Some initial conditions do not converge to the attractor. For instance, any x
belonging to an unstable periodic orbit will not converge to the attractor. Unstable
orbits are, by definition, not attractors, so that almost any orbit near an unstable
periodic orbit will diverge from it and head toward some attractor.

leaves the unit interval, then it will run off to infinity and never return (provided $\lambda \geq 1$). Further, when studying attractors we can limit our attention to values of $\lambda \in [1, 4]$. If $0 < \lambda < 1$ then the only attractor is a stable fixed point at zero, and if $\lambda > 4$ there are no attractors.[4]

For all these reasons, a bifurcation diagram is a particularly powerful method for studying the attractors in the quadratic map. Recall that a bifurcation diagram is a plot of an asymptotic solution on the vertical axis and a control parameter on the horizontal axis. To construct a bifurcation diagram for the quadratic map only requires some simple modifications of our previous program for iterating the quadratic map. As seen below, the new algorithm consists of the following steps:

1. Set $\lambda = 1$, and $x_0 = 0.1$ (almost any x_0 will do);

2. Iterate the quadratic map 200 times to remove the transient solution, and then print λ and x_n for the next 200 points, which are presumably part of the attractor;

3. Increment λ by a small amount, and set x_0 to the last value of x_n;

4. Repeat steps 2 and 3 until $\lambda = 4$.

A C program implementing this algorithm is as follows.

```
/* bifquad.c:  calculate bifurcation diagram for the quadratic map.
   input:      (none)
   output:     l1    x200
               l1    x201
               etc.,
               l2    x200
               l2    x201
               etc.
*/
#include <stdio.h>

main()
{
    int n;
    float lambda, x_n;
```

---

[4]Technically, the phase space of the quadratic map, **R**, can be compactified thereby making the point at infinity a valid attractor.

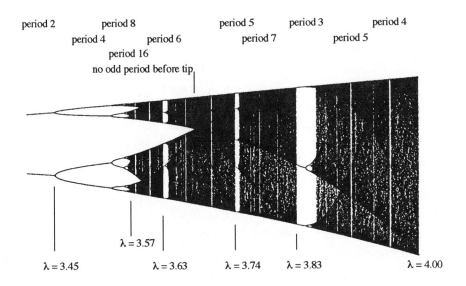

period 2   period 8   period 5   period 3   period 4

period 4   period 6   period 7   period 5

period 16

no odd period before tip

λ = 3.57

λ = 3.45   λ = 3.63   λ = 3.74   λ = 3.83   λ = 4.00

Figure 2.14: Bifurcation diagram for the quadratic map.

```
x_n = 0.1;
for(lambda = 1; lambda <= 4; lambda += 0.01) {
    for(n = 0; n <= 400; ++n) {
        x_n = lambda * x_n * (1 - x_n);
        if(n > 199)
            printf("%f %f\n", lambda, x_n);
    }
}
}
```

When plotted in Figure 2.14 (for $3.4 \leq \lambda \leq 4$), the output of our simple program produces a bifurcation diagram of stunning complexity. Above the diagram we provide comments on the type of attractor observed, and on the horizontal axis significant parameter values are indicated. This bifurcation diagram shows many qualitative similarities to bifurcation diagrams from the bouncing ball system (compare to Figure 1.16). Both exhibit the period doubling route to chaos. For the quadratic map an infinite number of period doublings occur for $1 < \lambda < 3.57$. Both also show periodic windows (white bands) within the chaotic regions. For the quadratic map a period three window begins at $\lambda \approx 3.83$ and a period five window begins at $\lambda \approx 3.74$. Looking closely at the periodic windows we see that each branch of these periodic windows also undergoes a period doubling cascade.

Bifurcation diagrams showing only attracting solutions can be somewhat misleading. Much of the structure in the bifurcation diagram can only be understood by keeping track of both the stable attracting solutions and the unstable repelling solutions, as we did in constructing the orbit stability diagram (Figure 2.13). Just as there are stable periodic orbits and chaotic attractors, there are also unstable periodic orbits and chaotic repellers. In Figure 2.13, for instance, we indicate the existence of an unstable period one orbit by the dashed line beginning at the first period doubling bifurcation. As we show in section 2.7.2, this unstable period one orbit is simply the continuation of the stable period one orbit that exists before this period doubling bifurcation (see Figures 2.14 and 2.22).

2.7 Local Bifurcation Theory

Poincaré used the term bifurcation to describe the "splitting" of asymptotic states of a dynamical system. Figure 2.14 shows a bifurcation diagram for the quadratic map (see Plate 2 for a color version of a quadratic map bifurcation diagram). As we examine Figure 2.14, we see that several different types of changes can occur. We would like to analyze and classify these bifurcations. At a bifurcation value, the qualitative nature of the solution changes. It can change to, or from, an equilibrium, periodic, or chaotic state. It can change from one type of periodic state to another, or from one type of chaotic state to another.

For instance, in the bouncing ball system we are initially in an equilibrium state, with the ball moving in unison with the table. As we turn up the table amplitude we first find sticking solutions. As we increase the amplitude further, we find a critical parameter value at which the ball switches from the sticking behavior to bouncing in a period one orbit. Such a change from an equilibrium state to a periodic state is an example of a saddle-node bifurcation. As we further turn up the table amplitude, we find that there is a second critical table amplitude at which the ball switches from a period one to a period

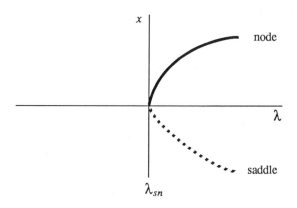

Figure 2.15: Saddle-node bifurcation diagram.

two orbit. The analogous period doubling bifurcation in the quadratic map occurs at $\lambda = 3$. Both the saddle-node and the period doubling bifurcations are examples of local bifurcations. At their birth (or death) all the orbits participating are localized in phase space; that is, they all start out close together. Global bifurcations can also occur, although typically these are more difficult to analyze since they can give birth to an infinite number of periodic orbits. In this section we analyze three simple types of local bifurcations that commonly occur in nonlinear systems. These are the saddle-node, period doubling, and transcritical bifurcations [6].

2.7.1   Saddle-node

In a *saddle-node bifurcation* a pair of periodic orbits are created "out of nothing." One of the periodic orbits is always unstable (the saddle), while the other periodic orbit is always stable (the node). The basic bifurcation diagram for a saddle-node bifurcation looks like that shown in Figure 2.15. The saddle-node bifurcation is fundamental to the study of nonlinear systems since it is one of the most basic processes by which periodic orbits are created.

A saddle-node bifurcation is also referred to as a *tangent bifurcation* because of the mechanism by which the orbits are born. Consider the nth composite of some mapping function f_λ which is near to a tangency

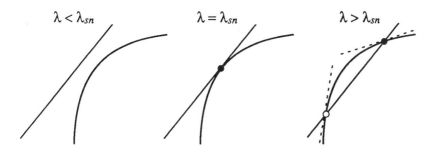

Figure 2.16: Tangency mechanism for a saddle-node bifurcation.

with the line $y = x$. Let λ_{sn} be the value at which a saddle-node bifurcation occurs. Notice in Figure 2.16 that f_λ^n is tangent to the line $y = x$ at λ_{sn}. For $\lambda < \lambda_{sn}$, no period n orbits exist in this neighborhood, but for $\lambda > \lambda_{sn}$ two orbits are born. The local stability of a point of a map is determined by $(f_\lambda^n)'$. Since $f_\lambda^n(x)$ is tangent to $y = x$ at a bifurcation, it follows that at λ_{sn},[5]

$$(f_{\lambda_{sn}}^n)'(x^*) = +1. \qquad (2.23)$$

Tangent bifurcations abound in the quadratic map. For instance, a pair of period three orbits are created by a tangent bifurcation in the quadratic map when $\lambda = 1 + \sqrt{8} \approx 3.828$. As illustrated in Figure 2.17 for $\lambda > 3.83$, there are eight points of intersection. Two of the intersection points belong to the period one orbits, while the remaining six make up a pair of period three orbits. Near to tangency, the absolute value of the slope at three of these points is greater than one—this is the unstable period three orbit. The remaining three points form the stable periodic orbit. The birth of this stable period three orbit is clearly visible as the period three window in our numerically constructed bifurcation diagram of the quadratic map, Figure 2.14. In fact, all the odd-period orbits of the quadratic map are created by some sort of tangent bifurcation.

---

[5]See reference [5] for more details.

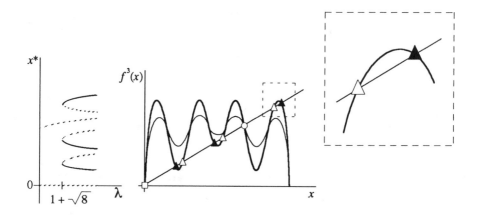

Figure 2.17: A pair of period three orbits created by a tangent bifurcation in the quadratic map (shown with the unstable period one orbits).

2.7.2   Period Doubling

Period doubling bifurcations are evident when we consider an even number of compositions of the quadratic map. In Figure 2.18 we show the second iteration of the quadratic map near a tangency. Below the period doubling bifurcation, a single stable period one orbit exists. As λ is increased, the period one orbit becomes unstable, and a stable period two orbit is born. This information is summarized in the bifurcation

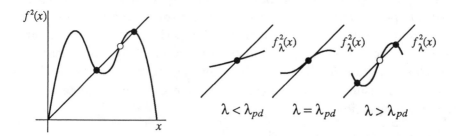

Figure 2.18: Second iterate of the quadratic map near a tangency.

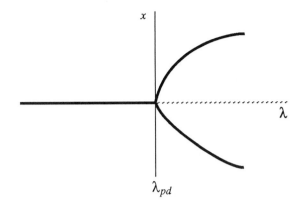

Figure 2.19: Period doubling (flip) bifurcation diagram.

diagram presented in Figure 2.19. Let λ_{pd} be the parameter value at which the period doubling bifurcation occurs. At this parameter value the period one and the nascent period two orbit coincide. As illustrated in Figure 2.18, $f^2_{\lambda_{pd}}(x^*)$ is tangent to $y = x$ so that $(f^2_{\lambda_{pd}})'(x^*) = +1$. However, $(f^n_{\lambda_{pd}})'(x^*) = -1$; that is, at a period doubling bifurcation the function determining the local stability of the periodic orbit is always -1. Figure 2.20 shows f'_λ just after period doubling. In general, for a period n to period $2n$ bifurcation, $(f^{2n}_{\lambda_{pd}})'(x^*) = +1$ and

$$(f^n_{\lambda_{pd}})'(x^*) = -1. \tag{2.24}$$

A period doubling bifurcation is also known as a *flip bifurcation*. In the period one to period two bifurcation, the period two orbit flips from side to side about its period one parent orbit. This is because $f'_{\lambda_{pd}}(x^*) = -1$ (see Prob. 2.14). The first flip bifurcation in the quadratic map occurs at $\lambda = 3$ and was analyzed in sections 2.5.2–2.5.4, where we considered the location and stability of the period one and period two orbits in the quadratic map.

2.7.3 Transcritical

The last bifurcation we illustrate with the quadratic map is a *transcritical bifurcation*, in which an unstable and stable periodic orbit collide and exchange stability. A transcritical bifurcation occurs in the

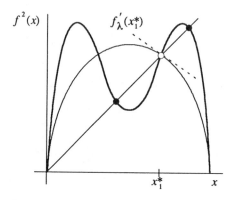

Figure 2.20: Tangency mechanism near a period doubling (flip) bifurcation.

quadratic map when $\lambda = \lambda_{tc} = 1$. As in a saddle-node bifurcation, $f'_{\lambda_{tc}} = +1$ at a transcritical bifurcation. However, a transcritical bifurcation also has an additional constraint not found in a saddle-node bifurcation, namely,

$$f_{\lambda_{tc}}(x^*) = 0. \tag{2.25}$$

For the quadratic map this fixed point is just the period one orbit at the origin, $x_0^* = 0$, found from equation (2.10).

A summary of these three types of bifurcations is presented in Figure 2.21. Other types of local bifurcations are possible; a more complete theory for both maps and flows is given in reference [7].

2.8   Period Doubling Ad Infinitum

A view of the bifurcation diagram for the quadratic map for λ between 2.95 and 4.0 is presented in Figure 2.22. This diagram reveals not one, but rather an infinite number of period doubling bifurcations. As λ is increased a period two orbit becomes a period four orbit, and this in turn becomes a period eight orbit, and so on. This sequence of period doubling bifurcations is known as a *period doubling cascade*. This process appears to converge at a finite value of λ around 3.57, beyond which a nonperiodic motion appears to exist. This period doubling

Bifurcation Diagram   **Mechanism**

Saddle-node (tangent)
Derivative:

$(f_{\lambda_{sn}}^n)'(x^*) = +1$

Birth of a pair of orbits,
one stable and one unstable.

Period doubling (flip)
Derivative:

$(f_{\lambda_{pd}}^n)'(x^*) = -1$

Birth of a stable period $2n$
orbit from a period n orbit.
The period n orbit becomes
unstable.

Transcritical
Derivative:

$f_{\lambda_{tc}}'(x^*) = +1$

$f_{\lambda_{tc}}'(x^*) = 0$

Collision of two periodic
orbits and exchange of
stability.

fixed point

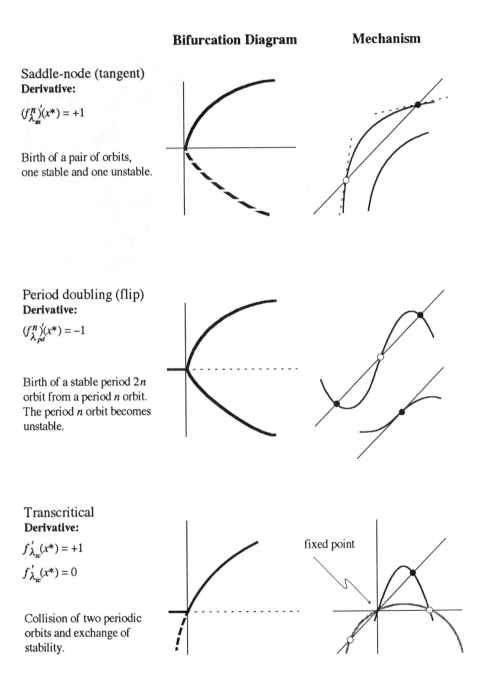

Figure 2.21: Summary of bifurcations.

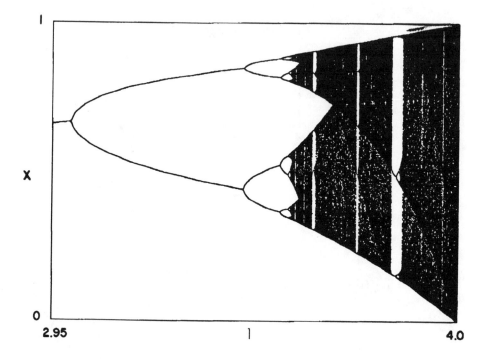

Figure 2.22:   Bifurcation diagram showing period doubling in the quadratic map.

cascade often occurs in nonlinear systems. For instance, a similar period doubling cascade occurs in the bouncing ball system (Figure 1.16). The period doubling route is one common way, but certainly not the only way, by which a nonlinear system can progress from a simple behavior (one or a few periodic orbits) to a complex behavior (chaotic motion and the existence of an infinity of unstable periodic orbits).

In 1976, Feigenbaum began to wonder about this period doubling cascade. He started playing some numerical games with the quadratic map using his HP65 hand-held calculator. His wondering soon led to a remarkable discovery. At the time, Feigenbaum knew that this period doubling cascade occurred in one-dimensional maps of the unit interval. He also had some evidence that it occurred in simple systems of nonlinear differential equations that model, for instance, the motion of a forced pendulum. In addition to looking at the qualitative similarities

$\lambda_1 = 3.0$	$\lambda_5 = 3.568759\ldots$
$\lambda_2 = 3.449490\ldots$	$\lambda_6 = 3.569692\ldots$
$\lambda_3 = 3.544090\ldots$	$\lambda_7 = 3.569891\ldots$
$\lambda_4 = 3.564407\ldots$	$\lambda_8 = 3.569934\ldots$

Table 2.1: Period doubling bifurcation values for the quadratic map.

between these systems, he began to ask if there might be some *quantitative* similarity—that is, some numbers that might be the same in all these different systems exhibiting period doubling. If these numbers could be found, they would be "universal" in the sense that they would not depend on the specific details of the system.

Feigenbaum was inspired in his search, in part, by a very successful theory of universal numbers for second-order phase transitions in physics.[6] A phase transition takes place in a system when a change of state occurs. During the 1970s it was discovered that there were quantitative measurements characterizing phase transitions that did not depend on the details of the substance used. Moreover, these universal numbers in the theory of phase transitions were successfully measured in countless experiments throughout the world. Feigenbaum wondered if there might be some similar universality theory for dissipative nonlinear systems [8].

By definition, such universal numbers are dimensionless; the specific mechanical details of the system must be scaled out of the problem. Feigenbaum began his search for universal numbers by examining the period doubling cascade in the quadratic map. He recorded, with the help of his calculator, the values of λ at which the first few period doubling bifurcations occur. We have listed the first eight values (orbits up to period 2^8) in Table 2.1. While staring at this sequence of bifurcation points, Feigenbaum was immediately struck by the rapid convergence

---

[6]Feigenbaum introduced the renormalization group approach of critical phenomena to the study of nonlinear dynamical systems. Additional early contributions to these ideas came from Cvitanović, and also Collet, Coullet, Eckmann, Lanford, and Tresser. The geometric convergence of the quadratic map was noted as early as 1958 by Myrberg (see C. Mira, *Chaotic dynamics* (World Scientific: New Jersey, 1987)), and also by Grossmann and Thomae in 1977.

of this series. Indeed, he recognized that the convergence appears to follow that of a *geometric series*, similar to the one we saw in equation (1.35) when we studied the sticking solutions of the bouncing ball.

Let λ_n be the value of the nth period doubling bifurcation, and define λ_∞ as $\lim_{n\to\infty} \lambda_n$. Based on his inspiration, Feigenbaum guessed that this sequence obeys a geometric convergence,[7]; that is,

$$\lambda_\infty - \lambda_n = c/\delta^n \ (n \to \infty), \tag{2.26}$$

where c is a constant, and δ is a constant greater than one. Using equation (2.26) and a little algebra it follows that if we define δ by

$$\delta = \lim_{n\to\infty} \frac{\lambda_n - \lambda_{n-1}}{\lambda_{n+1} - \lambda_n}, \tag{2.27}$$

then δ is a dimensionless number characterizing the rate of convergence of the period doubling cascade.

The three constants in this discussion have been calculated as

$$\lambda_\infty = 3.5699456..., \ \delta = 4.669202..., \ \text{and} \ c = 2.637.... \tag{2.28}$$

The constant δ is now called "Feigenbaum's delta," because Feigenbaum went on to show that this number is universal in that it arises in a wide class of dissipative nonlinear systems that are close to the single-humped map. This number has been measured in experiments with chicken hearts, electronic circuits, lasers, chemical reactions, and liquids in their approach to a turbulent state, as well as the bouncing ball system [9].

To experimentally estimate Feigenbaum's delta all one needs to do is measure the parameter values of the first few period doublings, and then substitute these numbers into equation (2.27). The geometric convergence of δ is a mixed blessing for the experimentalist. In practice it means that δ_n converges very rapidly to δ_∞, so that only the first few λ_n's are needed to get a good estimate of Feigenbaum's delta. It also means that only the first few δ_n's can be experimentally measured with any accuracy, since the higher δ_n's bunch up too quickly to δ_∞.

---

[7]For a review of geometric series see any introductory calculus text, such as C. Edwards and D. Penny, *Calculus and analytic geometry* (Prentice-Hall: Englewood Cliffs, NJ, 1982), p. 549.

To continue with more technical details of this story, see Rasband's account of renormalization theory for the quadratic map [6].

Feigenbaum's result is remarkable in two respects. Mathematically, he discovered a simple universal property occurring in a wide class of dynamical systems. Feigenbaum's discovery is so simple and fundamental that it could have been made in 1930, or in 1830 for that matter. Still, he had some help from his calculator. It took a lot of numerical work to develop the intuition that led Feigenbaum to his discovery, and it seems unlikely that the computational work needed would have occurred without help from some sort of computational device such as a calculator or computer. Physically, Feigenbaum's result is remarkable because it points the way toward a theory of nonlinear systems in which complicated differential equations, which even the fastest computers cannot solve, are replaced by simple models—such as the quadratic map—which capture the essence of a nonlinear problem, including its solution. The latter part of this story is still ongoing, and there are surely other gems to be discovered with some inspiration, perspiration, and maybe even a workstation.

2.9 Sarkovskii's Theorem

In the previous section we saw that for $3 < \lambda < 3.57$ an infinite number of periodic orbits with period 2^n are born in the quadratic map. In a period doubling cascade we know the sequence in which these periodic orbits are born. A period one orbit is born first, followed by a period two orbit, a period four orbit, a period eight orbit, and so on. For higher values of λ, additional periodic orbits come into existence. For instance, a period three orbit is born when $\lambda = 1 + \sqrt{8} \approx 3.828$, as we showed in section 2.7.1. In this section, we will explicitly show that all possible periodic orbits exist for $\lambda \geq 4$. One of the goals of bifurcation theory is to understand the different mechanisms for the birth and death of these periodic orbits. Pinning down all the details of an individual problem is usually very difficult, often impossible. However, there is one qualitative result due to Sarkovskii of great beauty that applies to any continuous mapping of the real line to itself.

The positive integers are usually listed in increasing order

$1, 2, 3, 4, \ldots$. However, let us consider an alternative enumeration that reflects the order in which a sequence of period n orbits is created. For instance, we might list the sequence of integers of the form 2^n as

$$2^n \triangleright \cdots \triangleright 2^4 \triangleright 2^3 \triangleright 2^2 \triangleright 2^1 \triangleright 2^0,$$

where the symbol $\triangleright$ means "implies." In the quadratic map system this ordering says that the existence of a period 2^n orbit implies the existence of all periodic orbits of period 2^i for $i < n$. We saw this ordering in the period doubling cascade. A period eight orbit thus implies the existence of both period four and period two orbits. This ordering diagram says nothing about the stability of any of these orbits, nor does it tell us how many periodic orbits there are of any given period.

Consider the ordering of *all* the integers given by

$$3 \triangleright 5 \triangleright 7 \triangleright 9 \triangleright \ldots \triangleright 2 \cdot 3 \triangleright 2 \cdot 5 \triangleright 2 \cdot 7 \triangleright 2 \cdot 9 \triangleright \ldots$$
$$\triangleright 2^n \cdot 3 \triangleright 2^n \cdot 5 \triangleright 2^n \cdot 7 \triangleright 2^n \cdot 9 \triangleright \ldots$$
$$\triangleright 2^n \triangleright \ldots \triangleright 16 \triangleright 8 \triangleright 4 \triangleright 2 \triangleright 1, \quad (2.29)$$

with $n \to \infty$. Sarkovskii's theorem says that the ordering found in equation (2.29) holds, in the sense of the 2^n ordering above, for any continuous map of the real line R to itself—the existence of a period i orbit implies the existence of all periodic orbits of period j where j follows i in the ordering. Sarkovskii's theorem is remarkable for its lack of hypotheses (it assumes only that f is continuous). It is of great help in understanding the structure of one-dimensional maps.

In particular, this ordering holds for the quadratic map. For instance, the existence of a period seven orbit implies the existence of all periodic orbits except a period five and a period three orbit. And the existence of a single period three orbit implies the existence of periodic orbits of all possible periods for the one-dimensional map. Sarkovskii's theorem forces the existence of period doubling cascades in one-dimensional maps. It is also the basis of the famous statement of Li and Yorke that "period three implies chaos," where chaos loosely means the existence of all possible periodic orbits.[8] An elementary

---

[8]In section 4.6.2 we show there exists a close connection between the existence

proof of Sarkovskii's theorem, as well as a fuller mathematical treatment of maps as dynamical systems, is given by Devaney in his book *An Introduction to Chaotic Dynamical Systems* [10].

Sarkovskii's theorem holds only for mappings of the real line to itself. It does not hold in the bouncing ball system because it is a map in two dimensions. It does not hold for mappings of the circle, S^1, to itself. Still, Sarkovskii's theorem is a lovely result, and it does point the way to what might be called "qualitative universality," that is, general statements, usually topological in nature, that are expected to hold for a large class of dynamical systems.

2.10 Sensitive Dependence

In section 1.4.5 we saw how a measurement of finite precision in the bouncing ball system has little predictive value in the long term. Such behavior is typical of motion on a chaotic attractor. We called such behavior sensitive dependence on initial conditions. For the special value $\lambda = 4$ in the quadratic map we can analyze this behavior in some detail.

Consider the transformation $x_n = \sin^2(\pi\theta_n)$ applied to the quadratic map when $\lambda = 4$. Making use of the identity

$$\sin 2\alpha = 2\sin\alpha\cos\alpha,$$

we find

$$
\begin{aligned}
x_{n+1} &= 4x_n(1 - x_n) \Longrightarrow \\
\sin^2(\pi\theta_{n+1}) &= 4\sin^2(\pi\theta_n)(1 - \sin^2(\pi\theta_n)) \\
&= 4\sin^2(\pi\theta_n)\cos^2(\pi\theta_n) \\
&= (2\sin(\pi\theta_n)\cos(\pi\theta_n))^2 \\
&= (\sin(2\pi\theta_n))^2 \Longrightarrow \\
\theta_{n+1} &= 2\theta_n \bmod 1. \quad (2.30)
\end{aligned}
$$

of an infinity of periodic orbits and the existence of a chaotic invariant set, not necessarily an attractor. The term "chaos" in nonlinear dynamics is due to Li and Yorke, although the current usage differs somewhat from their original definition (see T. Y. Li and J. A. Yorke, Period three implies chaos, Am. Math. Monthly **82**, 985 (1975)).

This last linear difference equation has the explicit solution

$$\theta_n = 2^n \theta_0 \bmod 1. \tag{2.31}$$

Sensitive dependence on initial conditions is easy to see in this example when we express the initial condition as a binary number,

$$\theta_0 = \frac{b_0}{2} + \frac{b_1}{4} + \frac{b_2}{8} + \cdots = \sum_{i=0}^{\infty} \frac{b_i}{2^{i+1}}, \quad b_i \in \{0,1\}. \tag{2.32}$$

Now the action of equation (2.31) on an initial condition θ_0 is a shift map. At each iteration we multiply the previous iterate by two (10 in binary), which is a left shift, and then apply the mod function, which erases the integer part. For example, if $\theta_0 = 0.10110101\ldots$ in binary, then

$$
\begin{aligned}
\theta_0 &= 0.10110101\ldots \\
\theta_1 &= (10 \cdot 0.10110101\ldots) \bmod 1 \\
&= 0.0110101\ldots \qquad \text{(shift left and drop the integer part)} \\
\theta_2 &= 0.110101\ldots \\
\theta_3 &= 0.10101\ldots \\
\theta_4 &= 0.0101\ldots \\
&\vdots
\end{aligned}
$$

and we see the precision of our initial measurement evaporating before our eyes.

We can even quantify the amount of sensitive dependence the system exhibits, that is, the rate at which an initial error grows. Assuming that our initial condition has some small error ϵ, the growth rate of the error is

$$f^n(\theta_0) - f^n(\theta_0 + \epsilon) = 2^n \theta_0 - 2^n(\theta_0 + \epsilon) = 2^n \epsilon = \epsilon e^{n(\ln 2)}.$$

If we think of n as time, then the previous equation is of the form e^{at} with $a = \ln 2$. In this example the error grows at a constant *exponential rate* of $\ln 2$. The exponential growth rate of an initial error is the defining characteristic of motion on a chaotic attractor. This rate of growth is called the *Lyapunov exponent*. A strictly positive Lyapunov exponent, such as we just found, is an indicator of chaotic motion.

The Lyapunov exponent is never strictly positive for a stable periodic motion.[9]

2.11 Fully Developed Chaos

The global dynamics of the quadratic map are well understood for $0 < \lambda < 3$, namely, almost all orbits beginning on the unit interval are asymptotic to a period one fixed point. We will next show that the orbit structure is also well understood for $\lambda > 4$. This is known as the hyperbolic regime. This parameter regime is "fully developed" in the sense that all of the possible periodic orbits exist and they are all unstable.[10] No chaotic attractor exists in this parameter regime, but rather a *chaotic repeller*. Almost all initial conditions eventually leave, or are repelled from, the unit interval. However, a small set remains. This remaining invariant set is an example of a fractal.

The analysis found in this book is based substantially on sections 1.5 to 1.8 of Devaney's *An Introduction to Chaotic Dynamical Systems* [10]. This section is more advanced mathematically than previous sections. The reader should consult Devaney's book for a complete treatment. Section 2.12 contains a more pragmatic description of the symbolic dynamics of the quadratic map and can be read independently of the current section.

2.11.1 Hyperbolic Invariant Sets

We begin with some definitions.

Definition. A set or region Γ is said to be *invariant* under the map f if for any $x_0 \in \Gamma$ we have $f^n(x_0) \in \Gamma$ for all n.

The simplest example of an invariant set is the collection of points

---

[9]For a well-illustrated exploration of the Lyapunov exponent in the quadratic map system see A. K. Dewdney, Leaping into Lyapunov space, Sci. Am. **265** (3), pp. 178–180 (1991).

[10]Technically, the system is "structurally stable." See section 1.9 of Devaney's book for more details. Hyperbolicity and structural stability usually go hand-in-hand.

forming a periodic orbit. But, as we will see shortly, there are more complex examples, such as strange invariant sets, which are candidates for chaotic attractors or repellers.

Definition. For mappings of $\mathbf{R} \longrightarrow \mathbf{R}$, a set $\Gamma \subset \mathbf{R}$ is a repelling (resp., attracting) *hyperbolic set* for f if Γ is closed, bounded, and invariant under f and there exists an $N > 0$ such that $|(f^n)'(x)| > 1$ (resp., < 1) for all $n \geq N$ and all $x \in \Gamma$ [10].

This definition says that none of the derivatives of points in the invariant set are exactly equal to one. A simple example of a hyperbolic invariant set is a periodic orbit that is either repelling or attracting, but not neutral. In higher dimensions a similar definition of hyperbolicity holds, namely, all the points in the invariant set are saddles.

The existence of both a simple periodic regime and a complicated fully developed chaotic (yet well understood) hyperbolic regime turns out to be quite common in low-dimensional nonlinear systems. In Chapter 5 we will show how information about the hyperbolic regime, which we can often analyze in detail using symbolic dynamics, can be exploited to determine useful physical information about a nonlinear system.

In examining the dynamics of the quadratic map for $\lambda > 4$ we proceed in two steps: first, we examine the invariant set, and second, we describe how orbits meander on this invariant set. The set itself is a fractal Cantor set [11], and to describe the dynamics on this fractal set we employ the method of symbolic dynamics.

Since $f(1/2) > 1$ for $\lambda > 4$ there exists an open interval centered at $1/2$ with points that leave the unit interval after one iteration, never to return. Call this open set A_0 (see Figure 2.23). These are the points in A_0 whose image under f is greater than one. On the second iteration, more points leave the unit interval. In fact, these are the points that get mapped to A_0 after the first iteration: $A_1 = \{x \in I | f(x) \in A_0\}$. Inductively, define $A_n = \{x \in I | f^i(x) \in I \text{ for } i \leq n \text{ but } f^{n+1}(x) \notin I\}$; that is, A_n consists of all points that escape from I at the $n + 1$st iteration. Clearly, the invariant limit set, call it Λ, consists of all the

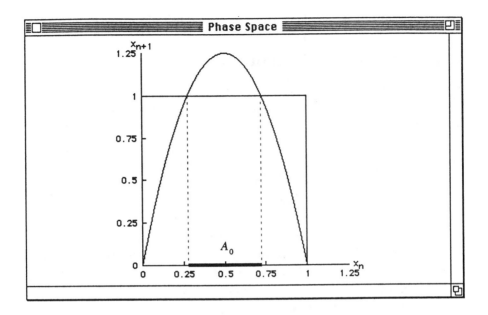

Figure 2.23: Quadratic map for $\lambda > 4$. (Generated by the Quadratic Map program.)

remaining points

$$\Lambda = I - \left(\bigcup_{n=0}^{\infty} A_n \right). \tag{2.33}$$

What does Λ look like? First, note that A_n consists of 2^n disjoint open intervals, so $\Lambda_n = I - (A_0 \cup .... \cup A_n)$ is 2^{n+1} disjoint closed intervals. Second, f^{n+1} monotonically maps each of these intervals onto I. The graph of f^{n+1} is a polynomial with 2^n humps. The maximal sections of the humps are the collection of intervals A_n that get mapped out of I, but more importantly this polynomial intersects the $y = x$ line 2^{n+1} times. Thus, Λ_n has 2^{n+1} periodic points in I.

The set Λ is a *Cantor set* if Λ is a closed, totally disconnected, perfect subset. A set is disconnected if it contains no intervals; a set is perfect if every point is a limit point. It is not too hard to show that the invariant set defined by equation (2.33) is a Cantor set [10]. Thus, we see that the invariant limit set arising from the quadratic map for

$\lambda > 4$ is a fractal Cantor set with a countable infinity of periodic orbits.

2.11.2   Symbolic Dynamics

Our next goal is to unravel the dynamics on Λ. In beginning this task it is useful to think how the unit interval gets stretched and folded with each iteration. The transformation of the unit interval under the first three iterations for $\lambda = 4$ is illustrated in Figure 2.24. This diagram shows that the essential ingredients that go into making a chaotic limit set are stretching and folding. The technique of symbolic dynamics is a bookkeeping procedure that allows us to systematically follow this stretching and folding process. For one-dimensional maps the complete symbolic theory is also known as *kneading theory* [10].

We begin by defining a symbol space for symbolic dynamics. Let $\Sigma_2 = \{\mathbf{s} = (s_0 s_1 s_2 ...) | s_j = 0 \text{ or } 1\}$. Σ_2 is known as the *sequence space* on the symbols 0 and 1. We sometimes use the symbols L (Left) and R (Right) to denote the symbols 0 and 1 (see Figure 2.24). If we define the distance between two sequences $\mathbf{s}$ and $\mathbf{t}$ by

$$d[\mathbf{s}, \mathbf{t}] = \sum_{i=0}^{\infty} \frac{|s_i - t_i|}{2^i}, \qquad (2.34)$$

then Σ_2 is a metric space. The metric $d[\mathbf{s}, \mathbf{t}]$ induces a topology on Σ_2 so we have a notion of open and closed sets in Σ_2. For instance, if $\mathbf{s} = (0100101...)$ and $\mathbf{t} = (001011...)$, then the metric $d[\mathbf{s}, \mathbf{t}] = 1/2 + 1/4 + 1/32 + \cdots$.

A dynamic on the space Σ_2 is given by the shift map $\sigma : \Sigma_2 \longrightarrow \Sigma_2$ defined by $\sigma(s_0 s_1 s_2 ...) = (s_1 s_2 s_3 ...)$. That is, the shift map drops the first entry and moves all the other symbols one place to the left. The shift map is continuous. Briefly, for any $\epsilon > 0$, pick n such that $1/2^n < \epsilon$, and let $\delta = 1/2^{n+1}$. Then the usual $\epsilon - \delta$ proof goes through when we use the metric given by equation (2.34) [10].

What do the orbits in Σ_2 look like? Periodic points are identified with exactly repeating sequences, $\mathbf{s} = (s_0 ... s_{n-1}, s_0 ... s_{n-1}, ...)$. For instance, there are two distinct period one orbits, given by $(0000000...)$ and $(111111...)$. The period two orbit takes the form $(01010101...)$ and $(10101010...)$, and one of the period three orbits looks like

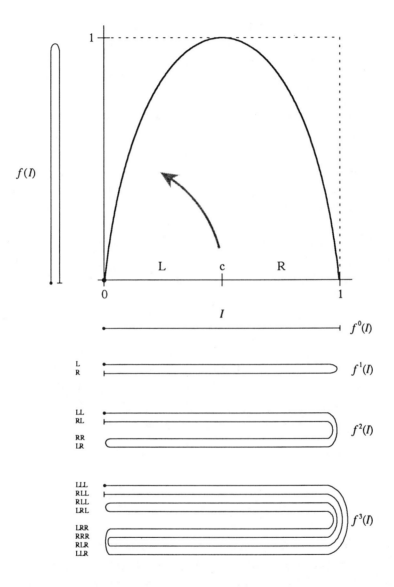

Figure 2.24: Keeping track of the stretching and folding of the quadratic map with symbolic dynamics.

(001001001...), (010010010...), and (100100100...), and so on. Evidently, there are 2^n periodic points of period n, although some of these points are of a lower period. But there is more. The periodic points are dense in Σ_2; that is, any nonperiodic point can be represented as the limit of some periodic sequence. Moreover, the nonperiodic points greatly outnumber the periodic points.

What does this have to do with the quadratic map, or more exactly the map f_λ restricted to the invariant set Λ? We now show that it is the "same" map, and thus to understand the orbit structure and dynamics of f_λ on Λ we need only understand the shift map, σ, on the space of two symbols, Σ_2. We can get a rough idea of the behavior of an orbit by keeping track of whether it falls to the left (L or 0) or right (R or 1) at the nth iteration. See Figure 2.25 for a picture of this partition. That is, the symbols 0 and 1 tell us the fold which the orbit lies on at the nth iteration.

Accordingly, define the *itinerary* of x as the sequence $S(x) = s_0 s_1 s_2 ...$ where $s_j = 0$ if $f_\lambda^j(x) < 1/2$ and $s_j = 1$ if $f_\lambda^j(x) > 1/2$. Thus, the itinerary of x is an infinite sequence of 0s and 1s: it "lives" in Σ_2. Further, we think of S as a map from Λ to Σ_2. If $\lambda > 4$, then it can be shown that $S : \Lambda \longrightarrow \Sigma_2$ is a *homeomorphism* (a map is a homeomorphism if it is a bijection and both f and f^{-1} are continuous). This last result says that the two sets Λ and Σ_2 are the same. To show the equivalence between the dynamics of f_λ on Λ and σ on Σ_2, we need the following theorem, which is quoted from Devaney.

Theorem. If $\lambda > 2 + \sqrt{5}$, then $S : \Lambda \longrightarrow \Sigma_2$ is a homeomorphism and $S \circ f_\lambda = \sigma \circ S$.

Proof. See section 1.7 in Devaney's book [10]. This theorem holds for all $\lambda > 4$, but the proof is more subtle.

As we show in the next section, the essential idea in this proof is to keep track of the preimages of points not mapped out of the unit interval. The symbolic dynamics of the invariant set gives us a way to uniquely name the orbits in the quadratic map that do not run off to infinity. In particular, the itinerary of an orbit allows us to name, and to find the relative location of, all the periodic points in the quadratic map. Symbolic dynamics is powerful because it is easy to keep track of the orbits in the symbol space. It is next to impossible to do this using only the quadratic map since it would involve solving polynomials of arbitrarily high order.

2.11.3   Topological Conjugacy

This last example suggests the following notion of equivalence of dynamical systems, which was originally put forth by Smale [12] and is fundamental to dynamical systems theory.

Definition. Let $f : A \longrightarrow A$ and $g : B \longrightarrow B$ be two maps. The functions f and g are said to be *topologically conjugate* if there exists a homeomorphism $h : A \longrightarrow B$ such that $h \circ f = g \circ h$.

The homeomorphism is called a topological conjugacy, and is more commonly defined by simply stating that the following diagram commutes:

$$
\begin{array}{ccc}
 & f & \\
A & \longrightarrow & A \\
h \downarrow & & \downarrow h \\
 & g & \\
B & \longrightarrow & B
\end{array}
$$

Using the theorem of the previous section, we know that if $\lambda > 2 + \sqrt{5}$ then f_λ (the quadratic map) is topologically conjugate to σ (the shift map). Topologically conjugate systems are the same system insofar as there is a one-to-one correspondence between the orbits of each system. Sometimes this is too restrictive and we only require that the mapping between orbits be many-to-one. In this latter case we say the two dynamical systems are *semiconjugate*.

In nonlinear dynamics, it is often advantageous to establish a conjugacy or a semiconjugacy between the dynamical system in question and the dynamics on some symbol space. The properties of the dynamical system are usually easy to see in the symbol space and, by the conjugacy or semiconjugacy, these properties must also exist in the original dynamical system. For instance, the following properties are easy to show in Σ_2 and must also hold in Λ, namely:

1. The cardinality of the set of periodic points (often written as $\text{Per}_n(f_\lambda)$) is 2^n.

2. $\text{Per}(f_\lambda)$ is dense in Λ.

3. f_λ has a dense orbit in Λ [10].

Although there is no universally accepted definition of chaos, most definitions incorporate some notion of sensitive dependence on initial conditions. Our notions of topological conjugacy and symbolic dynamics give us a promising way to analyze chaotic behavior in a specific dynamical system.

In the context of one-dimensional maps, we say that a map $f : I \longrightarrow I$ possesses sensitive dependence on initial conditions if there exists a $\delta > 0$ such that, for any $x \in I$ and any neighborhood N of x, there exist $y \in N$ and $n \geq 0$ such that $|f^n(x) - f^n(y)| > \delta$. This says that small errors due either to measurement or round-off errors become magnified upon iteration—they cannot be ignored.

Let S^1 denote the unit circle. Here we will think of the members of S^1 as being normalized to the range $[0, 1)$. A simple example of a map that is chaotic in the above sense is given by $g : S^1 \longrightarrow S^1$ defined by $g(\theta) = 2\theta$. As we saw in section 2.10, when θ is written in base two, $g(\theta)$ is simply a shift map on the unit circle. In ergodic theory the above shift map is known as a Bernoulli process. If we think of each symbol 0 as a Tail (T), and each symbol 1 as a Head (H), then the above shift map is topologically conjugate to a coin toss, our intuitive model of a random process. Each shift represents a toss of the coin. We now show that the shift map is essentially the same as the quadratic map for $\lambda = 4$; that is, the quadratic map (a fully deterministic process) can be as random as a coin toss.

If $f_4(x) = 4x(1 - x)$, then the limit set is the whole unit interval $I = [0, 1]$ since the maximum $f_4(1/2) = 1$; that is, the map is strictly *onto* (the map is measure-preserving and is roughly analogous to a Hamiltonian system that conserves energy). To continue with the analysis, define $h_1 : S^1 \longrightarrow [-1, 1]$ by $h_1 = \cos(\theta)$. Also define

$q(x) = 2x^2 - 1$. Then

$$
\begin{aligned}
h_1 \circ g(\theta) &= \cos(2\theta) \\
&= 2\cos^2(\theta) - 1 \\
&= q \circ h_1(\theta)
\end{aligned}
$$

so h_1 conjugates g and q. Note, however, that h_1 is two-to-one at most points so that we only have a semiconjugacy. To go further, if we define $h_2 : [-1, 1] \longrightarrow [0, 1]$ by $h_2(t) = \frac{1}{2}(1 - t)$, then $f_4 \circ h_2 = h_2 \circ q$. Then $h_3 = h_2 \circ h_1$ is a topological semiconjugacy between g and f_4; we have established the semiconjugacy between the chaotic linear circle map and the quadratic map when $\lambda = 4$. The reader is invited to work through a few examples to see how the orbits of the quadratic map, the linear circle map, and a coin toss can all be mapped onto one another.

2.12 Symbolic Coordinates

In the previous section we showed that when $\lambda > 4$, the dynamics of the quadratic map restricted to the invariant set Λ are "the same" as those given by the shift map σ on the sequence space on two symbols, Σ_2. We established this correspondence by partitioning the unit interval into two halves about the maximum point of the quadratic map, $x = 1/2$. The left half of the unit interval is labeled 0 while the right half is labeled 1, as illustrated in Figure 2.25. To any orbit of the quadratic map $f^n(x_0)$ we assign a sequence of symbols $\mathbf{s} = (s_0 s_1 s_2 \ldots)$—for example, 101001...—called the *itinerary*, or symbolic future, of the orbit. Each s_i represents the half of the unit interval in which the ith iteration of the map falls.

In part, the theorem of section 2.11.2 says that knowing an orbit's initial condition is exactly equivalent to knowing an orbit's itinerary. Indeed, if we imagine that the itinerary is simply an expression for some binary number, then perhaps the correspondence is not so surprising. That is, the mapping f_λ^n takes some initial coordinate number x_0 and translates it to a binary number $\beta_0 = \beta(s_0, s_1, \ldots)$ constructed from the symbolic future, which can be thought of as a "symbolic coordinate."

From a practical point of view, the renaming scheme described by symbolic dynamics is very useful in at least two ways:

1. Symbolic dynamics provides a good way to label all the periodic orbits.

2. The symbolic itinerary of an orbit provides the location of the orbit in phase space to any desired degree of resolution.

We will explain these two points further and in the process show the correspondence between β and x_0.

In practical applications we shall be most concerned with keeping track of the periodic orbits. Symbolic itineraries of periodic orbits are repeating finite strings, which can be written in various forms, such as

$$\left(s_0 s_1 \ldots s_{n-1}, s_0 s_1 \ldots s_{n-1}, \ldots\right) = \left(s_0 s_1 \ldots s_{n-1}\right)^\infty = \overline{s_0 s_1 \ldots s_{n-1}}.$$

To see the usefulness of the symbolic description, let us consider the following problem: for $\lambda > 4$, find the approximate location of all the periodic orbits in the quadratic map.

2.12.1   What's in a name? Location.

As discussed in section 2.5, the exact location of a period n orbit is determined by the roots of the fixed point equation, $f^n(x) = x$. This naive method of locating the periodic points is impractical in general because it requires finding the roots of an arbitrarily high-order polynomial. We now show that the problem is easy to solve using symbolic dynamics if we ask not for the exact location, but only for the location relative to all the other periodic orbits.

If $\lambda > 4$, then there exists an interval centered about $x = 1/2$ for which $f(x) > 1$. Call this interval A_0 (see Figure 2.23). Clearly, no periodic orbit exists in A_0 since all points in A_0 leave the unit interval I at the first iteration, and thereafter escape to $-\infty$. As we argued in section 2.11.1, the periodic points must be part of the invariant set, those points that are *never* mapped into A_0.

As shown in Figure 2.25, the points in the invariant set can be constructed by considering the *preimages* of the unit interval found from the inverse map f_λ^{-1}. The first iteration of f_λ^{-1} produces two

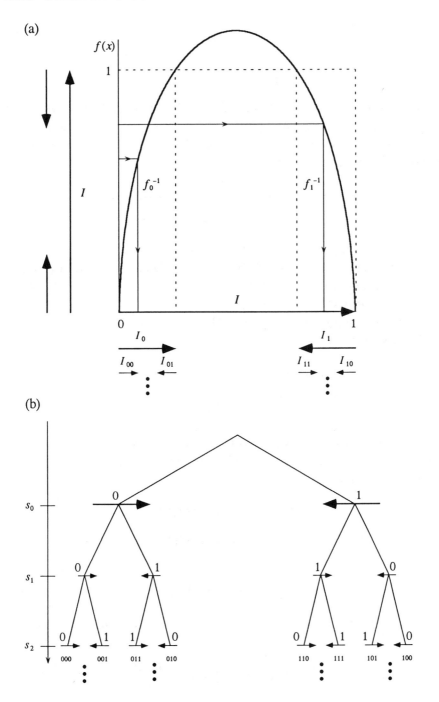

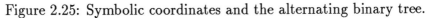

Figure 2.25: Symbolic coordinates and the alternating binary tree.

disjoint intervals,

which are labeled I_0 (the left interval) and I_1 (the right interval). As indicated by the arrows in Figure 2.25, I_0 preserves orientation, while I_1 reverses orientation. The orientation of the interval is simply determined by the slope (derivative) of $f_\lambda(x)$,

$$f_\lambda'(x) > 0 \qquad \text{if } x < 1/2, \text{ preserves orientation};$$
$$f_\lambda'(x) < 0 \qquad \text{if } x > 1/2, \text{ reverses orientation}.$$

We view $f^{-1}(I)$ as a first-level approximation to the invariant set Λ. In particular, $f^{-1}(I)$ gives us a very rough idea as to the location of both period one orbits, one of which is located somewhere in I_0, while the other is located somewhere in I_1.

To further refine the location of these periodic orbits, consider the application of f_λ^{-1} to both I_0 and I_1,

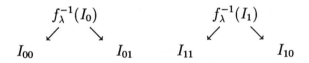

The second iteration gives rise to four disjoint intervals. Two of these contain the distinct period one orbits, and the remaining two intervals contain the period two orbit,

$$
\begin{aligned}
I_{00} &= I_0 \cap f_\lambda^{-1}(I_0), \\
I_{01} &= I_0 \cap f_\lambda^{-1}(I_1), \\
I_{11} &= I_1 \cap f_\lambda^{-1}(I_1), \\
I_{10} &= I_1 \cap f_\lambda^{-1}(I_0).
\end{aligned}
$$

In general we can define 2^n disjoint intervals at the nth level of refinement by

$$I_{s_0 s_1 \ldots s_{n-1}} = I_{s_0} \cap f_\lambda^{-1}(I_{s_1}) \cap \cdots \cap f_\lambda^{-(n-1)}(I_{s_{n-1}}). \qquad (2.35)$$

With each new refinement, we hone in closer and closer to the periodic orbits.

The one-to-one correspondence between x_0 and s is easy to see geometrically by observing that, as $n \to \infty$,

$$\bigcap_{n \geq 0} I_{s_0 s_1 \ldots s_n}$$

forms an infinite intersection of nested nonempty closed intervals that converges to a unique point in the unit interval.[11] The invariant limit set is the collection of all such limit points, and the periodic points are all those limit points indexed by periodic symbolic strings.

2.12.2 Alternating Binary Tree

We must keep track of two pieces of information to find the location of the orbits at the nth level: the relative location of the interval $I_{s_0 s_1 \ldots s_{n-1}}$ and its orientation. A very convenient way to encode both pieces of data is through the construction of a binary tree that keeps track of all the intervals generated by the inverse function, f_λ^{-1}. The quadratic map gives rise to the "alternating binary tree" illustrated in Figure 2.25(b) [13].

The nth level of the alternating binary tree has 2^n nodes, which are labeled from left to right by the sequence

$$n\text{th level}: \overbrace{0\ 1\ 1\ 0\ 0\ 1\ 1\ \ldots\ 0\ 0\ 1\ 1\ 0\ 0\ 1\ 1\ 0}^{2^n},$$

This sequence starts at the left with a zero. It is followed by a pair of ones, then a pair of zeros, and so on until 2^n digits are written down. To form the alternating binary tree, we construct the above list of 0s and 1s from level one to level n and then draw in the pair of lines from each $i - 1$st level node to the adjacent nodes at the ith level.

Now, to find the symbolic name for the interval at the nth level, $I_{s_0 s_1 \ldots s_{n-1}}$, we start at the topmost node, s_0, and follow the path down

---

[11]A partition of phase space that generates a one-to-one correspondence between points in the limit set and points in the original phase space is known in ergodic theory as a *generating partition*. Physicists loosely call such a generating partition a "good partition" of phase space.

N	B	$s_0 s_1 s_2$
x position	binary x position	symbolic name
0	000	000
1	001	001
2	010	011
3	011	010
4	100	110
5	101	111
6	110	101
7	111	100

Table 2.2: Symbolic coordinate for $n = 3$ from the alternating binary tree.

the alternating binary tree to the nth level, reading off the appropriate symbol name at each level along the way. By construction, we see that the symbolic name read off at the nth level of the tree mimics the location of the interval containing a period n orbit.

More formally, we identify the set of repeating sequences of period n in Σ_2 with the set of finite strings $s_0 s_1 \ldots s_{n-1}$. Let $\beta(s_0, s_1, \ldots, s_{n-1})$ denote the fraction between 0 and $(2^n - 1)/2^n$ giving the order, from left to right, generated by the alternating binary tree. Further, let $N(s_0, s_1, \ldots, s_{n-1})$ denote the integer position between 0 and $2^n - 1$ and let B denote N in binary form. It is not too difficult to show that $B(s_0, s_1, \ldots, s_{n-1}) = b_0 b_1 \ldots b_{n-1}$, where $b_i = 0$ or 1, and

$$\beta(s_0, s_1, \ldots, s_{n-1}) = \frac{b_0}{2} + \frac{b_1}{4} + \ldots + \frac{b_{n-1}}{2^n} \tag{2.36}$$

$$N(s_0, s_1, \ldots, s_{n-1}) = b_0 \cdot 2^{n-1} + b_1 \cdot 2^{n-2} + \ldots + b_{n-1} \cdot 2^0 \tag{2.37}$$

$$b_i = \sum_{j=0}^{i} s_j \bmod 2. \tag{2.38}$$

An application of the ordering relation can be read directly off of Figure 2.25(b) for $n = 3$ and is presented in Table 2.2. As expected, the left-most orbit is the string $\overline{0}$, which corresponds to the period one orbit at the origin, x_0. Less obvious is the position of the other period one orbit, x_1, which occupies the fifth position at the third level.

The itinerary of a periodic orbit is generated by a shift on the repeating string $(s_0 s_1 \ldots s_{n-1})^{\infty}$:

$$\sigma(s_0 s_1 \ldots s_{n-1}) = (s_1 s_2 \ldots s_{n-1} s_0). \tag{2.39}$$

In this case, the shift is equivalent to a cyclic permutation of the symbolic string. For instance, there are two period three orbits shown in Table 2.2; their itineraries and positions are

$$
\begin{array}{rccccc}
s_0 s_1 s_2 : & 001 & \longrightarrow & 010 & \longrightarrow & 100 \\
N : & 1 & \longrightarrow & 3 & \longrightarrow & 7
\end{array}
$$

and

$$
\begin{array}{rccccc}
s_0 s_1 s_2 : & 011 & \longrightarrow & 110 & \longrightarrow & 101 \\
N : & 2 & \longrightarrow & 4 & \longrightarrow & 6.
\end{array}
$$

So the itinerary of a periodic orbit is generated by cyclic permutations of the symbolic name.

2.12.3 Topological Entropy

To name a periodic orbit, we need only choose one of its cyclic permutations. The number of distinct periodic orbits grows rapidly with the length of the period. The symbolic names for all periodic orbits up to period eight are presented in Table 2.3. A simple indicator of the complexity of a dynamical system is its *topological entropy*. In the one-dimensional setting, the topological entropy, which we denote by h, is a measure of the growth of the number of periodic cycles as a function of the symbol string length (period),

$$h = \lim_{n \to \infty} \frac{\ln N_n}{n}, \tag{2.40}$$

where N_n is the number of distinct periodic orbits of length n. For instance, for the fully developed quadratic map, N_n is of order 2^n, so

$$h = \ln 2 \approx 0.6931 \ldots.$$

0	01011	0000011	0101011	00010011	00011111
1	01111	0000101	0011111	00010101	00101111
01	000001	0001001	0101111	00011001	00110111
001	000011	0000111	0110111	00100101	00111011
011	000101	0001011	0111111	00001111	00111101
0001	000111	0001101	00000001	00010111	01010111
0011	001011	0010011	00000011	00011011	01011011
0111	001101	0010101	00000101	00011101	00111111
00001	001111	0001111	00001001	00100111	01011111
00011	010111	0010111	00000111	00101011	01101111
00101	011111	0011011	00001011	00101101	01111111
00111	0000001	0011101	00001101	00110101	

Table 2.3: Symbolic names for all periodic orbits up to period eight occurring in the quadratic map for $\lambda > 4$. All names related by a cyclic permutation are equivalent.

The topological entropy is zero in the quadratic map for any value of λ below the accumulation point of the first period doubling cascade because N_n is of order $2n$ in this regime. The topological entropy is a continuous, monotonically increasing function between these two parameter values. The topological entropy increases as periodic orbits are born by different bifurcation mechanisms. A strictly positive value for the topological entropy is sometimes taken as an indicator for the amount of "topological chaos."

In addition to its theoretical importance, symbolic dynamics will also be useful experimentally. It will help us to locate and organize the periodic orbit structure arising in real experiments. In Chapter 5 we will show how periodic orbits can be extracted and identified from experimental data. We will further describe how to construct a periodic orbit's symbolic name directly from experiments and how to compare this with the symbolic name found from a model, such as the quadratic map. Reference [14] describes an additional refinement of symbolic dynamics called kneading theory, which is useful for analyzing nonhyperbolic parameter regions, such as occur in the quadratic map for $1 < \lambda < 4$.

Notice that the ordering relation described by the alternating binary tree between the periodic orbits does not change for any $\lambda > 4$. A simple observation, which will nevertheless be very important from an experimental viewpoint, is the following: *this ordering relation, which is easy to calculate in the hyperbolic regime, is often maintained in the nonhyperbolic regime.* This is the case, for instance, in the quadratic map for all $\lambda > 1$. This observation is useful experimentally because it will give us a way to name and locate periodic orbits in an experimental system at parameter values where a nonhyperbolic strange attractor exists. That is, we can name and identify periodic orbits in a hyperbolic regime, where the system can be analyzed analytically, and then carry over the symbolic name for the periodic orbit from the hyperbolic regime to the nonhyperbolic regime, where the system is more difficult to study rigorously. Symbolic dynamics and periodic orbits will be our "breach through which we may attempt to penetrate an area hitherto deemed inaccessible" [15].

The reader might notice that our symbolic description of the quadratic map used very little that was specific to this map. The same description holds, in fact, for any single-humped (*unimodal*) map of the interval. Indeed, the topological techniques we described here in terms of binary trees extend naturally to k symbols on k-ary trees when a map with many humps is encountered.

This concludes our introduction to the quadratic map. There are still many mysteries in this simple map that we have not yet begun to explore, such as the organization of the periodic window structure, but at least we can now continue our journey into nonlinear lands with words and pictures to describe what we might see [16].

Usage of *Mathematica*

In this section, we illustrate how *Mathematica*[12] can be used for research or course work in nonlinear dynamics. *Mathematica* is a complete system for doing symbolic, graphical, and numerical manipulations on symbols and data, and is commonly available on a wide range of machines from microcomputers (386s and Macs) to mainframes and supercomputers.

Mathematica is strong in two- and three-dimensional graphics, and, where appropriate, we would encourage its use in a nonlinear dynamics course. It can serve at least three important functions: (1) a means of generating complex graphical representations of data from experiments or simulations; (2) a method for double-checking complex algebraic manipulations first done by hand; and (3) a general system for writing both numerical and symbolic routines for the analysis of nonlinear equations.

What follows is text from a typical *Mathematica* session, typeset for legibility, used to produce some of the graphics and to double-check some of the algebraic results presented in this chapter. Of course, a real *Mathematica* session would not be so heavily commented.

(* This is a *Mathematica* comment statement. *Mathematica* ignores everything between star parentheses. *)

(* To try out this *Mathematica* session yourself, type everything in bold that is not enclosed in the comment statements. The following line is *Mathematica*'s answer typed in italic. *Mathematica*'s output from graphical commands are not printed here, but are left for the reader to discover. *)

(* This notebook is written on a Macintosh. 7/22/90 nbt. *)

(* In this notebook we will analytically solve for the period one and period two orbits in the quadratic map and plot their locations as a function of the control parameter, lambda. *)

---

[12] *Mathematica* is a trademark of Wolfram Research Inc. For a brief introduction to *Mathematica* see S. Wolfram, *Mathematica, a system for doing mathematics by computer* (Addison-Wesley: New York, 1988), pp. 1–23.

(* First we define the quadratic function with the *Mathematica*
Function[{x,y, ...}, f(x,y, ...)] command, which takes two
arguments, the first of which is a list of variables, and the second
of which is the function. The basic data type in *Mathematica*
is a list, and all lists are enclosed between braces,
{x1, x2, x3, ...}. Note that all arguments and functions in
Mathematica are enclosed in square brackets f□, which differs from
the standard mathematical notation of parentheses f(). This is, in
part, because square brackets are easier to reach on the keyboard.
Also, note that in defining a variable one must always put a space
around it, so xy is equal to a single variable named "xy", while
x y with a space between is equal to two variables, x and y. *)

(* To evaluate a *Mathematica* expression tap the enter key, not the
return key. Now to our first *Mathematica* command: *)

f = Function[{lambda, x}, lambda x (1 - x)]
Function[{lambda, x}, lambda x (1 − x)]

(* *Mathematica* should respond by saying Out[1], which tells us that
Mathematica successfully processed the first command and has put the
result in the variable Out[1], as well as the variable we created, f.
To evaluate the quadratic map, we now can feed f two arguments,
the first of which is lambda, and the second of which is x. *)

f[4, 1/2]
1

(* *Mathematica* should respond with the function Out[2], which
contains the quadratic map evaluated at lambda = 4 and x = 1/2. To
evaluate a list of values for x we could let the x variable be a
list, i.e., a series of numbers enclosed in braces {x1, x2, x3, ...}.
To try this command, type: *)

f[4, {0, 0.25, 0.5, 0.75, 1}]
{0, 0.75, 1., 0.75, 0}

(* To plot the quadratic map we use the *Mathematica* plot command,
Plot[f, {x, xmin, xmax}]. For instance a plot of the quadratic map
for lambda = 4 is given by: *)

Plot[f[4,x], {x, 0, 1}]

(* It is as easy as pie to take a composite function in *Mathematica*,
just type f[f[x]], so to plot f(f(x)) for the quadratic map we
simply type: *)

Plot[f[4,f[4,x]], {x, 0, 1}]

(* Now let's find the locations of the period one orbits, given by
the roots of f(x) = x. To find the roots, we use the *Mathematica*
command Solve[eqns, vars]. Notice that we are going to rename the
parameter "lambda" to "a" just to save some space when printing the
answer. The double equals "==" in *Mathematica* is equivalent to the
single equal "=" of mathematics. *)

Solve[f[a, x1] == x1, x1]
$\{\{x1 \longrightarrow 1 - a^{-1}\}, \{x1 \longrightarrow 0\}\}$

(* As expected, *Mathematica* finds two roots. Let's make a function
out of the first root so that we can plot it later using Plot. To do
this we need the following sequence of somewhat cryptic commands: *)

r1 = %[[1]]
$\{x1 \longrightarrow 1 - a^{-1}\}$

(* The roots are saved in a list of two items. To pull out the
first item we used the % command, a *Mathematica* variable that always
holds the value of the last expression. In this case it holds the
list of two roots, and the double square bracket notation %[[1]]
tells *Mathematica* we want the first item in the list of two items.
Now we must pull out the last part of the expression, 1 - 1/a, with
the replacement command Replace[expr, rules]: *)

x1 = Replace[x1, r1]
$1 - a^{-1}$

(* To plot the location of the period one orbit we just use the Plot
command again. *)

Plot[x1, {a, 0.9, 4}]

(* To find the location of the period two orbit,
we solve for f(f(x)) = x. *)

Solve[f[a,f[a, x2]] == x2, x2]
$\{\{x2 \longrightarrow 0\}, \{x2 \longrightarrow 1 - a^{-1}\},$
$\{x2 \longrightarrow (1 + ((-1 - a^{-1})^2 - (4(1 + a^{-1}))/a)^{1/2} + a^{-1})/2\},$
$\{x2 \longrightarrow (1 - ((-1 - a^{-1})^2 - (4(1 + a^{-1}))/a)^{1/2} + a^{-1})/2\}\}$

(* We find four roots, as expected. Before proceeding further, it's
a good idea to try and simplify the algebra for the last two new
roots by applying the Simplify command to the last expression. *)

rt = Simplify[%]
$\{\{x2 \longrightarrow 0\}, \{x2 \longrightarrow 1 - a^{-1}\},$
$\{x2 \longrightarrow (1 + (1 - 3/a^2 - 2/a)^{1/2} + a^{-1})/2\},$
$\{x2 \longrightarrow (1 - (1 - 3/a^2 - 2/a)^{1/2} + a^{-1})/2\}\}$

(* And we can now pull out the positive and negative roots of the
period two orbit. *)

x2plus = Replace[x2, rt[[3]]]
x2minus = Replace[x2, rt[[4]]]
$(1 - (1 - 3/a^2 - 2/a)^{1/2} + a^{-1})/2$

(* As a last step, we can plot the location of the period one orbit
and both branches of the period two orbit by making a list of
functions to be plotted. *)

Plot[{x1, x2plus, x2minus}, {a, 1, 4}]

```
(* This is how we originally plotted the orbit stability diagram
in the text.  Mathematica can go on to find higher-order periodic
orbits by numerically finding the roots of the nth composite of f,
if no exact solution exists.  *)
```

References and Notes

[1]   There are several excellent reviews of the quadratic map. Some of the oldest
      are still the best, and all of the following are quite accessible to undergraduates:
      E. N. Lorenz, The problem of deducing the climate from the governing equa-
      tions, Tellus **16**, 1–11 (1964); E. N. Lorenz, Deterministic nonperiodic flow, J.
      Atmos. Sci. **20**, 130–141 (1963); R. M. May, Simple mathematical models with
      very complicated dynamics, Nature **261**, 459–467 (1976); M. J. Feigenbaum,
      Universal behavior in nonlinear systems, Los Alamos Science **1**, 4–27 (1980).
      These last two articles are reprinted in the book, P. Cvitanović, ed., *Universal-*
      ity in Chaos (Adam Hidgler Ltd: Bristol, 1984). All four are models of good
      expository writing. A simple circuit providing an analog simulation of the
      quadratic map suitable for an undergraduate lab is described by T. Mishina,
      T. Kohmoto, and T. Hashi, Simple electronic circuit for the demonstration of
      chaotic phenomena, Am. J. Phys. **53** (4), 332–334 (1985).

[2]   A good review of the dynamics of the quadratic map from a dynamical systems
      perspective is given by R. L. Devaney, Dynamics of simple maps, in *Chaos and*
      fractals: The mathematics behind the computer graphics, Proc. Symp. Applied
      Math. **39**, edited by R. L. Devaney and L. Keen (AMS: Rhode Island,1989).

[3]   For an elementary proof, see E. A. Jackson, *Perspectives in nonlinear dynamics*,
      Vol. 1 (Cambridge University Press: New York, 1989), pp. 152–153.

[4]   Counting the number of periodic orbits of period n is a very pretty combina-
      torial problem. See Hao B.-L., *Elementary symbolic dynamics and chaos in*
      dissipative systems (World Scientific: New Jersey, 1989), pp. 196–201. An up-
      dated version for some results in this book can be found in Zheng W.-M. and
      Hao B.-L., Applied symbolic dynamics, in *Experimental study and characteri-*
      zation of chaos, edited by Hao B.-L. (World Scientific: New Jersey, 1990). Also
      see the original analysis published in the article by M. Metropolis, M. L. Stein,
      and P. R. Stein, On the finite limit sets for transformations of the unit interval,
      J. Comb. Theory **15**, 25–44 (1973). Also reprinted in Cvitanović, reference [1].

[5]   For a discussion concerning the existence of a single stable attractor in the quadratic map and its relation to the Schwarzian derivative, see Jackson, reference [3], pp. 148–149, and Appendix D, pp. 396–399.

[6]   A more complete mathematical account of local bifurcation theory is presented by S. N. Rasband, *Chaotic dynamics of nonlinear systems* (John Wiley & Sons: New York, 1990), pp. 25–31 and pp. 108–109. Chapter 3 deals with universality theory from the quadratic map.

[7]   G. Iooss and D. D. Joseph, *Elementary stability and bifurcation theory* (Springer-Verlag: New York, 1981).

[8]   For more about this tale see the chapter "Universality" of J. Gleick, *Chaos: Making a new science*, (Viking: New York, 1987).

[9]   See the introduction of P. Cvitanović, ed., *Universality in Chaos* (Adam Hidgler Ltd: Bristol, 1984).

[10]   R. L. Devaney, *An introduction to chaotic dynamical systems*, second edition (Addison-Wesley: New York, 1989). Section 1.10 covers Sarkovskii's Theorem and section 1.18 covers kneading theory. Another elementary proof of Sarkovskii's Theorem can be found in H. Kaplan, A cartoon-assisted proof of Sarkovskii's Theorem, Am. J. Phys. **55**, 1023–1032 (1987).

[11]   K. Falconer, *Fractal geometry* (John Wiley & Sons: New York, 1990).

[12]   S. Smale, *The mathematics of time. Essays on dynamical systems, economic processes, and related topics* (Springer-Verlag: New York, 1980).

[13]   P. Cvitanović, G. H. Gunaratne, and I. Procaccia, Topological and metric properties of Hénon-type strange attractors, Phys. Rev. A **38** (3), 1503–1520 (1988).

[14]   Some papers providing a hands-on approach to kneading theory include: P. Grassberger, On symbolic dynamics of one-humped maps of the interval, Z. Naturforsch. **43a**, 671–680 (1988); J.-P. Allouche and M. Cosnard, Itérations de fonctions unimodales et suites engendrées par automates, C. R. Acad. Sc. Paris Série I **296**, 159–162 (1983); and reference [13]. Also see section 1.18 of Devaney's book, reference [10], for a nice mathematical account of kneading theory. The classic reference in kneading theory is J. Milnor and W. Thurston, On iterated maps of the interval, Lect. Notes in Math. 1342, in Dynamical Systems Proceedings, University of Maryland 1986–87, edited by J. C. Alexander (Springer-Verlag: Berlin, 1988), pp. 465–563. The early mathematical contributions of Fatou, Julia, and Myrberg to the dynamics of maps, symbolic dynamics, and kneading theory are emphasized by C. Mira, *Chaotic dynamics: From the one-dimensional endomorphism to the two-dimensional diffeomorphism* (World Scientific: New Jersey, 1987).

[15] H. Poincaré, *Les méthodes nouvelles de la mécanique céleste*, Vol. 1–3 (Gauthier-Villars: Paris, 1899); reprinted by Dover, 1957. English translation: New methods of celestial mechanics (NASA Technical Translations, 1967). See Vol. 1, section 36 for the quote.

[16] See sections 2.1.1 and 3.6.4 of Hao B.-L., reference [4], for some results on periodic windows in the quadratic map.

Problems

Problems for section 2.1.

2.1. For the quadratic map (eq. (2.1)), show that the interval $[1 - \lambda/4, \lambda/4]$ is a trapping region for all x in the unit interval and all $\lambda \in (2, 4]$.

2.2. Use the transformation

$$x = (\lambda/4 - 1/2)y + 1/2, \quad \mu = \lambda(\lambda/4 - 1/2)$$

to show that the quadratic map (eq. (2.1)) can be written as

$$y_{n+1} = 1 - \mu y_n^2, \quad y \in [-1, +1], \quad \mu \in (0, 2], \tag{2.41}$$

or (using a different x-transformation) as

$$z_{n+1} = \mu - z_n^2, \quad z \in [-\mu, +\mu], \quad \mu \in (0, 2]. \tag{2.42}$$

Specify the ranges to which x and λ are restricted under these transformations.

2.3. Read the Tellus article by Lorenz mentioned in reference [1].

Section 2.2.

2.4. Write a program to calculate the iterates of the quadratic map.

Section 2.3.

2.5. For $f(x) = 4x(1-x)$ and $x_0 = 0.25$, calculate $f^6(x_0)$ by the graphical method described in section 2.3.

2.6. Show by graphical analysis that if $x_0 \notin [0, 1]$ and $\lambda > 1$ in the quadratic map, then as $n \to \infty$, $f^n(x_0) \to -\infty$. Further, show that if $0 < \lambda < 1$, then the fixed point at the origin is an attractor.

Section 2.4.

2.7. Find all the attractors and basins of attraction for the map $f(x) = mx^2$ where m is a constant.

2.8. The *tent map* (see sections 2.1–2.2 of Rasband, reference [6]) is defined by

$$\Delta_\mu = \mu(1 - 2|x - \frac{1}{2}|) = 2\mu \begin{cases} x & \text{if } 0 \le x \le 1/2, \\ 1 - x & \text{if } 1/2 \le x \le 1. \end{cases}$$

(a) Sketch the graph of the tent map for $\mu = 3/4$. Why is it called the tent map?

(b) Show that the fixed points for the tent map are

$$x_0^* = 0 \text{ and, for } \mu > 1/2, \ x_1^* = 2\mu/(1 + 2\mu).$$

(c) Show that x_1^* is always repelling and that x_0^* is attracting when $\mu \in (0, 1/2)$.

(d) For a one-dimensional map the Lyapunov exponent is defined by

$$\lambda(x_0) = \lim_{n \to \infty} \frac{1}{n} \ln \left| \frac{d}{dx} f^n(x) \right|_{x=x_0}. \tag{2.43}$$

Show that for $\mu = 1$, the Lyapunov exponent for the tent map is $\lambda = \ln 2$. Hint: For the tent map use the chain rule of differentiation to show that

$$\lambda(x_0) = \lim_{n \to \infty} \frac{1}{n} \sum_{i=0}^{n-1} \ln |f'(x_i)|. \tag{2.44}$$

2.9. Determine the local stability of orbits with $0 < f'(x^*) < 1$ and $1 < f'(x^*) < \infty$ using graphical analysis as in Figure 2.9.

Section 2.5.

2.10. Determine the parameter value(s) for which the quadratic map intersects the line $y = x$ just once.

2.11. Consider a period two orbit of a one-dimensional map,

$$f^2(x_s^*) = f(f(x_s^*)) = x_s^*.$$

(a) Use the chain rule for differentiation to show that $(f^2)'(x_{01}^*) = (f^2)'(x_{10}^*)$.

(b) Show that the period two orbit is stable if $|f'(x_{01}^*) \ f'(x_{10}^*)| < 1$, and unstable if $|f'(x_{01}^*) \ f'(x_{10}^*)| > 1$.

2.12. For $\lambda = 4$, use *Mathematica* to find the locations of both period three orbits in the quadratic map.

2.13. Consider a p-cycle of a one-dimensional map. If this p-cycle is periodic then it is a fixed point of f^p. Let $x_0, x_1, \ldots, x_{p-1}$ represent one orbit of period p. Show that this periodic orbit is stable if

$$\prod_{j=0}^{p-1} |f'(x_j)| < 1.$$

Hint: Show by the chain rule for differentiation that for any orbit (not necessarily periodic),

$$\left. \frac{df^p}{dx} \right|_{x=x_0} = f'(x_0)f'(x_1)\cdots f'(x_{p-1}). \tag{2.45}$$

Now assume the orbit is periodic. Then for any two points of a period p orbit, x_i and x_j, note that $(f^p)'(x_i) = (f^p)'(x_j)$.

2.14. Consider a seed near a period two orbit, $x_0 = x^* + \epsilon$, with slope near to $f'(x^*) = -1$. Show by graphical analysis that $f^n(x_0)$ "flips" back and forth between the two points on the period two orbit. Show by graphical construction that this period two orbit is stable if $|(f^2)'(x_0)| < 1$, and unstable if $|(f^2)'(x_0)| > 1$.

2.15. For $\lambda \geq 4$, show that the quadratic map $f_\lambda^n(x)$ intersects the $y = x$ line 2^n times, but only some of these intersection points belong to new periodic orbits of period n. For $n = 1$ to 10, build a table showing the number of different orbits of f_λ of period n.

2.16. Prove (see Prob. 2.15) that for prime n, $(2^n - 2)/n$ is an integer (this result is a special case of the Simple Theorem of Fermat).

2.17. **(a)** Derive the equation of the graph shown in Figure 2.12.

(b) Verify that the two intersection points shown in the figure occur at 3 and $1 + \sqrt{6}$.

2.18. Equation (2.21) follows directly from equation (2.12) and the chain rule discussion in Problem 2.13. Derive it using only equations (2.13), (2.19), and (2.20).

Section 2.6.

2.19. Write a program to generate a plot of the bifurcation diagram for the quadratic map.

Section 2.7.

2.20. Show that the period two orbit in the quadratic map loses stability at $\lambda = 1 + \sqrt{6}$; i.e., $(f^2)'(x^*) = -1$ at this value of λ.

2.21. Show that the "equispaced" orbits of the bouncing ball system (see Prob. 1.2) are born in a saddle-node bifurcation. Note that equation (1.41) gives two impact phases for each n. For $n = 1$ and for $n = 2$ in equation (1.41), which orbit is a saddle near birth, and which orbit is a node? What is the impact phase of the saddle? What is the impact phase of the node?

2.22. Use *Mathematica* (or another computer program) to show that a pair of period three orbits are born in the quadratic map by a tangent bifurcation at $\lambda = 1 + \sqrt{8}$.

2.23. Show that the absolute value of the slope of f^n evaluated at all points of a period n orbit have the same value at a bifurcation point.

2.24. In Figure 2.17, the two sets of triangles ("open"—white interior and "closed"—black interior) represent two different period three orbits. Show that the open triangles represent an unstable orbit.

Section 2.8.

2.25. Using Table 2.1 and equations (2.26 and 2.27), estimate δ, c, and λ_∞.

2.26. Use the bifurcation diagram of the quadratic map (Fig. 2.22) and a ruler to measure the first few period doubling bifurcation values, λ_n. It may be helpful to use a photocopying machine to expand the figure before doing the measurements. Based on these measurements, estimate "Feigenbaum's delta" with equation (2.27). Do the same thing for the bifurcation diagram for the bouncing ball system found in Figure 1.16. How do these two values compare?

Section 2.9.

2.27. Find a one-dimensional circle map $g : S^1 \to S^1$ for which Sarkovskii's ordering does not hold. Hint: Find a map that has no period one solutions by using a discontinuous map.

2.28. Consider the periodic orbits of periods $1, 2, 3, 4, 5, 6, 7, 8, 9$, and 10. Order these points according to Sarkovskii's ordering, equation (2.29). Show that Sarkovskii's ordering uniquely orders all the positive integers.

Section 2.11.

2.29. For $\lambda > 4$, determine the interval A_0, as shown in Figure 2.23, as a function of λ. Verify that points in this interval leave the unit interval and never return.

2.30. (a) Using the metric of equation (2.34), calculate the distance between the two period two points, $\overline{01}$ and $\overline{10}$.

(b) Create a table showing the distances between all six period three points: $\overline{001}, \overline{010}, \overline{100}, \overline{011}, \overline{101},$ and $\overline{110}$.

2.31. Establish an isomorphism between the unit interval and the sequence space Σ_2 of section 2.11.2. Hint: See Devaney [10], section 1.6.

2.32. (a) Define $f : \mathbf{R} \to \mathbf{R}$ by $f(x) = mx + b$ and define $g : \mathbf{R} \to \mathbf{R}$ by $g(x) = mx + nb$, where m, b, and $n \in \mathbf{R}$. Show that f and g are topologically conjugate.

(b) Define $f : S^1 \to S^1$ by $f(x) = x + \pi/2$. Define $g : [0,1] \to [0,1]$ by $g(x) = 1 - x$. Show that f and g are topologically semiconjugate.

(c) Find a set of functions f, g, and h that satisfies the definition of topological conjugacy.

Section 2.12.

2.33. Construct the *binary tree* up to the fourth level where the nth level is defined by the rule

$$n\text{th level}: \overbrace{0\,1\,0\,1\,0\,1\,0\,\ldots\,1\,0\,1\,0\,1\,0\,1\,0\,1}^{2^n}.$$

(a) Construct the sixteen symbolic coordinates $s_0 s_1 s_2 s_3$ at the fourth level of this binary tree, and show that the ordering from left to right at the nth level is given by

$$N(s_0, s_1, \ldots, s_{n-1}) = s_0 \cdot 2^{n-1} + s_1 \cdot 2^{n-2} + \cdots + s_{n-1} \cdot 2^0. \qquad (2.46)$$

Why is it called the "binary tree"?

(b) Show that the fractional ordering β is given by

$$\beta(s_0, s_1, \ldots, s_{n-1}) = \frac{s_0}{2} + \frac{s_1}{4} + \cdots + \frac{s_{n-1}}{2^n}. \qquad (2.47)$$

(c) Give an example of a one-dimensional map (not necessarily continuous) on the unit interval giving rise to the binary tree.

2.34. Construct the alternating binary tree up to the fourth level and calculate the symbolic coordinate and position of each of the sixteen points (2^4). Present this information in a table.

2.35. Show that the order on the x-axis of two points x_0 and y_0 in the quadratic map with $\lambda = 4$ is determined by their itineraries $\{a_k\}$ and $\{b_k\}$ as follows: suppose $a_1 = b_1, a_2 = b_2, \ldots, a_k = b_k$, and that $a_{k+1} = 0$ and $b_{k+1} = 1$ (i.e., the itineraries of each initial condition are identical up to the kth iteration, and differ for the first time at the $k+1$st iteration). Then

$$x_0 < y_0 \iff \sum_{i=1}^{k} a_i \bmod 2 = 0. \tag{2.48}$$

Hint: See Appendix A of reference [13] and theorem 18.10 on page 145 of Devaney, reference [10].

Chapter 3

String

3.1 Introduction

Like a jump rope, a string tends to swing in an ellipse, a fact well known
to children. When holding both ends of a rope or string, it is difficult to
shake it so that motion is confined to a single transverse plane. Instead
of remaining confined to planar oscillations, strings appear to prefer
elliptical or whirling motions like those found when playing jump rope.
Borrowing terminology from optics, we would say that a string prefers
circular polarization to planar polarization. In addition to whirling,
other phenomena are easily observed in forced strings including bifurcations between planar and nonplanar periodic motions, transitions
to chaotic motions, sudden jumps between different periodic motions,
hysteresis, and periodic and aperiodic cycling between large and small
vibrations.

In this chapter we will begin to explore the dynamics of an elastic
string by examining a single-mode model for string vibrations. In the
process, several new types of nonlinear phenomena will be discovered,
including a new type of attractor, the torus, arising from quasiperiodic
motions, and a new route to chaos, via torus doubling. We will also
show how power spectra and Poincaré sections are used in experiments
to identify different types of nonlinear attractors. In this way we will
continue building the vocabulary used in studying nonlinear phenomena
[1].

In addition to its intrinsic interest, understanding the dynamics of a string can also be important for musicians, instrument makers, and acoustical engineers. For instance, nonlinearity leads to the modulation and complex tonal structure of sounds from a cello or guitar. Whirling motions account for the rattling heard when a string is strongly plucked [2]. Linear theory provides the basic outline for the science of the production of musical sounds; its real richness, though, comes from nonlinear elements.

When a string vibrates, the length of the string must fluctuate. These fluctuations can be along the direction of the string, longitudinal vibrations, or up and down, vibrations transverse to the string. The longitudinal oscillations occur at about twice the frequency of the transverse vibrations. The modulation of a string's length is the essential source of a string's nonlinearity and its rich dynamical behavior. The coupling between the transverse and longitudinal motions is an example of a *parametric oscillation*. An oscillation is said to be parametric when some parameter of a system is modulated, in this case the string's length. Linear theory predicts that a string's free transverse oscillation frequency is independent of the string's vibration amplitude. Experimental measurements, on the other hand, show that the resonance frequency depends on the amplitude. Thus the linear theory has a restricted range of applicability.

Think of a guitar string. A string under a greater tension has a higher pitch (fundamental frequency). Whenever a string vibrates it gets stretched a little more, so its pitch increases slightly as its vibration amplitude increases.

We begin this chapter by describing the experimental apparatus we've used to study the string (section 3.2). In section 3.3 we model our experiment mathematically. Sections 3.4 to 3.6 examine a special case of string behavior, planar motion, which gives rise to the Duffing equation. Section 3.7 looks at the more general case, nonplanar motion. Finally, in section 3.8 we present experimental techniques used by nonlinear dynamicists. These experimental methods are illustrated in the string experiment.

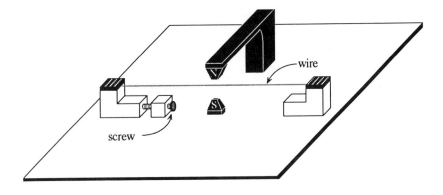

Figure 3.1: Schematic of the apparatus used to study the vibrations of a wire (string).

3.2 Experimental Apparatus

An experimental apparatus to study the vibrations of a string can be constructed by mounting a wire between two heavy brass anchors [3]. As shown in Figure (3.1), a screw is used to adjust the position of the anchors, and hence the tension in the wire (string). An alternating sinusoidal current passed through the wire excites vibrations; this current is usually supplied directly from a function generator. An electromagnet, or large permanent magnet, is placed at the wire's midpoint. The interaction between this magnetic field and the magnetic field generated by the wire's alternating current causes a periodic force to be applied at the wire's midpoint.[1] If a nonmagnetic wire is used, such as tungsten, then both planar and nonplanar whirling motions are easy to observe. On the other hand, if a magnetic wire is used, such as steel, then the motion always remains restricted to a single plane [4]. The use of a magnetic wire introduces an asymmetry into the system that causes the damping rate to depend strongly on the direction of oscillation.

---

[1]From the *Lorentz force law*, a wire carrying a current I, in a magnetic field of strength B, is acted on by a magnetic force $F_{mag} = \int (I \times B) dl$. If the current I in the wire varies sinusoidally, then so does the force on the wire. See D. J. Griffiths, *An introduction to electrodynamics* (Prentice-Hall: Englewood Cliffs, NJ, 1981), pp. 174–181.

A similar asymmetry is seen in the decay rates of violin, guitar, and piano strings. In these musical instruments the string runs over a bridge, which helps to hold the string in place. The bridge damps the motion of the string; however, the damping force applied by the bridge is different in the horizontal and vertical directions [5]. The clamps holding the string in our apparatus are designed to be symmetric, and it is easy to check experimentally that the decay rates in different directions (in the absence of a magnetic field) show no significant variation. Our experimental apparatus can be thought of as the inverse of that found in an electric guitar. There, a magnetic coil is used to detect the motion of a string. In our apparatus, an alternating magnetic field is used to excite motions in a wire.

The horizontal and vertical string displacements are monitored with a pair of inexpensive slotted optical sensors consisting of an LED (light-emitting diode) and a phototransistor in a U-shaped plastic housing [6]. Two optical detectors, one for the horizontal motion and one for the vertical motion, are mounted together in a holder that is fastened to a micropositioner allowing exact placement of the detectors relative to the string. The detectors are typically positioned near the string mounts. This is because the detector's sensitivity is restricted to a small-amplitude range, and the string displacement is minimal close to the string mounts. As shown in Figure 3.2, the string is positioned to obstruct the light from the LED and hence casts a shadow on the surface of the phototransistor. For a small range of the string displacements, the size of this shadow is linearly proportional to the position of the string, and hence also to the output voltage from the photodetector. This voltage is then monitored on an oscilloscope, digitized with a microcomputer, or further processed electronically to construct an experimental Poincaré section as described in section 3.8.1.

Care must be taken to isolate the rig mechanically and acoustically. In our case, we mounted the apparatus on a floating optical table. We also constructed a plastic cover to provide acoustical isolation. The string apparatus is small and easily fits on a desktop. Typical experimental parameters are listed in Table 3.1.

Most of our theoretical analysis will be concerned with single-mode oscillations of a string. If we pluck the string near its center, it tends to oscillate in a sinusoidal manner with most of its energy at some pri-

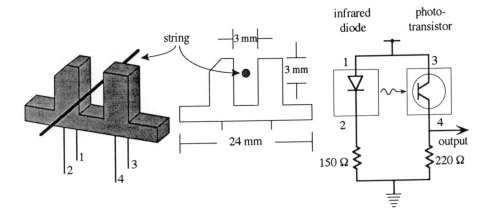

Figure 3.2: Module used to detect string displacements. (Adapted from Hanson [6].)

Parameter	Typical experimental value
Length	80 mm
Mass per unit length	0.59 g/m
Diameter	0.2 mm
Primary resonance	1 kHz
Range of hysteresis	300 Hz
Magnetic field strength	0.2 T
Current	0–2 A
Maximum displacement	3 mm
Damping	0.067

Table 3.1: Parameters for the string apparatus.

mary frequency called the *fundamental*. A similar plucking effect can be achieved by exciting wire vibrations with the current and the stationary magnetic field. To pluck the string, we switch off the current after we get a large-amplitude string vibration going. The fundamental frequency is recognizable by us as the characteristic pitch we hear when the string is plucked. A large-amplitude (resonant) response is expected when the forcing frequency applied to a string is near to this fundamental. This is the primary resonance of the string, and it is defined by the linear theory as

$$\omega_0 = \frac{\pi}{l}\left(\frac{T}{\mu}\right)^{1/2},\tag{3.1}$$

where $\mu = m/l$ is the mass per unit length and T is the tension in the string.[2]

The primary assumption of the linear theory is that the equilibrium length of the string, l, remains unchanged as the string vibrates, that is, $l(t) = l$, where $l(t)$ is the instantaneous length. In other words, the linear theory assumes that there are no longitudinal oscillations. In developing a simple nonlinear model for the vibrations of a string we must begin to take into account these longitudinal oscillations and the dependence of the string's length on the vibration amplitude.

3.3 Single-Mode Model

A model of a string oscillating in its fundamental mode is presented in Figure 3.3 and consists of a single mass fastened to the central axis by a pair of linearly elastic springs [7]. Although the springs provide a linear restoring force, the resulting force toward the origin is nonlinear because of the geometric configuration. The ends of the massless springs are fixed a distance l apart where the relaxed length of the spring is l_0 and the spring constant is k. In the center a mass is attached that is free to make oscillations in the x–y plane centered at the origin. The motion in the two transverse directions, x and y, is coupled directly,

---

[2]For a review of the linear theory for the vibrations of a stretched string see A. P. French, *Vibrations and waves* (W. W. Norton: New York, 1971), pp. 161-170.

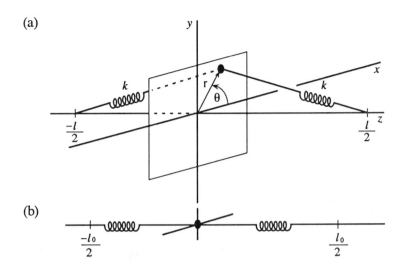

Figure 3.3: Single-mode model for nonlinear string vibrations. String vibrations are assumed to be in the fundamental mode and are measured in the transverse x–y plane by the polar coordinates (r, θ). (a) Equilibrium length; (b) relaxed length.

and also indirectly, via the longitudinal motion of the spring. Both of these coupling mechanisms are nonlinear. The multimode extension of this single-mode model would consist of n masses hooked together by $n + 1$ springs.

The restoring force on the mass shown in Figure 3.3 is

$$F = -2kr\left(1 - \frac{l_0}{\sqrt{l^2 + 4r^2}}\right),\qquad(3.2)$$

where the position of the mass is given by polar coordinates (r, θ) of the transverse plane (see Prob. 3.8). Expanding the right-hand side of equation (3.2) in a Taylor series ($2r < l$), we find that

$$F = -2kr(l - l_0)(\frac{r}{l}) - 4kl_0[(\frac{r}{l})^3 - 3(\frac{r}{l})^5 + \cdots].$$

The force can be written as

$$F = m\ddot{r},$$

so

$$mr \approx -2k(l - l_0)\left(\frac{r}{l}\right)\left[1 + \frac{2l_0}{(l - l_0)}\left(\frac{r}{l}\right)^2\right]. \tag{3.3}$$

Note the cubic restoring force. Also note that nonlinearity dominates when $l \approx l_0$. That is, the nonlinear effects are accentuated when the string's tension is low.

Define

$$\omega_0^2 = \frac{2k}{m}\frac{(l - l_0)}{l} \tag{3.4}$$

and

$$K = \frac{2l_0}{l^2(l - l_0)}. \tag{3.5}$$

Then from equation (3.3) we get, because of symmetry in the angular coordinate, the vector equation for $\mathbf{r} = (x(t), y(t))$,

$$\ddot{\mathbf{r}} + \omega_0^2\mathbf{r}(1 + K\mathbf{r}^2) = 0, \tag{3.6}$$

which is the equation of motion for a two-dimensional conservative cubic oscillator.[3] The behavior of equation (3.6) depends critically upon the ratio (l_0/l). If $l_0 < l$, the coefficient of the nonlinear term, K, is positive, the equilibrium point at $r = 0$ is stable, and we have a model for a string vibrating primarily in its fundamental mode. On the other hand, if $l_0 > l$, then K is negative, the origin is an unstable equilibrium point, and two stable equilibrium points exist at approximately $r = \pm l$. This latter case models the motions of a single-mode elastic beam [8]. For our purpose we will mostly be concerned with the case $l_0 < l$, or $K > 0$.

In general, we will want to consider damping and forcing, so equation (3.6) is modified to read

$$\ddot{\mathbf{r}} + \lambda\dot{\mathbf{r}} + \omega_0^2(1 + K\mathbf{r}^2)\mathbf{r} = \mathbf{f}(t), \tag{3.7}$$

where $\mathbf{f}(t)$ is a periodic forcing term and λ is the damping coefficient. Usually, the forcing term is just a sinusoidal function applied in one radial direction, so that it takes the form $\mathbf{f}(t) = (A\cos(\omega t), 0)$. For

---

[3]The term $\mathbf{r}^2$ in equation (3.6) is a typical physicist's notation meaning the dot product of the vector, $\mathbf{r}^2 = (\mathbf{r} \cdot \mathbf{r})^2 = x^2 + y^2 = r^2$.

simplicity, we have assumed that the energy losses are linearly proportional to the radial velocity of the string, $\dot{\mathbf{r}}$. We also assumed that the ends of the string are symmetrically fixed, so that λ is a scalar. In general, the damping rate depends on the radial direction, so the damping term is a vector function. This is the case, for instance, when a string is strung over a bridge that breaks the symmetry of the damping term.

Equation (3.7) was also derived by Gough [2] and Elliot [9], both of whom related ω_0 and K to actual string parameters that arise in experiments. For instance, Gough showed that the natural frequency is given by

$$\omega_0 = \frac{c\pi}{l} \tag{3.8}$$

and the strength of the nonlinearity is

$$K = \frac{1}{\epsilon l}(\frac{\pi}{2})^2, \tag{3.9}$$

where ϵ is the longitudinal extension of a string of equilibrium length l, ω_0 is the low-amplitude angular frequency of free vibration, and c is the transverse wave velocity. Again, we see that the nonlinearity parameter, K, increases as the longitudinal extension, ϵ, approaches zero. That is, the nonlinearity is enhanced when the longitudinal extension—and hence the tension—is small. Nonlinear effects are also amplified when the overall string length is shortened, and they are easily observable in common musical instruments. For a viola D-string with a vibration amplitude of 1 mm, typical values of the string parameters showing nonlinear effects are: $l = 27.5$ cm, $\omega_0 = 60$ Hz, $\epsilon = 0.079$ mm, $K = 0.128$ mm^{-2} [2].

Equation (3.7) constitutes our single-mode model for nonlinear string vibrations and is the central result of this section. For some calculations it will be advantageous to write equation (3.7) in a dimensionless form. To this end consider the transformation

$$\tau = \omega_0 t, \quad \mathbf{s} = \frac{\mathbf{r}}{l_0}, \tag{3.10}$$

which gives

$$\mathbf{s}'' + \alpha \mathbf{s}' + [1 + \beta \mathbf{s}^2]\mathbf{s} = \mathbf{g}(\gamma\tau), \tag{3.11}$$

where the prime denotes differentiation with respect to τ and

$$\alpha \equiv \frac{\lambda}{\omega_0}, \ \beta \equiv K l_0^2, \ \mathbf{g} \equiv \frac{\mathbf{f}}{l_0 \omega_0^2}, \ \text{and} \ \gamma \equiv \frac{\omega}{\omega_0} . \qquad (3.12)$$

Before we begin a systematic investigation of the single-mode model it is useful to consider the unforced linear problem, $\mathbf{f}(t) = (0,0)$. If the nonlinearity parameter K is zero, then equation (3.7) is simply a two-degree of freedom linear harmonic oscillator with damping that admits solutions of the form

$$\mathbf{r} = (X_0 \cos \omega_0 t, Y_0 \sin \omega_0 t) e^{-\lambda t/2} , \qquad (3.13)$$

where X_0 and Y_0 are the initial amplitudes in the x and y directions. Equation (3.13) is a solution of (3.7) if we discard second-order terms in λ. In the conservative limit ($\lambda = 0$), the orbits are ellipses centered about the z-axis. As we show in section 3.7, one effect of the nonlinearity is to cause these elliptical orbits to precess. The trajectories of these precessing orbits resemble Lissajous figures, and these precessing orbits will be one of our first examples of quasiperiodic motion on a torus attractor.

3.4   Planar Vibrations: Duffing Equation

An external magnetic field surrounding a magnetic wire restricts the forced vibrations of a wire to a single plane. Alternatively, we could fasten the ends of the wire in such a way as to constrain the motion to planar oscillations. In either case, the nonlinear equation of motion governing the single-mode planar vibrations of a string is the *Duffing equation*,

$$\ddot{x} + \lambda \dot{x} + \omega_0^2 x (1 + K x^2) = A \cos(\omega t), \qquad (3.14)$$

where equation (3.14) is calculated from equation (3.7) by assuming that the string's motion is confined to the x–z plane in Figure 3.3. The forcing term in equation (3.7) is assumed to be a periodic excitation of the form

$$f(t) = A \cos(\omega t), \qquad (3.15)$$

where the constant A is the forcing amplitude and ω is the forcing frequency. The literature studying the Duffing equation is extensive, and it is well known that the solutions to equation (3.14) are already complicated enough to exhibit multiple periodic solutions, quasiperiodic orbits, and chaos. A good guide to the nonchaotic properties of the Duffing equation is the book by Nayfeh and Mook [10]. Highly recommended as a pioneering work in nonlinear dynamics is the book by Hayashi, which deals almost exclusively with the Duffing equation [11].

3.4.1 Equilibrium States

The first step in analyzing any nonlinear system is the identification of its equilibrium states. The equilibrium states are the stationary points of the system, that is, where the system comes to rest. For a system of differential equations, the equilibrium states are calculated by setting all the time derivatives equal to zero in the unforced system. Setting $\ddot{x} = 0$, $\dot{x} = 0$, and $A = 0$ in equation (3.14), we immediately find that the location of the equilibrium solutions is given by

$$\omega_0^2 x(1 + Kx^2) = 0, \tag{3.16}$$

which, in general, has three solutions:

$$x_0 = 0, \quad \text{and} \quad x_+ = +\sqrt{\frac{-1}{K}}, \quad x_- = -\sqrt{\frac{-1}{K}}. \tag{3.17}$$

Clearly, there is only one real solution if $K > 0$, x_0, since the other two solutions, $x_\pm$, are imaginary in this case. If $K < 0$, then there are three real solutions.

To understand the stability of the stationary points it is useful to recall the physical model that goes with equation (3.14). If $l < l_0$, then $K < 0$ (see eqs. (3.4 and 3.5)) and the Duffing equation (3.14) is a simple model for a beam under a compressive load. As illustrated in Figure 3.4, the solutions $x_\pm$ correspond to the two asymmetric stable beam configurations. The position x_0 corresponds to the symmetric unstable beam configuration—a small tap on the beam would immediately send it to one of the $x_\pm$ configurations. If $l > l_0$, then $K > 0$ and the Duffing equation is a simple model of a string or wire under tension, so there is only one symmetric stable configuration, x_0 (Fig. 3.5).

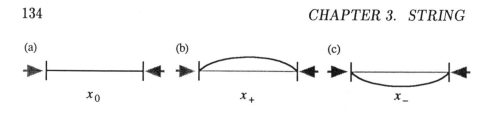

Figure 3.4: Equilibrium states of a beam under a compressive load.

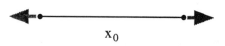

Figure 3.5: Equilibrium state of a wire under tension.

3.4.2 Unforced Phase Plane

Conservative Case

After identifying the equilibrium states, our next step is to understand the trajectories in phase space in a few limiting cases. In the unforced, conservative limit, a complete account of the orbit structure is given by integrating the equations of motion by using the chain rule in the form

$$
\ddot{x} \;=\; \dot{v} = \frac{d}{dt}v(x) = \frac{dv}{dx}\frac{dx}{dt}
$$

$$
\;=\; v\frac{dv}{dx}. \tag{3.18}
$$

Applying this identity to equation (3.14) with $\lambda = 0$ and $A = 0$ yields

$$
v\frac{dv}{dx} = -\omega_0^2 x(1 + Kx^2), \tag{3.19}
$$

which can be integrated to give

$$
\frac{1}{2}v^2 = h - \omega_0^2\left(\frac{x^2}{2} + K\frac{x^4}{4}\right), \tag{3.20}
$$

where h is the constant of integration.

The term on the left-hand side of equation (3.20) is proportional to the kinetic energy, while the term on the right-hand side,

$$
V(x) = \omega_0^2\left(\frac{x^2}{2} + K\frac{x^4}{4}\right), \tag{3.21}
$$

is proportional to the potential energy. Therefore, the constant h is proportional to the total energy of the system, as illustrated in Figure 3.6(a).

The phase space is a plot of the position x and the velocity v of all the orbits in the system. In this case, the phase space is a *phase plane*, and in the unforced conservative limit we find

$$v = \pm\sqrt{2}[h - V(x)]^{1/2}$$

$$= \pm\sqrt{2}\left[h - \omega_0^2\left(\frac{x^2}{2} + K\frac{x^4}{4}\right)\right]^{1/2}. \tag{3.22}$$

The last equation allows us to explicitly construct the *integral curves* (a plot of $v(t)$ vs. $x(t)$) in the phase plane. Each integral curve is labeled by a value of h, and the qualitative features of the phase plane depend critically on the signs of ω_0^2 and K.

If $l > l_0$, then both ω_0^2 and K are positive. A plot of equation (3.22) for several values of h is given in Figure 3.6(c). If $h = h_0$, then the integral curve consists of a single point called a *center*. When $h > h_0$, the orbits are closed, bounded, simply connected curves about the center. Each curve corresponds to a distinct periodic motion of the system. Going back to the string model again, we see that the center corresponds to the symmetric equilibrium state of the string, while the integral curves about the center correspond to finite-amplitude periodic oscillations about this equilibrium point.

If $l < l_0$, then K is negative. The phase plane has three stationary points. This parameter regime models a compressed beam. The left and right stationary points, $x_{\pm}$, are centers, but the unstable point at $x = 0$, labeled S in Figure 3.6(b), is a *saddle point* because it corresponds to a local maximum of $V(x)$.

Curves that pass through a saddle point are very important and are called *separatrices*. In Figure 3.6(d) we see that there are two integral curves approaching the saddle point S and two integral curves departing from S. These separatrices "separate" the phase plane into two distinct regions. Each integral curve inside the separatrices goes around one center, and hence corresponds to an asymmetric periodic oscillation about either the left or the right center, but not both. The integral curves outside the separatrices go around all three stationary

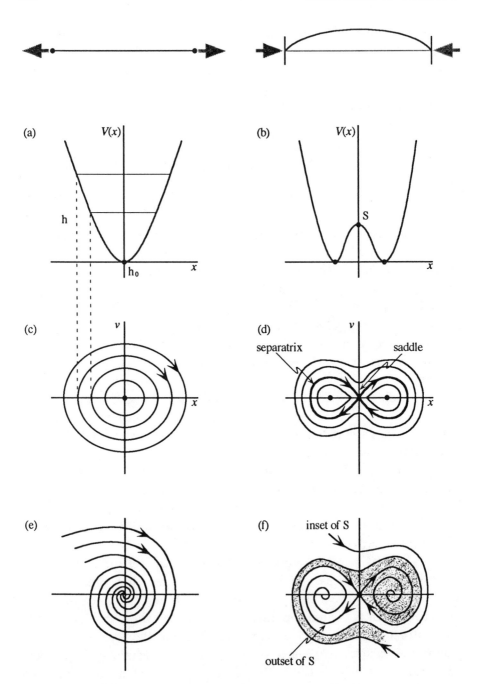

Figure 3.6: Potential and phase space for a single-mode string (a,c,e) and beam (b,d,f).

points and correspond to large-amplitude symmetric periodic orbits (Figure 3.6(d)). Thus, the separatrices act like barriers in phase space separating motions that are qualitatively different.

Dissipative Case

If damping is included in the system, then the phase plane changes to that shown in Figure 3.6(e). For the string, damping destroys all the periodic orbits, and all the motions are damped oscillations that converge to the *point attractor* at the origin. That is, if we pluck a string, the sound fades away. The string vibrates with a smaller and smaller amplitude until it comes to rest. Moreover, the basin of attraction for the point attractor is the entire phase plane. This particular point attractor is an example of a *sink*.

The phase plane for the oscillations of a damped beam is a bit more involved, as shown in Figure 3.6(f). The center points at $x_{\pm}$ become point attractors, while the stationary point at x_0 is a saddle. There are two separate basins of attraction, one for each point attractor (sink). The shaded region shows all the integral curves that head toward the right sink. Again we see the important role played by separatrices, since they separate the basins of attraction of the left and right attracting points. In the context of a dissipative system, the separatrix naturally divides into two parts: the *inset* consisting of all integral curves that approach the saddle point S, and the *outset* consisting of all points departing from S. Formally, the outset of S can be defined as all points that approach S as time runs backwards. That is, we simply reverse all the arrows in Figure 3.6(f).

The qualitative analysis of a dynamical system can usually be divided into two tasks: first, identify all the attractors and repellers of the system, and second, analyze their respective insets and outsets. Attractors and repellers are limit sets. Insets, outsets, and limit sets are all examples of invariant sets (see section 4.3.1). Thus, much of dynamical systems theory is concerned not simply with the analysis of attractors, but rather with the analysis of invariant sets of all kinds, attractors, repellers, insets, and outsets. For the unforced, damped beam the task is relatively easy. There are two attracting points and one saddle point. The inset and outset of the saddle point spiral around the two attracting points and completely determine the structure of the basins of attraction (see Figure 3.6(f)).

3.4.3 Extended Phase Space

To continue with the analysis of planar string vibrations, we now turn our attention to the forced Duffing equation in dimensionless variables (from eqs. (3.11 and 3.14)),

$$x'' + \alpha x' + (1 + \beta x^2)x = F\cos(\gamma\tau), \tag{3.23}$$

where F is the forcing amplitude and γ is the normalized forcing frequency.

It is often useful to rewrite an nth-order differential equation as a system of first-order equations, and to recall the geometric interpretation of a differential equation as a vector field. To this end, consider the change of variable $v = x'$, so that

$$\left.\begin{array}{rl} x' &= v, \\ v' &= aut(x,v) + g(\gamma\tau), \end{array}\right\} \quad (x,v) \in \mathbf{R}^2 \tag{3.24}$$

where $aut(x,v) = -[\alpha v + (1 + \beta x^2)x]$ is the *autonomous*, or time-independent, term of v' and $g(\gamma\tau) = F\cos(\gamma\tau)$ is the time-dependent term of v'. The phase space for the forced Duffing equation is topologically a plane, since each dependent variable is just a copy of $\mathbf{R}$, and the phase space is formally constructed from the Cartesian product of these two sets, $\mathbf{R} \times \mathbf{R} = \mathbf{R}^2$.

A vector field is obtained when to each point on the phase plane we assign a vector whose coordinate values are equal to the differential system evaluated at that point. The vector field for the unforced, undamped Duffing equation is shown in Figure 3.7(a). This vector field is static (time-independent). In contrast, the forced Duffing equation has a time-dependent vector field since the value of the vector field at (x,v) at τ is

$$(x', v') = (v, aut(x,v) + F\cos(\gamma\tau)).$$

In Figure 3.7(b) we show what the integral curves look like when plotted in the *extended phase space*, which is obtained by introducing a third variable,

$$z = \gamma\tau. \tag{3.25}$$

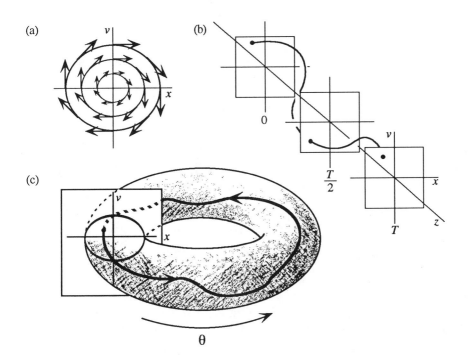

Figure 3.7: Extended phase space for the Duffing oscillator.

With this variable the differential system can be rewritten as

$$\left.\begin{array}{ll} x' & = v, \\ v' & = aut(x,v) + g(z), \\ z' & = \gamma. \end{array}\right\} \quad (x,v,z) \in \mathbf{R}^3 \qquad (3.26)$$

By increasing the number of dependent variables by one, we can formally change the forced (time-dependent) system into an autonomous (time-independent) system.

Moreover, since the vector field is a periodic function in z, it is sensible to introduce the further transformation

$$\theta = \gamma\tau \bmod 2\pi, \qquad (3.27)$$

thereby making the third variable topologically a circle, S^1. With this transformation the forced Duffing equation becomes

$$\left.\begin{array}{ll} x' & = v, \\ v' & = -[\alpha v + (1 + \beta x^2)x] + F\cos(\theta), \\ \theta' & = \gamma. \end{array}\right\} \quad (x, v, \theta) \in \mathbf{R}^2 \times S^1 \quad (3.28)$$

One last reduction is possible in the topology of the phase space of the Duffing equation. It is usually possible to find a trapping region that topologically is a disk, a circular subset $D \subset \mathbf{R}^2$. In this last instance, the topology of the phase space for the Duffing equation is simply $D \times S^1$, or a *solid torus* (see Figure 3.7(c)).

3.4.4 Global Cross Section

The global solution to a system of differential equations (the collection of all integral curves) is also known as a flow. A *flow* is a one-parameter family of diffeomorphisms of the phase space to itself (see section 4.2).

To visualize the flow in the Duffing equation, imagine the extended phase space as the solid torus illustrated in Figure 3.8. Each initial condition in the disk D, at $\theta = 0$, must return to D when $\theta = 2\pi$, because D is a trapping region and the variable θ is 2π-periodic. That is, the region D flows back to itself. An initial point in D labeled $(x_0, v_0, \theta_0 = 0)$ is carried by its integral curve back to some new point labeled $(x_1, v_1, \theta_1 = 2\pi)$ also in D. The Duffing equation satisfies the fundamental uniqueness and existence theorems in the theory of ordinary differential equations [12]. Hence, each initial point in D gets carried to a unique point back in D and no two integral curves can ever intersect in $D \times S^1$.

As originally observed by Poincaré, this unique dependence with respect to initial conditions, along with the existence of some region in phase space that is recurrent, allows one to naturally associate a map to any flow. The map he described is now called the Poincaré map. For the Duffing equation this map is constructed from the flow as follows. Define a *global cross section* Σ^{θ_0} of the vector field (eq. (3.28)) by

$$\Sigma^{\theta_0} = \{(x, v, \theta) \in D \times S^1 \mid \theta = \theta_0 \in [0, 2\pi)\}. \quad (3.29)$$

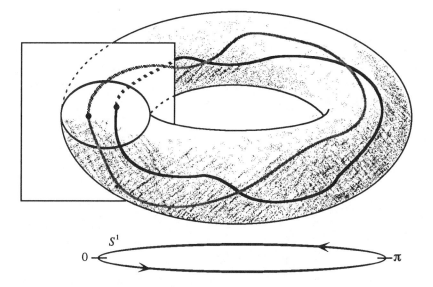

Figure 3.8: Phase space for the Duffing oscillator as a solid torus.

Next, define the Poincaré map of Σ^{θ_0} as

$$P_{\theta_0} : \Sigma^{\theta_0} \longrightarrow \Sigma^{\theta_0}, \quad x_0 \mapsto x_1, \quad v_0 \mapsto v_1, \qquad (3.30)$$

where (x_1, v_1) is the next intersection with Σ^{θ_0} of the integral curve emanating from (x_0, v_0). For the Duffing equation the Poincaré map is also known as a stroboscopic map since it samples, or strobes, the flow at a fixed time interval.

The dynamics of the Poincaré map are often easier to study than the dynamics in the original flow. By constructing the Poincaré map we reduce the dimension of the problem from three to two. This dimension reduction is important both for conceptual clarity as well as for graphical representations (both numerical and experimental) of the dynamics. For instance, a periodic orbit is a closed curve in the flow. The corresponding periodic orbit in the map is a collection of points in the map, so the fixed point theory for maps is easier to handle than the corresponding periodic orbit theory for flows.

The construction of a map from a flow via a cross section is generally unique. However, constructing a flow from a map is generally not unique. Such a construction is called a *suspension* of the map. Studies of maps and flows are intimately related—but they are not identical. For instance, a fixed point of a flow (an equilibrium point of the differential system) has no natural analog in the map setting.

A complete account of Poincaré maps along with a thorough case study of the Poincaré map for the harmonic oscillator is presented by Wiggins [13].

3.5   Resonance and Hysteresis

We now turn our attention to *resonance* in the Duffing oscillator. The notion of a resonance is a physical concept with no exact mathematical definition. Physically, a resonance is a large-amplitude response, or output, of a system that is subject to a fixed-amplitude input. The concept of a resonance is best described experimentally, and resonances are easy to see in the string apparatus described in section 3.2 by constructing a sort of experimental bifurcation diagram for forced string vibrations.

Imagine that the string apparatus is running with a small excitation amplitude (the amount of current in the wire is small) and a low forcing frequency (the frequency of the alternating current in the wire is much less than the natural frequency of free wire vibrations). To construct a resonance diagram we need to measure the response of the system, by measuring the maximum amplitude of the string vibrations as a function of the forcing frequency. To do this we slowly increase (scan through) the forcing frequency while recording the response of the string with the optical detectors. The results of this experiment depend on the forcing amplitude as well as where the frequency scan begins and ends. Decreasing frequency scans can produce different results from increasing frequency scans.

3.5.1   Linear Resonance

For a very small forcing amplitude the string responds with a *linear resonance*, such as that illustrated in Figure 3.9. According to linear theory, the response of the string is maximum when $\gamma = \omega/\omega_0 = 1$.

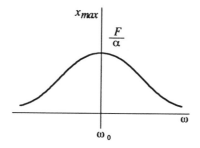

Figure 3.9: Response curve for a harmonic oscillator.

In other words, it is maximum when the forcing frequency ω exactly equals the natural frequency ω_0. A *primary* (or *main*) *resonance* exists when the natural frequency and the excitation frequency are close. The resonance diagram (Fig. 3.9) is called a linear response because it can be obtained by solving the periodically forced, linearly damped harmonic oscillator,

$$x'' + \alpha x' + x = F \cos(\gamma \tau), \tag{3.31}$$

which has a general solution of the form

$$\begin{aligned}x(\tau) \;=\; & x_0 e^{-\alpha \tau/2} \cos[(1-\alpha^2)\tau + \theta_0] + \\ & F[(1-\gamma^2)^2 + \alpha^2\gamma^2]^{-1/2} \cos(\gamma\tau + \delta).\end{aligned} \tag{3.32}$$

The constants x_0 and θ_0 are initial conditions. Equation (3.32) is a solution to equation (3.31) if we discard higher-order terms in α. The maximum amplitude of x, as a function of the driving frequency γ, is found from the asymptotic solution of equation (3.32),

$$\lim_{\tau \to \infty} x(\tau) \approx \frac{F \cos(\gamma\tau + \delta)}{[(1-\gamma^2)^2 + \alpha^2\gamma^2]^{1/2}}, \tag{3.33}$$

which produces the linear response diagram shown in Figure 3.9, since

$$a(\gamma) = x_{max}(\gamma) = \max\left[\lim_{\tau \to \infty} x(\tau)\right] = \frac{F}{[(1-\gamma^2)^2 + \alpha^2\gamma^2]^{1/2}}. \tag{3.34}$$

After the transient solution dies out, the steady-state response has the same frequency as the forcing term, but it is phase shifted by an amount

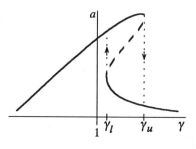

Figure 3.10: Schematic of the response curve for a cubic oscillator.

δ that depends on α, γ, and F. As with all damped linear systems, the steady-state response is independent of the initial conditions so that we can speak of *the* solution.

In the linear solution, motions of significant amplitude occur when F is large or when $\gamma \approx 1$. Under these circumstances the nonlinear term in equation (3.28) cannot be neglected. Thus, even for planar motion, a nonlinear model of string vibrations may be required when a resonance occurs or where the excitation amplitude is large.

3.5.2   Nonlinear Resonance

A nonlinear resonance curve is produced when the frequency is scanned with a moderate forcing amplitude, F. Figure 3.10 shows the results of both a backward and a forward scan, which can be constructed from a numerical solution of the Duffing oscillator, equation (3.28) (see Appendix C on *Ode* [14] for a description of the numerical methods). The two scans are identical except in the region marked by $\gamma_l < \gamma < \gamma_u$. Here, the forward scan produces the upper branch of the response curve. This upper branch makes a sudden jump to the lower branch at the frequency γ_u. Similarly, the backward (decreasing) scan makes a sudden jump to the upper branch at γ_l. In the region $\gamma_l < \gamma < \gamma_u$, at least two stable periodic orbits coexist. The sudden jump between these two orbits is indicated by the upward and downward arrows at γ_l and γ_u. This phenomenon is known as *hysteresis*.

The nonlinear response curve also reveals several other intriguing features. For instance, the maximum response amplitude no longer oc-

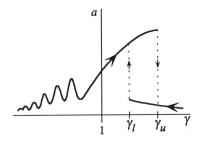

Figure 3.11: Nonlinear resonance curve showing secondary resonances in addition to the main resonance.

curs at $\gamma = 1$, but is shifted forward to the value γ_u. This is expected in the string because, as the string's vibration amplitude increases, its length increases, and this increase in length (and tension) is accompanied by a shift in the natural frequency of free oscillations.

Several *secondary resonances* are evident in Figure 3.11. These secondary resonances are the bumps in the amplitude resonance curve that occur away from the main resonance.

The main resonance and the secondary resonances are associated with periodic orbits in the system. The main resonance occurs near $\gamma = 1$ when the forcing amplitude is small and corresponds to the period one orbits in the system, those orbits whose period equals the forcing period. The secondary resonances are located near some rational fraction of the main resonance and are associated with periodic motions whose period is a rational fraction of γ. These periodic orbits (denoted by $\bar{x}$) can often be approximated to first order by a sinusoidal function of the form

$$\bar{x}_{m,n}(\tau) \approx A \cos(\frac{m}{n}\tau + \delta), \tag{3.35}$$

where A is the amplitude of the periodic orbit, (m/n) is its frequency, and δ is the phase shift. These periodic motions are classified by the integers m and n as follows ($m \neq 1, n \neq 1$):

$$
\begin{array}{llll}
\gamma & = & \omega/\omega_0 & = & m, & \text{an } ultraharmonic, \\
\gamma & = & \omega/\omega_0 & = & 1/n, & \text{a } subharmonic, \\
\gamma & = & \omega/\omega_0 & = & m/n, & \text{an } ultrasubharmonic.
\end{array}
$$

(a)   (b)   (c)

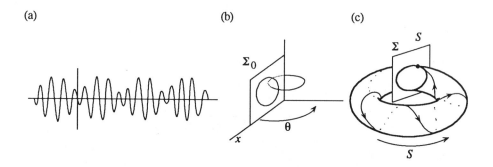

Figure 3.12: Amplitude-modulated, quasiperiodic motions on a torus.

Equation (3.35) is used as the starting point for the *method of harmonic balance*, a pragmatic technique that takes a trigonometric series as the basis for an approximate solution to the periodic orbits of a nonlinear system (see Prob. 3.12) [11]. It is also possible to have solutions to differential equations involving frequencies that are not rationally related. Such orbits resemble amplitude-modulated motions and are generally known as *quasiperiodic* motions (see Figure 3.12). A more complete account of nonlinear resonance theory is found in Nayfeh and Mook [10]. Parlitz and Lauterborn also provide several details about the nonlinear resonance structure of the Duffing oscillator [15].

3.5.3   Response Curve

In this section we focus on understanding the hysteresis found at the main resonance of the Duffing oscillator because hysteresis at the main resonance and at some secondary resonances is easy to observe experimentally. The results in this section can also be derived by the method of harmonic balance by taking $m = n = 1$ in equation (3.35); however, we will use a more general method that is computationally a little simpler.

We generally expect that in a nonlinear system the maximum response frequency will be detuned from its natural frequency. An estimate for this detuning in the undamped, free cubic oscillator,

$$x'' + (1 + \beta x^2)x = 0, \tag{3.36}$$

is obtained by studying this equation by the method of *slowly varying amplitude* [16]. Write[4]

$$x(\tau) = \frac{1}{2}[A(\tau)e^{i\gamma\tau} + A^*(\tau)e^{-i\gamma\tau}] \tag{3.37}$$

and substitute equation (3.37) into equation (3.36) while assuming $A(\tau)$ varies slowly in the sense that $|A''| << \gamma^2 A$. Then equation (3.36) is approximated by

$$(2i\gamma)A' + (1 - \gamma^2 + \frac{3\beta}{4}|A|^2)A = 0, \tag{3.38}$$

where we ignore all terms not at the driving frequency. Equation (3.38) has a steady-state solution in A, denoted by $a \in \mathbf{R}$. In this case,

$$\gamma^2 = 1 + \frac{3\beta}{4}|a|^2, \tag{3.39}$$

since $A' = 0$, and [17]

$$\bar{x}(\tau) = a\cos(\gamma\tau) . \tag{3.40}$$

To first order, the strength of the nonlinearity increases the normalized frequency by an amount depending on the amplitude of oscillation and the nonlinearity parameter. This approximate value for a Duffing oscillator is consistent with the results found in Appendix B where exact solutions for a cubic oscillator are presented.

Hysteresis is discovered when we apply the slowly varying amplitude approximation to the forced, damped Duffing equation,

$$x'' + \alpha x' + [1 + \beta x^2]x = F\cos(\gamma\tau). \tag{3.41}$$

Substituting equation (3.37) into equation (3.41), and again keeping only the terms at the first harmonic, we arrive at the complex amplitude equation

$$(\alpha + 2i\gamma)A' + \left(1 - \gamma^2 + i\alpha\gamma + \frac{3\beta}{4}|A|^2\right)A = F, \tag{3.42}$$

---

[4]If we write $A(\tau) = a(\tau) + ib(\tau)$, and $A^*(\tau) = a(\tau) - ib(\tau)$, then it is easy to show from Euler's identity that $x(\tau) = a(\tau)\cos(\gamma\tau) - b(\tau)\sin(\gamma\tau)$. Thus, the use of complex numbers is not essential in this calculation; it is merely a trick that simplifies some of the manipulations with the sinusoidal functions.

which in steady-state $(A' = 0)$ becomes

$$\left(1 - \gamma^2 + i\alpha\gamma + \frac{3\beta}{4}|\bar{A}|^2\right)\bar{A} = F. \tag{3.43}$$

To find the set of real equations for the steady state, write the complex amplitude in the form

$$\bar{A} = ae^{-i\delta} , \tag{3.44}$$

where both a and δ are real constants. Then equation (3.43) separates into two real equations,

$$\alpha\gamma a = F\sin\delta \tag{3.45}$$

and

$$(1 - \gamma^2 + \frac{3\beta}{4}a^2)a = F\cos\delta, \tag{3.46}$$

which collectively determine both the phase and the amplitude of the response. Squaring both equations (3.45) and (3.46) and then adding the results, we obtain a cubic equation in a^2,

$$[(\alpha\gamma)^2 + (1 - \gamma^2 + \frac{3\beta}{4}a^2)^2]a^2 = F^2, \tag{3.47}$$

illustrated in Figure 3.10, which is known as the *response curve*. In this approximation the steady-state response is given by

$$\bar{x} = a\cos(\gamma\tau - \delta), \tag{3.48}$$

where a is the maximum amplitude of the harmonic response determined from equation (3.47) and δ is the phase shift determined from equations (3.45 and 3.46).

3.5.4   Hysteresis

The response curve shown in Figure 3.10 is a plot of a versus γ calculated from equation (3.47), the nonlinear phase–amplitude relation. This curve shows that the string can exhibit hysteresis near a primary resonance; a slow scan of the variable γ (the so-called quasistatic approximation) results in a sudden jump between the two stable solutions indicated by the solid lines in Figure 3.10. The jump from the upper branch to the lower branch takes place at γ_u. The jump from the lower branch to the upper branch takes place at γ_l.

In the parameter regime $\gamma_l < \gamma < \gamma_u$, the response curve reveals the coexistence of three periodic orbits at the same frequency, but with different amplitudes. All these orbits are possible solutions to equation (3.47) for the values of a indicated in the diagram. All three orbits are harmonic responses (or period one orbits) since their frequency equals the forcing frequency. The middle solution, indicated by the dashed line in Figure 3.10, is an unstable periodic orbit.

3.5.5 Basins of Attraction

In a linear system with damping, the attracting periodic orbit is independent of the initial conditions. In contrast, the existence of two or more stable periodic orbits for the same parameter values in a nonlinear system indicates that the initial conditions play a critical role in determining the system's overall response. These attracting periodic orbits are called *limit cycles*, and their global stability is determined by constructing their basins of attraction. A very nice three-dimensional picture of the basins of attraction for the two stable periodic orbits found in the Duffing oscillator is presented by Abraham and Shaw [18]. However, this picture of the basins of attraction within the three-dimensional flow is very intricate. An equivalent picture of the basins of attraction constructed with a two-dimensional cross section and a Poincaré map is easier to understand.

A schematic for the basins of attraction in a Duffing oscillator with three coexisting orbits is portrayed in Figure 3.13. The cross section shows two stable orbits, P_1 and P_3, and one unstable orbit, P_2. In the region surrounding the inset of P_2, a small change in the initial conditions can produce a large change in the response of the system since initial conditions in this region can go to either attracting periodic orbit. The unstable periodic orbit indicated by P_2 is a saddle fixed point in the Poincaré map, and the stable periodic orbits are sinks in the Poincaré map.

The inset of the saddle is the collection of all points that approach P_2. This inset divides the Poincaré map into two distinct regions: the initial conditions that approach P_1 and the initial conditions that ap-

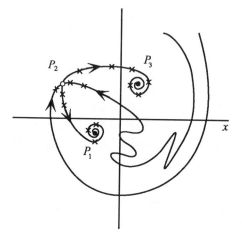

Figure 3.13: Schematic of the basins of attraction in the Duffing oscillator. (Adapted from Hayashi [11].)

proach P_3. That is, the inset to the saddle determines the boundary separating the two basins of attraction. Again, we see the importance of keeping track of the unstable solutions, as well as the stable solutions, when analyzing a nonlinear system. Figure 3.14(a) should be compared to—but not confused with—Figure 3.6(e), the phase plane for the unforced, damped Duffing oscillator. In the Poincaré map each fixed point represents an entire periodic orbit, not just an equilibrium point of the flow as in Figure 3.6. More importantly, in the Poincaré map, the *inset to the saddle point at P_2 is* not *a trajectory of the flow.* Rather, it is the collection of all initial conditions that converge to P_2. The approach of a single orbit toward P_2 is a sequence of discrete points, indicated by the crosses ($\times$) in Figure 3.13. In general, the inset and the outset of the saddle represent an infinite continuum of distinct orbits, all of which share a common property: namely, they arrive at or depart from a periodic point of the map.

(a) (b)

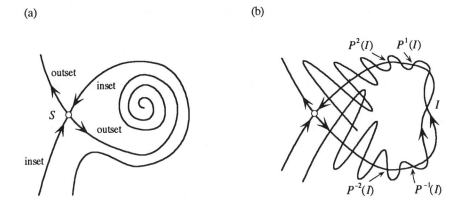

Figure 3.14: Schematic of a homoclinic tangle in the Duffing oscillator.

3.6 Homoclinic Tangles

Figure 3.13 shows the Poincaré map for the flow arising in the Duffing oscillator in a parameter region where hysteresis exists. We see that the inset to P_2 consists of two different curves. Similarly, the outset of P_2 also consists of two different curves. One branch of the outset approaches P_1, while the other branch approaches P_3. The inset and the outset of the saddle are not trajectories in the flow, so they can intersect without violating a fundamental theorem of ordinary differential equations, the unique dependence of an orbit with respect to initial conditions.

It is possible to find parameter values so that the inset and the outset of the saddle point at P_2 do indeed cross. A self-intersection of the inset of a saddle with its outset is illustrated in Figure 3.14(b) and, as originally observed by Poincaré, it always gives rise to wild oscillations about the saddle (see section 4.6.2).

Such a self-intersection of the inset of a saddle with its outset is called a *homoclinic intersection*, and it is a fundamental mechanism by which chaos is created in a nonlinear dynamical system. The reason is roughly the following. Consider a point at a crossing of the inset and the outset indicated by the point I in Figure 3.14(b). By definition, this point is part of an orbit that approaches the saddle by both its inset and its outset; that is, it is doubly asymptotic. Consider the next intersection point of I with the cross section, $P^1(I)$. This point must lie on the inset at a point closer to the saddle. Next, the second iterate of I under the Poincaré map, $P^2(I)$, must be even closer to the saddle. Similarly, the preimage of I approaches the saddle along the outset of the saddle. The outset and inset get bunched up near the saddle, creating an image known as a *homoclinic tangle*. Homoclinic tangles beat at the heart of chaos because, in the region of a homoclinic tangle, initial conditions are subject to a violent stretching and folding process, the two essential ingredients for chaos. A marvelous pictorial description of homoclinic tangles along with an explanation as to their importance in dynamical systems is presented by Abraham and Shaw [18].

Homoclinic tangles are often associated with the existence of strange sets in a system. Indeed, it is thought that in many instances a strange attractor is nothing but the closure[5] of the outset of some saddle when this outset is bunched up in a homoclinic tangle. Figure 3.15(a) shows the cross section for a strange attractor of the Duffing oscillator. Figure 3.15(b) shows the cross section of the outset of the period one saddle in this strange set for the exact same parameter values. The resemblance between these two structures is striking. Indeed, developing methods to dissect homoclinic tangles will be central to the study of chaos in low-dimensional nonlinear systems. In fact, one could call it <u>the</u> problem of low-dimensional chaos.

3.7   Nonplanar Motions

Additional dynamical possibilities arise when we consider nonplanar string vibrations. These vibrations are also easy to excite with the string apparatus described in section 3.2. When a nonmagnetic wire is used, out-of-plane motions are observed which are sometimes called ballooning or whirling motions. These nonplanar vibrations arise even when the excitation is only planar.

Indeed, ballooning motions are hard to avoid. Imagine scanning the forcing frequency of the string apparatus through a resonance. The re-

---

[5]The closure of a set X is the smallest closed set that is a superset of X.

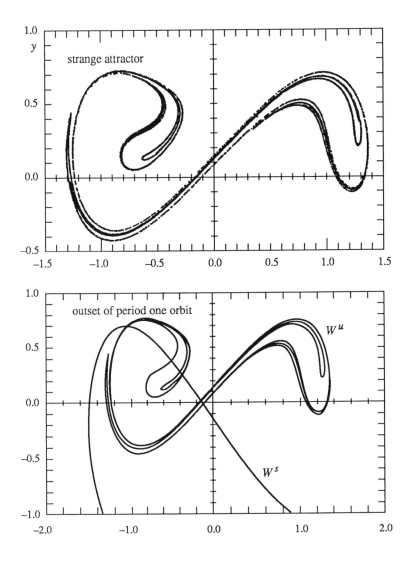

Figure 3.15: Comparison of a strange attractor and the outset of a period one saddle in the Duffing oscillator.

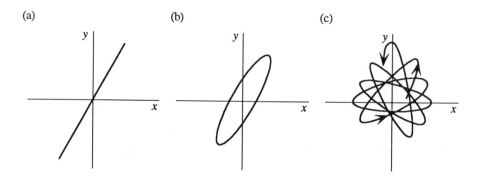

Figure 3.16: Planar periodic, elliptical (nonplanar) periodic, and precessing (quasiperiodic) motions of a string.

sponse of the string increases as the resonance frequency is approached, and the following behavior is typically observed. Well below the resonance frequency, the string responds with a planar, periodic oscillation (Fig. 3.16(a)). As the forcing frequency is increased, the amplitude of the response also grows until the string "pops out of the plane" and begins to move in a nonplanar, elliptical, periodic pattern (Fig. 3.16(b)). That is, the string undergoes a bifurcation from a planar to a nonplanar oscillation. At a still higher frequency the elliptical periodic orbit becomes unstable and begins to precess, as illustrated in Figure 3.16(c).

We will present a more complete qualitative account of these whirling motions in section 3.7.2, which is based on the recent work of Johnson and Bajaj [19], Miles [20], and O'Reilly [21]. Now, though, we turn our attention to the whirling motion that occurs when no forcing is present.

3.7.1   Free Whirling

If we pluck a string hard and look closely, we typically see the string whirling around in an elliptical pattern with a diminishing amplitude. Some understanding of these motions is obtained by considering the free planar oscillations of a string modeled by the two-dimensional equation

$$\ddot{\mathbf{r}} + \lambda \dot{\mathbf{r}} + \omega_0^2 (1 + K \mathbf{r}^2)\mathbf{r} = 0, \qquad (3.49)$$

which is equation (3.7) with no forcing term. We noted in section 3.3 that the linear approximation to equation (3.7) results in elliptical motion (eq. (3.13)). We shall use this observation to calculate an approximate solution to equation (3.49) using a procedure put forth by Gough [2]; similar results were obtained by Elliot [9].

Transform the problem of nonlinear free vibrations to a reference frame rotating with an angular frequency Ω, with Ω to be determined. In this rotating frame, equation (3.49) becomes

$$\ddot{\mathbf{u}} + \lambda\dot{\mathbf{u}} + 2\boldsymbol{\Omega} \times \dot{\mathbf{u}} + \lambda\boldsymbol{\Omega} \times \mathbf{u} - \Omega^2\mathbf{u} + \omega_0^2(1 + K\mathbf{u}^2)\mathbf{u} = 0, \qquad (3.50)$$

where $\mathbf{u}$ is the new radial displacement vector, subjected to the addition of Coriolis and centrifugal accelerations. Let us now look for a solution of the form

$$\mathbf{u}(t) = [x(t), y(t)] =$$
$$e^{-\lambda t/2}[X_1 \cos \tilde{\omega}t + X_3 \cos 3\tilde{\omega}t, \ Y_1 \sin \tilde{\omega}t + Y_3 \sin 3\tilde{\omega}t], \quad (3.51)$$

where X_3 and Y_3 are small compared to X_1 and Y_1. Looking at the x coordinate only, when we substitute equation (3.51) into equation (3.50) and discard appropriate higher-order terms, we get

$$(\omega_0^2 - \Omega^2 - \tilde{\omega}^2)X_1 \cos \tilde{\omega}t + (\omega_0^2 - \Omega^2 - 9\tilde{\omega}^2)X_3 \cos 3\tilde{\omega}t -$$
$$2\Omega\tilde{\omega}Y_1 \cos \tilde{\omega}t + \omega_0^2 K(X_1^2 \cos^2 \tilde{\omega}t + Y_1^2 \sin^2 \tilde{\omega}t)e^{-\lambda t}X_1 \cos \tilde{\omega}t = 0;$$

a similar relation holds for the y coordinate. On equating sinusoidal terms of the same frequency we—after considerable algebra—discover

$$\tilde{\omega}^2 = \omega_0^2 \left[1 + \frac{3K}{4}(X_1^2 + Y_1^2)e^{-\lambda t} \right] - \Omega^2, \qquad (3.52)$$

$$\frac{\tilde{\omega}\Omega}{\omega_0^2} = \frac{-K}{4}X_1 Y_1 e^{-\lambda t}, \qquad (3.53)$$

and

$$\frac{X_3}{X_1} = \frac{Y_3}{Y_1} = \left(\frac{K}{4} \right) \frac{\omega_0^2(X_1^2 - Y_1^2)e^{-\lambda t}}{(9\tilde{\omega}^2 - \omega_0^2 + \Omega^2)}. \qquad (3.54)$$

If there is no damping ($\lambda = 0$), then this approximate solution is periodic in the rotating reference frame and is slightly distorted from an elliptical orbit. The angular frequency $\tilde{\omega}$ is detuned from ω_0 by an amount proportional to the mean-square radius $X_1^2 + Y_1^2$. In the original stationary reference frame, equation (3.53) shows us that the orbit precesses at a rate Ω proportional to the orbital area $\pi X_1 Y_1$. The angular frequency $\tilde{\omega}$ in equations (3.52) to (3.54) is measured in the rotating reference frame. It is related to the angular frequency in the stationary reference frame ω by

$$\omega^2 = (\tilde{\omega} + \Omega)^2 = \omega_0^2 \left\{ 1 + \frac{K}{4}[3(X_1^2 + Y_1^2) - 2X_1 Y_1]e^{-\lambda t} \right\}. \qquad (3.55)$$

Thus, in the stationary reference frame, the undamped motion is quasiperiodic unless $\tilde{\omega}$ and Ω are accidentally commensurate, in which case the orbit is periodic. The damped oscillations are also elliptical in character and precess at a rate Ω. In both cases, the detuning given by equation (3.55) is due to two sources: the nonlinear planar motion detuning plus a detuning resulting from the precessional frequency.

3.7.2   Response Curve

The case of whirling motions subject to a planar excitation is described by the equation

$$\ddot{\mathbf{r}} + \lambda \dot{\mathbf{r}} + \omega_0^2(1 + K\mathbf{r}^2)\mathbf{r} = (A \cos \omega t, 0), \qquad (3.56)$$

where the phase space is four-dimensional: $(x, v_x, y, v_y) \in \mathbf{R}^4$. The extended phase space, when we add the forcing variable, is five-dimensional: $(x, v_x, y, v_y, \theta) \in \mathbf{R}^4 \times S^1$.

Equation (3.56) can be analyzed for periodic motions by a combination of averaging and algebraic techniques not unlike the harmonic balance method. Because our text is an experimental introduction to nonlinear dynamics, we present here a qualitative description of the results of this analysis. For further details see references [19], [20], and [21].

As mentioned in the introduction to this section, an experimental frequency scan that passes through a main resonance can result in the following sequence of motions:

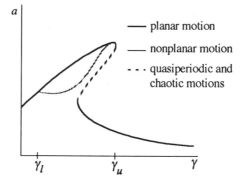

Figure 3.17: Response curve for planar and nonplanar motion. (Adapted from Johnson and Bajaj [19].)

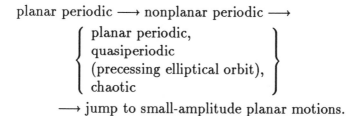

planar periodic $\longrightarrow$ nonplanar periodic $\longrightarrow$

$$\left\{ \begin{array}{l} \text{planar periodic,} \\ \text{quasiperiodic} \\ \text{(precessing elliptical orbit),} \\ \text{chaotic} \end{array} \right\}$$

$\longrightarrow$ jump to small-amplitude planar motions.

The basic features of these experimental observations agree with those predicted by equation (3.56), and are summarized in the response curve shown in Figure 3.17.

In the parameter range $\gamma_l < \gamma < \gamma_u$, the response curve indicates the coexistence of three planar periodic motions and one nonplanar periodic orbit. In this parameter regime, the planar periodic orbit becomes unstable; the string "pops out of the plane" and begins to execute a whirling motion. At some parameter value γ_q, the nonplanar periodic motion itself becomes unstable and the system may do any number of things depending on the exact system parameters and initial conditions. For instance, it may hop to the small-amplitude planar periodic orbit. Or the ballooning orbit itself may become unstable and begin to precess (quasiperiodic motion). In addition, chaotic motions can sometimes be observed in this parameter range. These various dynamical possibilities are illustrated schematically in Figure 3.17. We repeat, the motion observed depends on the exact system parameters and the initial conditions, because there can be many coexisting attractors with complicated basins of attraction in this region. In particular, the chaotic motions are difficult to isolate (and observe experimentally) without a thorough understanding of the system.

3.7.3 Torus Attractor

To construct a cross section for nonplanar periodic motion we could imagine a plot of the position of the center of the string, and the forcing phase, $(x, y, \theta) \in \mathbf{R}^2 \times S^1$ (Fig. 3.18). A map can be associated to an orbit in $\mathbf{R}^2 \times S^1$ by recording the position of the orbit once each forcing period. Although this map is not a true Poincaré map, it is easy to obtain experimentally and will be useful in explaining the notion of a torus attractor (see section 3.8.1).[6]

An elliptical periodic orbit in the flow is represented in this cross section by a discrete set of points that lie on a closed curve. This curve is topologically a circle, S^1 (Fig. 3.18(b)). Similarly, a precessing ellipse (quasiperiodic motion) can generate an infinite number of points; these points fill out this circle (Fig. 3.18(c)).

In the extended space (x, y, θ), this quasiperiodic motion represents a dense winding of a torus as shown in Figure 3.18(d). Topologically, a torus is a space constructed from the Cartesian product of two circles, $T^2 = S^1 \times S^1$. In general, an n torus is constructed from n copies of a circle,

$$T^n = \overbrace{S^1 \times S^1 \cdots \times S^1}^{n},$$

and a *torus attractor* naturally arises whenever quasiperiodic motion is encountered in a dissipative dynamical system.[7] The torus is an

---

[6]A proper cross section would be a manifold transverse to the flow in $\mathbf{R}^4 \times S^1$, i.e., a four manifold such as $\Sigma = \{(x, v_x, y, v_y, \theta) \mid \theta = 0\}$. The torus attractor arises from a Hopf bifurcation—a bifurcation from a fixed point to an invariant curve. In a cross section, the limit cycle is represented by a fixed point. At the transition to quasiperiodic motion, this fixed point loses stability and gives birth to an invariant circle, which is a cross section of the torus in the flow.

[7]The attracting torus for nonplanar string vibrations is actually a four torus, T^4.

(a)

(b)

(c)

(d)

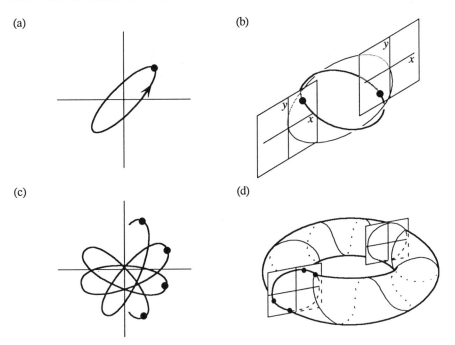

Figure 3.18: Experimental cross section for the nonplanar string vibrations.

attractor because it is an invariant set and an attracting limit set. This is illustrated in Figure 3.19, which shows how orbits are attracted to a torus. A graph of a quasiperiodic orbit on a torus attractor is an amplitude-modulated time series (Fig. 3.12).

3.7.4 Circle Map

We found that the single-humped map of the interval, $f_\lambda : I \longrightarrow I$, was a good model for some aspects of the dynamics of the bouncing ball system. Similarly, insight about motion near a torus attractor can be gained by studying a *circle map*,

$$g : S^1 \longrightarrow S^1,$$

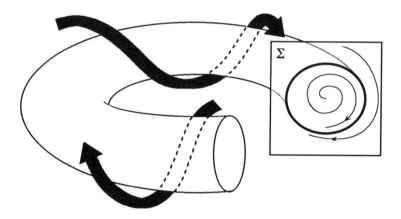

Figure 3.19: Torus attractor.

where the mapping g most often studied is a two-parameter map of the form

$$\theta_{n+1} = g(\theta_n) = \theta_n + \Omega - \frac{K}{2\pi}\sin(2\pi\theta_n), \theta_n \in [0,1). \qquad (3.57)$$

This map has a linear term θ_n, a constant bias term Ω, and a nonlinear term whose strength is determined by the constant K.

The frequency of the circle map is monitored by the *winding number*

$$W = \lim_{n\to\infty}\frac{g_{\Omega,K}^n(\theta) - \theta}{n}. \qquad (3.58)$$

If the nonlinear term K equals zero, then $W = \Omega$. The winding number measures the average increase in the angle θ per unit time (Fig. 3.20). An orbit of the circle map is periodic if, after q iterations, $\theta_{n+q} = \theta_n + p$, for integers p and q. The winding number for a periodic orbit is $W = p/q$. A quasiperiodic orbit has an irrational winding number [22]. Circle maps have a devilish dynamical structure, which is explored in reference [22].

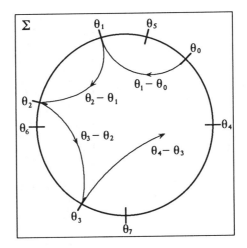

Figure 3.20: The winding number measures the average increase in the angle of the circle map.

3.7.5 Torus Doubling

A nonlinear system can make a transition from quasiperiodic motion directly to chaos. This is known as the *quasiperiodic route to chaos*. It is of great practical and historical importance since it was one of the first proposed mechanisms leading to the formation of a strange attractor [23].

There are, in fact, many routes to chaos even from a humble T^2 torus attractor. For instance, when the T^2 attractor loses stability, a stable higher-dimensional torus attractor sometimes forms. Another possibility in the string system is the formation of a doubled torus, illustrated schematically in Figure 3.21. In the *torus doubling route to chaos*, our original torus (which is a closed curve in cross section) appears to split into two circles at the torus doubling bifurcation point [24]. The torus doubling route to chaos is reminiscent of the period doubling route to chaos. However, it differs in at least two significant ways. First, in most experimental systems, there are only a finite number of torus doublings before the onset of chaotic motion. In fact, no more than two torus doublings have ever been observed in the string experiment. Second, the torus doubling route to chaos is a higher-dimensional phenomenon,

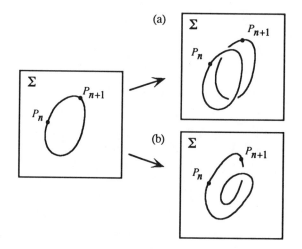

Figure 3.21: Schematic of a torus doubling bifurcation.

requiring at least a four-dimensional flow, or a three-dimensional map. It is not observed in one-dimensional maps, unlike the period doubling route to chaos.

Now that we have reviewed some of the more salient dynamical features of a string's motions, let's turn our attention to assembling the tools required to view these motions in a real string experiment.

3.8 Experimental Techniques

The dynamics of a forced string raises experimental challenges common to a variety of nonlinear systems. In this section we describe a few of the experimental diagnostics that help with the visualization and identification of different attractors, such as:

    equilibrium points

    limit cycles (periodic orbits)

    invariant tori (quasiperiodic orbits)

    strange attractors (chaotic orbits)

The main tools used in the *real-time* identification of an attractor are Fourier power spectra and real-time Poincaré maps. In addition, the correlation dimension calculated from an experimental time series helps to confirm the existence of a strange attractor, as well as providing a measure of its fractal structure.

The terms "strange attractor" and "chaotic attractor" are not always interchangeable. Specifically, a *strange attractor* is an attractor that is a fractal. That is, the term strange refers to a static geometric property of the attractor. The term *chaotic attractor* refers to an attractor whose motions exhibit sensitive dependence on initial conditions. That is, the term chaotic refers to the dynamics on the attractor. We mention this distinction because it is possible for an attractor to be strange (a fractal), but not chaotic (exhibit sensitive dependence on initial conditions) [25]. Experimental methods are available for quantifying both the geometric structure of an attractor (fractal dimensions) and the dynamic properties of orbits on an attractor (Lyapunov exponents).

3.8.1 Experimental Cross Section

Variables measured directly in the string apparatus include the forcing phase, $\theta(t)$, and the displacement amplitudes of the string, $x(t)$ and $y(t)$. The phase is measured directly from the function generator that provides the sinusoidal current to the wire. The string displacement is measured from the optical detectors that record the horizontal and vertical displacement at a fixed point along the wire. In addition, the wire's velocity can be measured by sending the amplitude displacement signal through a differentiator, a circuit that takes the analog derivative of an input signal.

Many approaches are possible for constructing an experimental Poincaré section. The particular approach taken depends on both the type of system and the equipment at hand. Here, we assume that the lab is stocked with a dual-trace storage oscilloscope and some basic electronic components. A different approach might be taken, for instance, if we have access to a digital oscilloscope either in the form of a commercial instrument or a plug-in board to a microcomputer.

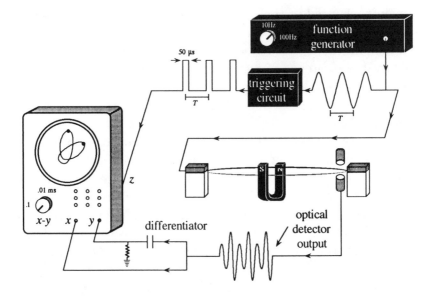

Figure 3.22: Schematic for the construction of an experimental Poincaré map for the string apparatus.

Planar Cross Section

We will record the Poincaré section on the storage oscilloscope. Our first step is to adjust the optical detectors so that the axis for a purely planar vibration is well aligned with one of the optical detectors. Next, the signal from this optical detector is sent to one of the input channels of the oscilloscope. The other channel is used to record the velocity, via the differentiator, of this same signal. Lastly, the time-base on the oscilloscope must be set to X-Y mode, thereby allowing both the horizontal and vertical oscilloscope sweeps to be controlled by the external signals.

The result, as shown schematically in the oscilloscope in Figure 3.22, is an experimental rendering of the phase space trajectory for a string. To construct a Poincaré map, we need to sample this trajectory once each forcing period. That is, instead of recording the entire trajectory, we only want to record a sequence of points on this trajectory. This can be accomplished by turning the oscilloscope's beam intensity on for a brief moment once each forcing period. This is easy to do because on the back of most oscilloscopes is an analog input line labeled "z" that controls the oscilloscope's beam intensity. Finally, we need a *triggering* circuit that takes as input the sinusoidal forcing signal and generates as output a clock pulse, which is used to briefly turn on the oscilloscope's beam once each forcing period.

Triggering Circuit

A simple triggering circuit can be constructed from a monostable vibrator and a Schmitt trigger. Here, though, we describe a slightly more sophisticated approach based on a phase-locked loop (PLL). The phase-locked loop circuit has the advantage that it allows us to trigger more than once each forcing period. This feature will be useful when we come to digitizing a signal because the phase-locked loop circuit can be used to trigger a digitizer an integer number k times each forcing period, thereby giving us k samples of the trajectory each cycle.

A good account of all things electronic, including phase-locked loops, is presented in the book by Horowitz and Hill, *The art of electronics*. For our purposes, a phase-locked loop chip contains a phase detector, an amplifier, and a voltage-controlled oscillator (VCO), in one package. A PLL, when used in conjunction with a stage counter, generates a clock signal (or triggering pulse) that is ideal for constructing a Poincaré section. A schematic of the triggering circuit used with the string apparatus is presented in Figure 3.23, and is constructed from two off-the-shelf chips: a CMOS 4046 PLL and a 4040 stage counter. This circuit, with small adjustments, is useful for generating a triggering signal in any forced system. The counter is set to one for a Poincaré section; that is, it generates one pulse each period. It can be readjusted to produce k pulses per period when digitizing.

Nonplanar Cross Section

To generate an experimental cross section for nonplanar motions we replace the $v_x(t)$ input to the oscilloscope with the output $y(t)$ from the second displacement detector. The resulting plot on the oscilloscope is proportional to the actual horizontal and vertical displacement of the string, thus providing us with a magnified view of the string's whirling.

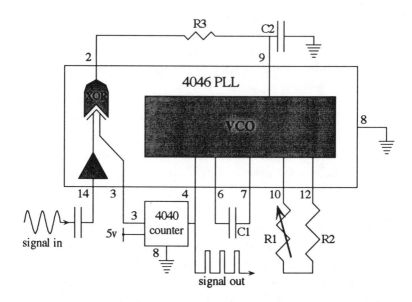

Figure 3.23: Triggering circuit used to construct a Poincaré map in the string apparatus. (Courtesy of K. Adams and T. C. A. Molteno.)

3.8.2   Embedding

In the string system there is no difficulty in specifying and measuring the major system variables. Usually, though, we are not so lucky. Imagine an experimental dynamical system as a black box that generates a time series, $x(t)$. In practice, we may know little about the process inside the black box. Therefore, the experimental construction of a phase space trajectory, or a Poincaré map, seems very problematic.

One approach to this problem—the experimental reconstruction of a phase space trajectory—is as follows. We start out by assuming that the time series is produced by a deterministic dynamical system that can be modeled by some nth-order ordinary differential equation. For this particular example we assume that the system is modeled by a third-order differential system, as is the case for planar string vibrations. Then a given trajectory of the system is uniquely specified by the value of the time series and its first and second derivatives at time $t_0 = 0$:

$$x(t_0),\ \dot{x}(t_0),\ \ddot{x}(t_0).$$

This suggests that in reconstructing the phase space we can begin with our measured time series, and then use $x(t)$ to calculate two new phase space variables $y(x(t))$ and $z(x(t))$ defined by

$$y(t) = \frac{d}{dt}x(t),$$

$$z(t) = \frac{d}{dt}y(t).$$

In estimating the pointwise derivatives of $x(t)$ in an experiment we can proceed in at least two ways: first, we can process the original signal through a differentiator, and then record (digitize) the original signal along with the differentiated signals; or second, we could digitize the signal, and then compute the derivatives numerically. While both techniques are feasible, each is fraught with experimental difficulties because differentiation is an inherently noisy process. This is because approximating a derivative often involves taking the difference of two numbers that are close in value. To see this, consider the numerical derivative of a digitized time series $\{x(t_i)\}$ defined by

$$y_i = \frac{x_i - x_{i-1}}{t_i - t_{i-1}}. \tag{3.59}$$

For instance, let $x_{i-1} = 1.0 \pm 0.1$, $x_i = 1.1 \pm 0.1$, and $t_i - t_{i-1} = 1$. Then $y_i = 0.1 \pm 0.2$; that is, the value of the first derivative is already buried in the noise, and the problem just gets worse when taking higher-order derivatives.

However, let's look at equation (3.59) again. If the sampling time is evenly spaced, then $t_i - t_{i-1}$ is constant, so

$$y_i \propto x_i - x_{i-1}.$$

That is, almost all the information about the derivative is contained in a variable constructed by taking the difference of two points in the original time series $\{x(t_i)\}$. This idea can be generalized as follows. Instead of defining the new variables for the reconstructed phase space in terms of the derivatives, we can recover almost all of the same information about an orbit from the *embedded variables* defined by [26]:

$$y_i = x_{i-r}, \tag{3.60}$$

$$z_i = x_{i-s}, \quad s \neq r, \tag{3.61}$$

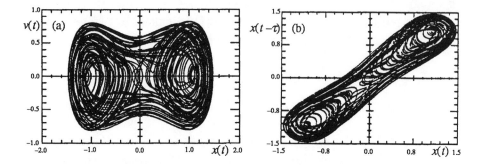

Figure 3.24: Trajectory of the Duffing oscillator in (a) phase space and (b) the embedded phase space with delay time $\tau = 0.8$.

where r and s are integers. Each new embedded variable is defined by taking a time delay of the original time series. Clearly, any number of embedded variables can be created in this way, and this method can be used to reconstruct a phase space of dimension far greater than three.

There are many technical issues associated with the construction of an embedded phase space. An in-depth discussion of these issues can be found in reference [1]. The first concern is determining a good choice for the *delay times*, r and s. One rule of thumb is to take r to be small, say 3 or 4. In fact, we could define the second variable as

$$v_i = x_i - x_{i-r}, \tag{3.62}$$

and think of it as a velocity variable. The second embedding time should be much larger than r, but not too large. To be more specific, consider the planar oscillations of a string again. In this example a natural cycle time is given by the period of the forcing term. A good choice for s is some sizable fraction of the cycle time. For instance, let's say our digitizer is set to sample the signal 64 times each period. Then a sensible choice for r might be 4, and for s might be 16, or one-quarter of the forcing period. Figure 3.24 shows a plot of a chaotic trajectory in the Duffing oscillator in both the original phase space and the phase space reconstructed from the embedding variables. The similarity of the two representations lends support to the claim that a trajectory in the embedded phase space provides a faithful representation of the dynamics.

Constructing a real-time two-dimensional embedded phase space is straightforward. An embedded signal is obtained by sending the original signal through a delay line. The current signal and the delayed signal are sent to the oscilloscope, thereby giving us a real-time representation of the phase space dynamics from our black box.

3.8.3 Power Spectrum

A signal from a nonlinear process is a function of time which we will call $F(t)$ in this section. It is possible to develop signatures for periodic, quasiperiodic, and chaotic signals by analyzing the periodic properties of $F(t)$. These signatures, which are based on Fourier analysis, are valuable experimental aids in identifying different types of attractors.

Fourier Series

To analyze the periodic properties of $F(t)$ it is useful to uncover a functional representation of the signal in terms of the orthogonal functions $\cos(2\pi t/L)$ and $\sin(2\pi t/L)$ of period L. To determine the *Fourier series* for $F(t)$ we must find the constants a_k and b_k so that the following identity holds [27]:

$$F(t) = \frac{a_0}{2} \ + \ a_1 \cos\frac{2\pi}{L}t + b_1 \sin\frac{2\pi}{L}t$$
$$+ \ a_2 \cos\frac{2\pi}{L}2t + b_2 \sin\frac{2\pi}{L}2t + \cdots. \qquad (3.63)$$

Fourier showed that the constants a_k and b_k in equation (3.63) are computed from $F(t)$ by means of the integral formulas

$$a_k = \frac{2}{L}\int_0^L F(t) \cos\left(\frac{2\pi}{L}kt\right) dt \quad k = 0, 1, 2, 3, \ldots \qquad (3.64)$$

and

$$b_k = \frac{2}{L}\int_0^L F(t) \sin\left(\frac{2\pi}{L}kt\right) dt \quad k = 1, 2, 3, 4, \ldots. \qquad (3.65)$$

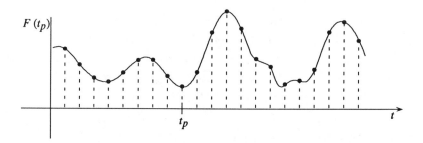

Figure 3.25: Digitized time series.

Finite Fourier Series

In equations (3.63–3.65) we are assuming that $F(t)$ is a continuous function of t. Equations (3.64) and (3.65) dictate that to calculate the kth constants, a_k and b_k in the Fourier series (eq. (3.63)), we substitute $F(t)$ into the previous equations and integrate. In experiments with digitized data, the signal we actually work with is an equally spaced discrete set of points, $F(t_p)$, measured at the set of times $\{t_p\}$ (Fig. 3.25). Therefore, we need to develop a discrete analog to the Fourier series, the *finite Fourier series*.

Let us consider an even number of points per period, $2N$. Then the $2N$ sample points are measured at

$$0, \ \frac{L}{2N}, \frac{2L}{2N}, \dots, \frac{(2N-1)L}{2N},$$

or more succinctly,

$$t_p = \frac{pL}{2N}, \quad p = 0, 1, 2, \dots, 2N-1. \tag{3.66}$$

The finite Fourier series for a function sampled at $F(t_p)$ is

$$F(t) = \frac{A_0}{2} + \sum_{k=1}^{N-1} \left(A_k \cos \frac{2\pi}{L} kt + B_k \sin \frac{2\pi}{L} kt \right) + \frac{A_N}{2} \cos \frac{2\pi}{L} Nt, \tag{3.67}$$

where

$$A_k = \frac{1}{N} \sum_{p=0}^{2N-1} F(t_p) \cos \frac{2\pi}{L} kt_p, \quad k = 0, 1, \dots, N \tag{3.68}$$

and

$$B_k = \frac{1}{N} \sum_{p=0}^{2N-1} F(t_p) \sin \frac{2\pi}{L} k t_p, \quad k = 1, 2, \ldots, N-1. \qquad (3.69)$$

In the above formulas, we assumed that the function is sampled at $2N$ points, that the value at the point $2N + 1$ is L, and that the endpoints 0 and L satisfy the periodic boundary condition,

$$F(0) = F(L).$$

If the latter assumption does not hold, the convention is to average the values at the endpoints so that the value at $F(0)$ is taken to be

$$\frac{F(0) + F(L)}{2}.$$

In this case the formulas for the coefficients are

$$A_k = \frac{1}{N} \left[\frac{F(0)}{2} + \sum_{p=1}^{2N-1} F(t_p) \cos \frac{2\pi}{L} k t_p + \frac{F(L)}{2} \right] \qquad (3.70)$$

and

$$B_k = \frac{1}{N} \sum_{p=1}^{2N-1} F(t_p) \sin \frac{2\pi}{L} k t_p. \qquad (3.71)$$

Equations (3.68 and 3.69) can be viewed as a transform or mapping. That is, given a set of numbers $F(t_p)$, these relations generate two new sets of numbers, $A(k)$ and $B(k)$. It is easy to program this transform. The code to calculate the discrete Fourier transform of $F(t_p)$ involves a double loop (see eqs. (3.68 and 3.69)): the inner loop cycles through the index k, and the outer loop covers the index p. Each loop has order N steps, so the total number of computations is of order N^2. The discrete Fourier transform is, therefore, quite slow for large N (say $N > 500$) (see Appendix D for programs to calculate discrete Fourier transforms). Fortunately, there exists an alternative method for calculating the Fourier transform called the fast Fourier transform or FFT. The FFT is an $N \ln N$ computation, which is much faster than the discrete Fourier for large data sets. A detailed explanation of the FFT along with C code examples is found in *Numerical Recipes* [28].

Power Spectrum

The amplitude coefficients A_k and B_k give us a measure of how well the signal fits the kth sinusoidal term. A plot of A_k versus k or B_k versus k is called a *frequency spectrum*. The (normalized) power spectrum amplitude is defined by

$$H_k = \sqrt{A_k^2 + B_k^2}. \tag{3.72}$$

A plot of H_k versus k is called the *power spectrum*. This graphical representation tells us how much of a given frequency is in the original signal.

Experimental power spectra can be obtained in at least three ways: (i) obtain a spectrum analyzer or signal analyzer, which is a commercial instrument dedicated to displaying real-time power spectra;[8] (ii) obtain a signal analyzer card for a microcomputer (this is usually less expensive than option (i)); or (iii) digitize your data and write an FFT for your computer (this option is the cheapest). Having successfully procured spectra capabilities, we now move on to describing how to use them.

Spectral Signatures

The spectral signatures for periodic, quasiperiodic, and chaotic motion are illustrated in Figure 3.26. The power spectrum of a period one orbit is dominated by one central peak, call it ω_1. The power spectrum of a period two orbit also has a sharp peak at ω_1, and additional peaks at the subharmonic $\omega/2$ and the ultrasubharmonic $3\omega/2$. These new spectral peaks are the "sidebands" about the primary frequency that come into existence through, say, a period doubling bifurcation of the period one orbit. More generally, the power spectrum of a period n orbit consists of a collection of discrete peaks showing the primary frequency and its overtones. In periodic motion, all the peaks are rationally related to the primary peak (resonance).

---

[8]All things are fair in love, war, and experimental physics. Methods to obtain a spectrum analyzer may include: (a) locating and "nationalizing" a spectrum analyzer from a nearby laboratory, or (b) locating and appropriating a spectrum analyzer on grounds of "national security." Another option is to use an older model spectrum analyzer, such as a Tektronix 1L5 spectrum analyzer which plugs into an older model Tek scope.

Time Series   **Power Spectrum**

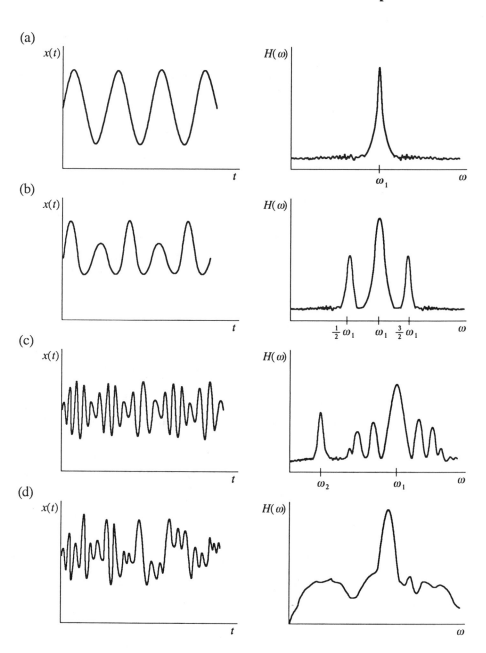

Figure 3.26: Time series and power spectra: (a) periodic (period one), (b) periodic (period two), (c) quasiperiodic, (d) chaotic.

Quasiperiodic motion is characterized by the coexistence of two incommensurate frequencies. Thus, the power spectrum for a quasiperiodic motion is made up from at least two primary peaks, ω_1 and ω_2, which are not rationally related. Additionally, each of the primary peaks can have a complicated overtone spectrum. The mixing of the overtone spectra from ω_1 and ω_2 usually allows one to distinguish periodic motion from quasiperiodic motion. A quick examination of the time series, or the Poincaré section, can also help to distinguish periodic motion from quasiperiodic motion.

The power spectrum of a chaotic motion is easy to distinguish from periodic or quasiperiodic motion. Chaotic motion has a broad-band power spectrum with a rich spectral structure. The broad-band nature of the chaotic power spectrum indicates the existence of a continuum of frequencies. A purely random or noisy process also has a broad-band power spectrum, so we need to develop methods to distinguish noise from chaos. In addition to its broad-band feature, a chaotic power spectrum can also have many broad peaks at the nonlinear resonances of the system. These nonlinear resonances are directly related to the unstable periodic orbits embedded within the chaotic attractor. So the power spectrum of a chaotic attractor does provide some limited information concerning the dynamics of the system, namely, the existence of unstable periodic orbits (nonlinear resonances) that strongly influence the recurrence properties of the chaotic orbit.

Additionally, in the periodic and quasiperiodic regimes, new humps and peaks can appear in a power spectrum whenever the system is near a bifurcation point. These spectral features are called *transient precursors* of a bifurcation. A detailed theory of these precursors with many practical applications has been developed by Wiesenfeld and coworkers [29].

3.8.4   Attractor Identification

So far we have discussed four measurements that allow us to visualize and identify the attractor coming from a nonlinear process:

time series

power spectra

phase space portrait, or reconstructed phase space

experimental Poincaré sections

All these qualitative techniques can be set up with instruments that are commonly available in any laboratory.

To get a time series we hook up the output signal from the nonlinear process to an input channel of the oscilloscope and use the time base of the scope to generate the temporal dimension of the plot. To obtain a power spectrum, we use a spectrum analyzer or digitize the data and use an FFT. An experimental phase space portrait can be plotted on an oscilloscope either by recording two system variables directly, such as (x, y) or $(x, \dot{x})$, or from the delayed variable $(x(t), x(t - \tau))$, where τ is the delay time. Lastly, in a forced system, the Poincaré section is obtained from the phase space trajectory by strobing it once each forcing period using the "z" blanking on the back of the oscilloscope.

Now to identify an attractor, we monitor these four diagnostic tools as we vary a system parameter. A bifurcation point is easy to identify by using these tools, and the existence of a particular bifurcation sequence, say a sequence of period doubling bifurcations, is a strong indicator for the possible existence of chaotic motion.

The use of these diagnostic tools is illustrated schematically in Figure 3.27 for the period doubling route to chaos. For the parameter values λ_1, λ_2, and λ_3, an examination of any one of these diagnostics is sufficient to identify the existence of a periodic motion, as well as its period. For $\lambda > \lambda_c$, these four diagnostics, as well as the fact that the strange Poincaré section arose from a sequence of bifurcations from a periodic state, support the claim that the motion is chaotic.

In particular, the Poincaré section is useful for distinguishing low-dimensional chaos from noise. The chaotic Poincaré section illustrated in Figure 3.27 resembles the familiar one-humped map studied in Chapter 2. In contrast, the Poincaré map for a noisy signal from a stochastic process fills the whole oscilloscope screen with a random collection of dots (see Fig. 3.28). Thus motion on a strange attractor has a strong spatial correlation not present in a purely random signal. In the next section we will quantify this observation by introducing the correlation dimension, a measure that allows us to distinguish whether the signal is coming from a low-dimensional strange attractor or from noise.

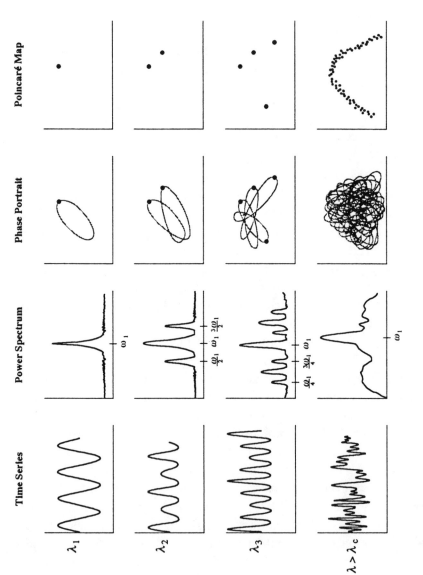

Figure 3.27: Period doubling route to chaos: (a) Period one, (b) period two, (c) period four, and (d) chaotic. (Adapted from Tredicce and Abraham [1].)

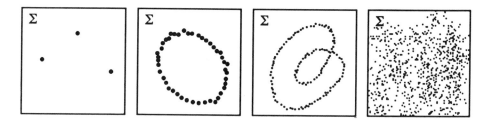

Figure 3.28: Poincaré maps from periodic, quasiperiodic, chaotic, and noisy (random) processes.

The quasiperiodic route to chaos is illustrated in Figure 3.29. In this case, the phase portrait for a quasiperiodic motion would resemble a Lissajous pattern, which may be hard to distinguish from a slowly evolving chaotic orbit. The Poincaré map, on the other hand, is useful in distinguishing between these two cases. The sequence of dots forming the Poincaré map in the quasiperiodic regime lie on a closed curve that is easy to distinguish from the spread of points in the Poincaré map for a strange attractor.

3.8.5 Correlation Dimension

One difference between a chaotic signal from a strange attractor and a signal from a noisy random process is that points on the chaotic attractor are spatially organized. One measure of this spatial organization is the *correlation integral*,

$$C(\epsilon) = \lim_{n \to \infty} \frac{1}{n^2} \times [\text{number of pairs } i, j \text{ whose distance } |\mathbf{y_i} - \mathbf{y_j}| < \epsilon],$$

where n is the total number of points in the time series. This correlation function can be written more formally by making use of the Heaviside function $H(z)$,

$$C(\epsilon) = \lim_{n \to \infty} \frac{1}{n^2} \sum_{i,j=1}^{n} H(\epsilon - |\mathbf{y_i} - \mathbf{y_j}|), \qquad (3.73)$$

where $H(z) = 1$ for positive z, and 0 otherwise. Typically, the vector $\mathbf{y_i}$ used in the correlation integral is a point in the embedded phase

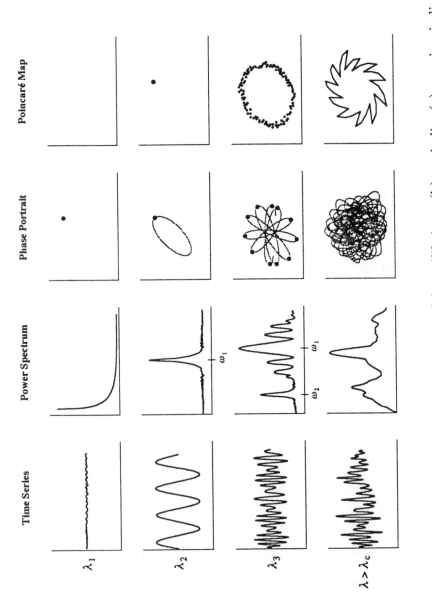

Figure 3.29: Quasiperiodic route to chaos: (a) equilibrium, (b) periodic, (c) quasiperiodic, and (d) chaotic. (Adapted from Tredicce and Abraham [1].)

space constructed from a single time series according to

$$\mathbf{y_i} = (x_i, x_{i+r}, x_{i+2r}, \ldots, x_{i+(m-1)r}), \quad i = 1, 2, \ldots. \qquad (3.74)$$

For a limited range of ϵ it is found that

$$C(\epsilon) \propto \epsilon^{\nu}, \qquad (3.75)$$

that is, the correlation integral is proportional to some power of ϵ [30]. This power ν is called the *correlation dimension*, and is a simple measure of the (possibly fractal) size of the attractor.

The correlation integral gives us an effective procedure for assigning a *fractal dimension* to a strange set. This fractal dimension is a simple way to distinguish a random signal from a signal generated by a strange (possibly chaotic) set. In principle, a random process has an "infinite" correlation dimension. Intuitively, this is because an orbit of a random process is not expected to have any spatial structure. In contrast, the correlation dimension for a closed curve (a periodic orbit) is 1, and for a two-dimensional surface (such as quasiperiodic motion on a torus) is 2. A strange (fractal) set can have a correlation dimension that is not an integer. For instance, the strange set arising at the end of the period doubling cascade found in the quadratic map has a correlation dimension of $0.583\ldots$, indicating that the dimension of this strange attractor is somewhere between that of a finite collection of points $(\nu = 0)$ and a curve $(\nu = 1)$.

There exist some technical issues associated with the calculation of a correlation dimension ν from a time series that have to do with the choice of the embedding dimension m and the delay time r. These issues are dealt with more fully in references [1] and [23]. There now exist several computer codes in the public domain that have, to a large extent, automated the calculation of ν from a single time series $x(t_i)$. Such a time series could come from a simulation or from an experiment. See, for example, the BINGO code by Albano [31], or the efficient algorithm discussed by Grassberger [32]. Thus, the correlation dimension is now a standard tool in a nonlinear dynamicist's toolbox that helps one distinguish between noise and low-dimensional chaos.

References and Notes

[1]   J. R. Tredicce and N. B. Abraham, Experimental measurements to identify and/or characterize chaotic signals, in *Lasers and quantum optics*, edited by L. M. Narducci, E. J. Quel, and J. R. Tredicce, CIF Series Vol. 13 (World Scientific: New Jersey, 1988); N. Gershenfeld, An experimentalist's introduction to the observation of dynamical systems, in *Directions in Chaos*, Vol. 2, edited by Hao B.-L. (World Scientific, New Jersey, 1988), pp. 310–382; N. B. Abraham, A. M. Albano, and N. B. Tufillaro, Introduction to measures of complexity and chaos, in NATO ASI Series B: Physics. Proceedings of the international workshop on Quantitative Measures of Dynamical Complexity in Nonlinear Systems, Bryn Mawr, PA, USA, June 22–24, 1989, edited by N. B. Abraham, A. M. Albano, T. Passamante, and P. Rapp, (Plenum Press: New York, 1990); J. Crutchfield, D. Farmer, N. Packard, R. Shaw, G. Jones, and R. Donnelly, Power spectral analysis of a dynamical system, Phys. Lett. **76A** (1), 1–4 (1980).

[2]   C. Gough, The nonlinear free vibration of a damped elastic string, J. Acoust. Soc. Am. **75**, 1770–1776 (1984).

[3]   The original version of this apparatus was constructed and examined for possible chaotic motions by T. C. A. Molteno at the University of Otago, Dunedin, New Zealand. For more details, see T. C. A. Molteno and N. B. Tufillaro, Torus doubling and chaotic string vibrations: Experimental results, J. Sound Vib. **137** (2), 327–330 (1990).

[4]   R. C. Cross, A simple measurement of string motion, Am. J. Phys. **56** (11), 1047–1048 (1988).

[5]   D. Martin, Decay rates of piano tones, J. Acoust. Soc. Am. **19** (4), 535–541 (1974).

[6]   R. Hanson, Optoelectronic detection of string vibrations, The Physics Teacher (March 1987), 165–166.

[7]   N. B. Tufillaro, Nonlinear and chaotic string vibrations, Am. J. Phys. **57** (5), 404–414 (1989); N. B. Tufillaro, Torsional parametric oscillations in wires, Eur. J. Phys. **11**, 122–124 (1990).

[8]   For theoretical and experimental details on chaos in an elastic beam, see F. C. Moon, *Chaotic vibrations: An introduction for applied scientists and engineers* (John Wiley & Sons: New York, 1987). In particular, Appendix C contains a detailed description of the construction of an inexpensive beam experiment showing chaos. Some discussion of the Duffing equation as it relates to the motion of a beam can also be found in J. M. Thompson and H. B. Steward, *Nonlinear dynamics and chaos* (John Wiley & Sons: New York, 1986).

[9] J. Elliot, Intrinsic nonlinear effects in vibrating strings, Am. J. Phys. **48** (6), 478–480 (1980); J. Elliot, Nonlinear resonance in vibrating strings, Am. J. Phys. **50** (12), 1148–1150 (1982).

[10] See Chapter 4 of A. Nayfeh and D. Mook, *Nonlinear oscillators* (Wiley-Interscience: New York, 1979). A more elementary account of perturbative solutions to the Duffing equation is presented by A. Nayfeh, *Introduction to perturbation techniques* (John Wiley & Sons: New York, 1981).

[11] A classic in nonlinear studies is C. Hayashi's *Nonlinear oscillations in physical systems* (Princeton University Press: Princeton, NJ, 1985). The method of harmonic balance is used extensively by Hayashi; see pages 28–30, 114–120, and 238–251 for a discussion and examples of this technique applied to the Duffing oscillator.

[12] A real gem of mathematical exposition is V. I. Arnold's *Ordinary differential equations* (MIT Press: Cambridge, MA, 1973). The basic theorems of ordinary differential equations are proved in Chapter 4. Arnold's text is the best introductory exposition to the geometric study of differential equations available. A must read for all students of dynamical systems.

[13] S. Wiggins, *Introduction to applied nonlinear dynamical systems and chaos* (Springer-Verlag: New York, 1990), pp. 64–82.

[14] N. B. Tufillaro and G. A. Ross, *Ode—A program for the numerical solution of ordinary differential equations*, Bell Laboratories Technical Memorandum 83-52321-39 (1983); program and documentation are in the public domain. Also see Appendix C.

[15] U. Parlitz and W. Lauterborn, Superstructure in the bifurcation set of the Duffing equation, Phys. Lett. **107A** (8), 351–355 (1985); U. Parlitz and W. Lauterborn, Resonance and torsion numbers of driven dissipative nonlinear oscillators, Z. Naturforsch. **41a**, 605–614 (1986).

[16] The method of slowly varying amplitude is also known as the averaging method of Krylov-Bogoliubov-Mitropolsky; see E. A. Jackson, *Perspectives in nonlinear dynamics*, Vol. 1 (Cambridge University Press: New York, 1989), pp. 264, 308–322. In the optics community, it is also known as the rotating-wave approximation; see P. W. Milonni, M.-L. Shih, and J. R. Ackerhalt, *Chaos in laser-matter interactions* (World Scientific: Singapore, 1987), pp. 58–70.

[17] This is a bit sneaky. $A = A^*$ here because there is no damping term causing a frequency shift in the approximation to first order. In general $A \neq A^*$, and then the slowly varying amplitude approximation results in both a detuning and a phase shift. Both data are encoded by $A \in \mathbf{C}$.

[18] A marvelous pictorial introduction to dynamical systems theory is the unique series of books by R. Abraham and C. Shaw, *Dynamics—The geometry of behavior*, Vol. 1–4 (Aerial Press: Santa Cruz, CA, 1988). Volume 1, pp. 143–151, contains a good description of the geometry of the phase space associated with hysteresis in the Duffing oscillator, and Volume 3, pp. 83–87, shows how homoclinic tangles are formed.

[19] J. M. Johnson and A. K. Bajaj, Amplitude modulated and chaotic dynamics in resonant motion of strings, J. Sound Vib. **128** (1), 87–107 (1989).

[20] J. Miles, Resonant, nonplanar motion of a stretched string, J. Acoust. Soc. Am. **75** (5), 1505–1510 (1984).

[21] O. M. O'Reilly, The chaotic vibration of a string, Ph.D. thesis, Cornell University, 1990.

[22] P. Bak, The devil's staircase, Physics Today **39** (12), 38–45 (1986); P. Bak, T. Bohr, and M. H. Jensen, Mode-locking and the transition to chaos in dissipative systems, Physica Scripta, Vol. T9, 50–58 (1985); P. Cvitanović, B. Shraiman, and B. Söderberg, Scaling laws for mode lockings in circle maps, Physica Scripta **32**, 263–270 (1985).

[23] For a fuller account of torus attractors and quasiperiodic motion in a nonlinear system, see Chapter 7 of P. Bergé, Y. Pomeau, and C. Vidal, *Order within chaos* (John Wiley & Sons: New York, 1984). An elementary discussion of fractal dimensions and the correlation integral is found on pages 146–154.

[24] A. Arnédo, P. Coullet, and E. Spiegel, Cascade of period doublings of tori, Phys. Lett. **94A** (1), 1–6 (1983); K. Kaneko, Doubling of torus, Prog. Theor. Phys. **69** (6), 1806–1810 (1983); F. Argoul and A. Arnédo, From quasiperiodicity to chaos: an unstable scenario via period doubling bifurcation or tori, J. de Mécanique Théorique et Appliquée, Numeréro spécial, 241–288 (1984).

[25] W. Ditto, M. Spano, H. Savage, S. Rauseo, J. Heagy, and E. Ott, Experimental observation of a strange nonchaotic attractor, Phys. Rev. Lett. **65** (5), 533–536 (1990).

[26] N. Packard, J. Crutchfield, J. Farmer, and R. Shaw, Geometry from a time series, Phys. Rev. Lett. **45**, 712 (1980). Also see Chapter 6 of D. Ruelle, *Chaotic evolution and strange attractors* (Cambridge University Press: New York, 1989). For a geometric approach to the embedding problem see Th. Buzug, T. Reimers, and G. Pfister, Optimal reconstruction of strange attractors from purely geometrical arguments, Europhys. Lett. **13** (7), 605–610 (1990).

[27] R. Hamming, *An introduction to applied numerical analysis* (McGraw-Hill: New York, 1971).

[28] W. Press, B. Flannery, S. Teukolsky, and W. Vetterling, *Numerical recipes in C* (Cambridge University Press: New York, 1988).

[29] K. Wiesenfeld, Virtual Hopf phenomenon: A new precursor of period doubling bifurcations, Phys. Rev. A **32**, 1744 (1985); K. Wiesenfeld, Noisy precursors of nonlinear instabilities, J. Stat. Phys. **38**, 1701 (1985); P. Bryant and K. Wiesenfeld, Suppression of period doubling and nonlinear parametric effects in periodically perturbed systems, Phys. Rev. A **33**, 2525–2543 (1986); B. McNamara and K. Wiesenfeld, Theory of stochastic resonance, Phys. Rev. A **39** (9), 4854 (1989); K. Wiesenfeld and N. B. Tufillaro, Suppression of period doubling in the dynamics of the bouncing ball, Physica **26D**, 321–335 (1987).

[30] P. Grassberger and I. Procaccia, Characterization of strange attractors, Phys. Rev. Lett. **50**, 346–349 (1983); P. Grassberger and I. Procaccia, Measuring the strangeness of strange attractors, Physica **9D**, 189 (1983).

[31] The BINGO program runs on the IBM-PC. For a copy contact A. Albano, Department of Physics, Bryn Mawr College, Bryn Mawr, PA 19010-2899.

[32] P. Grassberger, An optimized box-assisted algorithm for fractal dimensions, Phys. Lett. A **148**, 63–68 (1990). This brief paper provides a Fortran program for calculating the correlation integral. For another new approach to calculating the correlation integral see Xin-Jun Hou, Robert Gilmore, Gabriel B. Mindlin, and Hernán Solari, An efficient algorithm for fast $O(N * \ln(N))$ box counting, Phys. Lett. **151** (1,2), 43–46 (1990).

Problems

Problems for section 3.2.

3.1. Use Ampère's law to find the magnetic field at a radial distance r from a long straight current-carrying wire.

3.2. The Joule heating law says that the power dissipated by a current-carrying object is

$$P = VI = I^2R, \qquad (3.76)$$

where V is the voltage, I is the current, and R is the resistance. Furthermore, for small temperature changes $\Delta Temp$, the fractional change of length of a solid obeys

$$\frac{\Delta L}{L} = \kappa \Delta Temp, \qquad (3.77)$$

where κ is the linear coefficient of thermal expansion of the material. Discuss the relevance of these two physical laws on the string apparatus.

Section 3.3.

3.3. From equation (3.3), show that the location of the two equilibrium points is approximately given by $r_{\pm} \approx \pm l \sqrt{\frac{l_0 - l}{2 l_0}}$.

3.4. Solve the differential equation $\ddot{x} = -\lambda \dot{x} - \omega^2 x$ and graph $x(t)$ versus t for a few values of λ. Why is λ called the damping coefficient?

3.5. Solve the differential equation $\ddot{\mathbf{r}} = -\lambda \dot{\mathbf{r}}$, $\mathbf{r} = (x, y)$, $\lambda = (\lambda_x, \lambda_y)$. Draw a plot of $\mathbf{r}(t) = (x(t), y(t))$ for a fixed (λ_x, λ_y).

3.6. Verify that the variables in equation (3.11) are dimensionless.

3.7. Verify that equation (3.13) is a solution to equation (3.7) with $\mathbf{f}(t) = (0, 0)$ and $K = 0$ (discard second-order terms in λ). Draw the solution, $\mathbf{r}(t)$, in the x–y plane.

3.8. Derive equation (3.2) from Figure 3.3. Hint: To account for the factor of 2 realize that each spring makes a separate contribution to the restoring force.

Section 3.4.

3.9. **(a)** Show that the equation of motion for a simple pendulum in dimensionless variables is

$$\ddot{\phi} + \sin \phi = 0. \tag{3.78}$$

(b) Write this as a first-order system with $(\phi, v) \in S^1 \times \mathbf{R}$. Using equation (3.18), find the potential energy function and a few integral curves for a pendulum.

(c) Show that the pendulum has equilibrium points at $(0, 0)$ and $(\pm \pi, 0)$ and discuss the stability of these fixed points by relating them to the configurations of the physical pendulum. Are there any saddle points? Are there any centers?

(d) Draw a schematic of the phase plane. Identify the separatrix in this phase portrait. Orbits inside the separatrix are called *oscillations*. Why? Orbits outside the separatrix are called *rotations*. Why?

(e) Add a dissipative term $(\lambda \dot{\phi})$ to get a damped pendulum and discuss how this changes the phase portrait. In particular, discuss the relation of the insets and outsets of the equilibrium points with the basins of attraction. Are there any attractors? Are there any repellers?

(f) Now add a forcing term $f \cos(\omega t)$ and write the differential equations for the system in the extended phase space $(\phi, v, \theta) \in \mathbf{R} \times S^1 \times S^1$, where $\theta = \omega t$. Define a global cross section for the forced damped pendulum (see eq. (3.29)).

Section 3.5.

3.10. Verify that equation (3.32) is a solution to equation (3.31) (discard higher-order terms in α).

3.11. Plot the linear response curve $a(\gamma)$ (eq. (3.34)) for a few representative values of F and α.

3.12. This exercise illustrates the method of harmonic balance. Assume an approximate solution of the form

$$x_0 = X \sin \gamma \tau + Y \cos \gamma \tau$$

to the Duffing equation (3.23).

(a) Substitute x_0 into equation (3.23) and equate terms containing $\sin \gamma \tau$ and $\cos \gamma \tau$ separately to zero.

(b) Show that

$$AX + \gamma \alpha Y = 0, \quad \gamma \alpha X - AY = F,$$

where

$$A = \gamma^2 - 1 - \frac{3}{4}\beta R^2, \quad R^2 = X^2 + Y^2.$$

(c) Show that

$$(A^2 + \gamma^2 \alpha^2)R^2 = F^2.$$

(d) Plot the response (R^2 versus γ) for $\beta = 0.9$, $\alpha = 0.2$, and several values of F. Plot the amplitude characteristic (R^2 versus F) for $\beta = 0.9$, $\alpha = 0.2$, and $\gamma = 1$. Indicate the unstable solution with a dashed line. Hint: It is easier to plot R^2 as the independent variable.

Section 3.7.

3.13. Verify equations (3.52–3.54) for the free whirling motions of a string.

3.14. Write a program to iterate the circle map (eq. (3.57)) and explore its solutions for different values of K and Ω.

Section 3.8.

3.15. For a time series $\{x_i\}$, in which the x_i are sampled at evenly spaced times, write a transformation between the phase space variable $y_i = (x_i - x_{i-1})/(t_i - t_{i-1})$ and the embedded phase space variable $y_i = x_{i-r}$, with $r = 1$. Why is the embedded phase space rotated like it is in Figure 3.24? Why does one usually choose $r > 1$?

3.16. Write a program based on equations (3.67–3.71) to calculate the discrete Fourier amplitude coefficients and the power spectrum (eq., (3.72)) for a discrete time series. Test the program on some sample functions for periodic motion (e.g., $x(t) = \cos(\omega t) + \cos(3\omega t)$), and quasiperiodic motions (e.g., $x(t) = \cos(\omega t) + \cos(\sqrt{2}\omega t)$).

Chapter 4

Dynamical Systems Theory

4.1   Introduction

This chapter is an eclectic mix of standard results from the mathematical theory of dynamical systems along with practical results, terminology, and notation useful in the analysis of a low-dimensional dynamical system [1]. The examples in this chapter are usually confined to two-dimensional maps and three-dimensional flows.

In the first three chapters we presented nonlinear theory by way of examples. We now present the theory in a more general setting. Many of the fundamental ideas have already been illustrated in the one-dimensional setting. For example, in one-dimension we found that the stability of a fixed point is determined by the derivative at the fixed point. The same result holds in higher dimensions. However, the actual computational machinery needed is far more intricate because the derivative of an n-dimensional map is an $n \times n$ matrix.

Another key idea we have already introduced is hyperbolicity, hyperbolic sets, and symbolic analysis (see section 2.11). In this chapter we present a detailed study of the Smale horseshoe, which is the canonical example of a "hyperbolic chaotic invariant set." A thorough understanding of this example is essential to the analysis of a chaotic repeller or attractor. We conclude this chapter with a discussion of sensitive dependence on initial conditions. This brings us back to the Lyapunov exponent, which we define for a general n-dimensional dynamical system.

Flows   Maps

Lorenz Equation

$\phi(x, y, z) : \mathbf{R}^3 \to \mathbf{R}^3$

$$\left.\begin{array}{l} \dot{x} = \sigma(y - x) \\ \dot{y} = \rho x - y - xz \\ \dot{z} = -\beta z + xy \end{array}\right\}$$

parameters

σ, ρ, β

Duffing Equation

$\phi(x, v, \theta) : \mathbf{R}^2 \times S^1 \to \mathbf{R}^2 \times S^1$

$$\left.\begin{array}{l} \dot{x} = v \\ \dot{v} = -(x + \beta x^3 + \alpha v) + f\cos(\theta) \\ \dot{\theta} = \omega \end{array}\right\}$$

nonautonomous form

$\ddot{x} + \alpha\dot{x} + x + \beta x^3 = f\cos(\omega t)$

parameters

α, β, f, ω

Forced Damped Pendulum

$\phi(\theta, v, \varphi) : S^1 \times S^1 \times \mathbf{R} \to S^1 \times S^1 \times \mathbf{R}$

$$\left.\begin{array}{l} \dot{\theta} = v \\ \dot{v} = -[\alpha v + \beta\sin(\theta)] + f\cos(\varphi) \\ \dot{\varphi} = \omega \end{array}\right\}$$

parameters

α, β, f, ω

Modulated Laser

$\phi(u, z, \theta) : \mathbf{R}^2 \times S^1 \to \mathbf{R}^2 \times S^1$

$$\left.\begin{array}{l} \dot{u} = [z - f\cos(\theta)]u \\ \dot{z} = (1 - \alpha_1 z) - (1 + \alpha_2 z)u \\ \dot{\theta} = \omega \end{array}\right\}$$

parameters

$\alpha_1, \alpha_2, f, \omega$

Quadratic Map

$f(x) : \mathbf{R} \to \mathbf{R}$

$x_{n+1} = \lambda x_n(1 - x_n)$

parameter

λ

Sine Circle Map

$f(\theta) : S^1 \to S^1$

$\theta_{n+1} = \theta_n + \Omega + \frac{K}{2\pi}\sin(2\pi\theta_n)$

parameters

Ω, K

Hénon Map

$f(x, y) : \mathbf{R}^2 \to \mathbf{R}^2$

$$\left.\begin{array}{l} x_{n+1} = \alpha - x_n^2 + \beta y_n \\ y_{n+1} = x_n \end{array}\right\}$$

parameters

α, β

Baker's Map

$f(x, y) : I \times I \to I \times I$

$$\left.\begin{array}{l} x_{n+1} = 2x_n \bmod 1 \\ y_{n+1} = \begin{cases} \alpha y_n & \text{for } 0 \le x_n < 1/2 \\ 1/2 + \alpha y_n & \text{for } 1/2 \le x_n \le 1 \end{cases} \end{array}\right\}$$

parameter

$\alpha < 1/2$

IHJM Optical Map [2]

$f(z) : \mathbf{C} \to \mathbf{C}$

$z_{n+1} = \gamma + B z_n \exp\left[i\left(\kappa - \frac{\alpha}{1+|z_n|^2}\right)\right]$

parameters

$\alpha, \kappa, \gamma, B$

4.2 Flows and Maps

Most of the dynamical systems studied in this book are either three-dimensional flows or one- or two-dimensional maps. Common examples of maps and flows are listed on the previous page.

Flows are specified by differential equations (section 4.2.1). Similarly, maps are specified by difference equations:

$$\mathbf{x}_{n+1} = \mathbf{f}(\mathbf{x}_n; \mu). \tag{4.1}$$

Maps can also be written as $\mathbf{x} \mapsto \mathbf{f}(\mathbf{x}; \mu)$. The notation $\mapsto$ is read as "maps to." The *forward orbit* of $\mathbf{x}$ is $O^+(\mathbf{x}) = \{\mathbf{f}^n(\mathbf{x}) : n \geq 0\}$, where $\mathbf{f}^n = \mathbf{f} \circ \cdots \circ \mathbf{f}$ is the nth composite of $\mathbf{f}$, and f^0 is the identity function. If the inverse $\mathbf{f}^{-1}$ is well defined, then the *backward orbit* of $\mathbf{x}$ is $O^-(\mathbf{x}) = \{\mathbf{f}^{-n}(\mathbf{x}) : n \geq 0\}$. Finally, the *orbit* of $\mathbf{x}$ is the sequence of all positions visited by $\mathbf{x}$, $O(\mathbf{x}) = O^-(\mathbf{x}) \cup O^+(\mathbf{x})$.

4.2.1 Flows

A first-order system of differential equations is written as

$$\frac{d\mathbf{x}}{dt} = \mathbf{f}(\mathbf{x}, t; \mu), \tag{4.2}$$

where $\mathbf{x} = (x_1, x_2, \ldots, x_n)$ are the dependent variables, t is the independent variable time, and μ is the set of parameters for the system. Sometimes the parameter dependence is denoted by a subscript as in $\mathbf{f}_\mu(\mathbf{x}, t)$.

A *vector field* is formally defined by a map $\mathbf{F} : A \subset \mathbf{R}^n \to \mathbf{R}^n$ that assigns a vector $\mathbf{F}(\mathbf{x})$ to each point $\mathbf{x}$ in its domain A. More generally, a vector field on a manifold M is given by a map that assigns a vector to each point in M. For most of this chapter we only need to work with $\mathbf{R}^n$. A system governed by a time-independent vector field is called *autonomous*; otherwise it is called *nonautonomous*. As we saw in section 3.4.3, any nonautonomous vector field can be converted to an autonomous vector field of a higher dimension.

The *flow*, ϕ, of the vector field $\mathbf{F}$ is analytically defined by:

$$\frac{\partial}{\partial t}\phi(\mathbf{x}, t) = \mathbf{F}(\phi(\mathbf{x}, t)), \tag{4.3}$$

$$\phi(\mathbf{x}, 0) = \mathbf{x}. \tag{4.4}$$

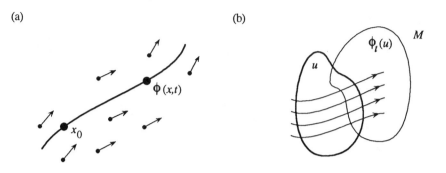

Figure 4.1: (a) The flow of a solution curve. (b) The flow of a collection of solution curves resulting in a continuous transformation of the manifold.

The position $\mathbf{x}$ is the *initial condition* or initial *state*. The initial condition is also written as $\mathbf{x}_0$ when it is specified at $t = 0$. A *solution curve*, *trajectory*, or *integral curve* of the flow is an individual solution of the above differential equation based at $\mathbf{x}_0$. We often explicitly note the time-dependence of the position (the solution curve to the above differential equation) by writing $\mathbf{x}(t)$ when we want to indicate the position of the trajectory at a time $t > 0$. The collection of all states of a dynamical system is called the *phase space*.

The term "flow" describing the evolution of the system in phase space comes by analogy from the motion of a real fluid flow. The flow $\phi(\mathbf{x}, t)$ is regarded as a function of the initial condition $\mathbf{x}$ and the single parameter time t. The flow ϕ tells us the position of the initial condition $\mathbf{x}$ after a time t. As illustrated in Figure 4.1(a), the position of the point on the flow line through $\mathbf{x}$ is carried or flows to the point $\phi(\mathbf{x}, t)$. A geometric description of a flow says that it is a one-parameter family of diffeomorphisms of a manifold. That is, the flow lines of the flow, which are solution curves of the differential equation, provide a continuous transformation of the manifold into itself (Fig. 4.1(b)).

The flow is often written as $\phi_t(\mathbf{x})$ to highlight its dependence on the single parameter t. The flow ϕ_t is also known as the *evolution operator*. Composition of the evolution operator is defined in a natural way: starting at the state $\mathbf{x}$ at time $s = 0$, it first flows to the point

$\phi_s = \phi(\mathbf{x}, s)$ and then onto the point $\phi_{t+s} = \phi(\phi(\mathbf{x}, s), t)$. The evolution operator satisfies the group properties

$$(\text{i}) \ \phi_0 = identity, \quad (\text{ii}) \ \phi_{t+s} = \phi_t \circ \phi_s, \qquad (4.5)$$

which are taken as the defining relations of a flow for an *abstract dynamical system*. If the system is nonreversible we speak of a *semiflow*. A semiflow flows forward in time, but not backward.

An explicit example of an evolution operator is given by solving the linear differential equation $\ddot{x} = -x$. Here the vector field is found from the equivalent first-order system $(\dot{x} = v, \dot{v} = -x)$ so that the vector field is

$$\mathbf{F}(x, v) = (v, -x).$$

This vector field generates a flow that is a simple rotation about the origin (see Fig. 0.3). The evolution operator is given by the rotation matrix[1]

$$\phi_t(\mathbf{x}) = \begin{bmatrix} x(t) \\ v(t) \end{bmatrix} = \begin{bmatrix} \cos(t) & \sin(t) \\ -\sin(t) & \cos(t) \end{bmatrix} \begin{bmatrix} x \\ v \end{bmatrix}.$$

4.2.2 Poincaré Map

A continuous flow can generate a discrete map in at least two ways: by a time-T map and by a Poincaré map. A *time-T map* results when a flow is sampled at a fixed time interval T. That is, the flow is sampled whenever $t = nT$ for $n = 0, 1, 2, 3$, and so on.

The more important way (as described, for instance, in Guckenheimer and Holmes [1]) in which a continuous flow generates a discrete map is via a *Poincaré map*. Let γ be an orbit of a flow ϕ_t in $\mathbf{R}^n$. As illustrated in Figure 4.2, it is often possible to find a *local cross section* $\Sigma \in \mathbf{R}^n$ about γ, which is of dimension $n - 1$. The cross section need not be planar; however, it must be *transverse* to the flow. All the orbits in the neighborhood of γ must pass through Σ. The technical requirement is that $\mathbf{F}(\mathbf{x}) \cdot \mathbf{N}(\mathbf{x}) \neq 0$ for all $\mathbf{x} \in \Sigma$, where $\mathbf{N}(\mathbf{x})$ is the

---

[1]The components of the map ϕ should be written as (ϕ^x, ϕ^v), where the superscript indicates the dependent coordinate. It is a common convention, though, to suppress the ϕ and mix the dependent variables with the coordinate functions so that $[\phi^x(t), \phi^v(t)] = [x(t), v(t)]$.

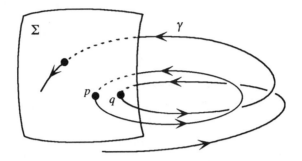

Figure 4.2: Construction of a Poincaré map from a local cross section.

unit normal vector to Σ at $\mathbf{x}$. Let $\mathbf{p}$ be a point where γ intersects Σ, and let $\mathbf{q} \in \Sigma$ be a point in the neighborhood of $\mathbf{p}$. Then the *Poincaré map* (or *first return map*) is defined by

$$P : \Sigma \to \Sigma, \quad P(\mathbf{q}) = \phi_\tau(\mathbf{q}), \tag{4.6}$$

where $\tau = \tau(\mathbf{q})$ is the time taken for an orbit starting at $\mathbf{q}$ to return to Σ. It is useful to define a Poincaré map in the neighborhood of a periodic orbit. If the orbit γ is periodic of period T, then $\tau(\mathbf{p}) = T$. A periodic orbit that returns directly to itself is a fixed point of the Poincaré map. Moreover, an orbit starting at $\mathbf{q}$ close to $\mathbf{p}$ will have a return time close to T.

For forced systems, such as the Duffing oscillator studied in section 3.4, a *global cross section* and Poincaré map are easy to define since the phase space topology is $\mathbf{R}^2 \times S^1$. All periodic orbits of a forced system have a period that is an integer multiple of the forcing period. In this situation, it is sensible to pick a planar global cross section that is transverse to S^1 (see Fig. 3.8). The return time for this cross section is independent of position and equals the forcing period. In this special case, the Poincaré map is equivalent to a time-T map.

The Poincaré map for the example of the rotational flow generated by $\mathbf{F}(x,v) = (v, -x)$ is particularly trivial. A good cross section is defined by the positive half of the x-axis, $\Sigma = \{(x,v)|v = 0 \text{ and } x > 0\}$. All the orbits of the flow are fixed points in the cross section, so the Poincaré map is just the identity map, $P(x) = x$.

(a)

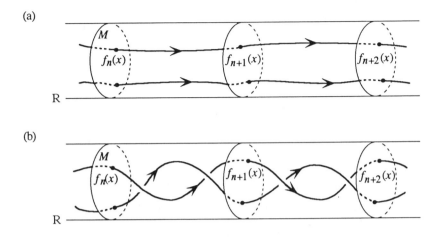

(b)

Figure 4.3: (a) Construction of a suspension of a map. (b) The suspension is not unique; an arbitrary number of twists can be added.

See Appendix E, Hénon's Trick, for a discussion of the numerical calculation of a Poincaré map from a cross section.

4.2.3   Suspension of a Map

A discrete map can also be used to generate a continuous flow. A canonical construction for this is the so-called *suspension of a map* [1], which is in a sense the inverse of a Poincaré map.[2] Given a discrete map f of an n-dimensional manifold M, it is always possible to construct a flow on an $n+1$-dimensional manifold formed by the Cartesian product $\mathbf{R}$ with the original manifold: $\phi : M \times \mathbf{R} \to M$. This suspension process is illustrated in Figure 4.3(a) where we show a mapping and the flow formed by "suspending" this map. Each sequence of points of the map becomes an orbit of the flow with the property that if $f_n(\mathbf{x}) = \phi_t(\mathbf{x})$ then $f_{n+1}(\mathbf{x}) = \phi_{T+t}(\mathbf{x})$. The original map is recovered from the suspended flow by a time-T map. The suspension construction is far from unique. For instance, as illustrated in Figure 4.3(b), we

---

[2]The suspension is defined globally, while the cross section for a Poincaré map is only defined locally. Thus, these two constructions are not completely complementary.

could add an arbitrary number of complete twists to this particular suspension and still get an identical time-T map. The number of full twists in the suspended flow is called the *global torsion*, and it is a topological invariant of the flow independent of the underlying map.

4.2.4   Creed and Quest

The close connection between maps and flows—Poincaré maps and suspensions—gives rise to "The Discrete Creed":

> Anything that happens in a flow also happens in a (lower-dimensional) discrete dynamical system (and conversely).

"The creed is stated in the mode of the sunshine patriot, leaving plenty of room to duck as necessary [3]." This creed is implicit in Poincaré's original work, but was first clearly enunciated by Smale.

The essence of dynamical systems studies is stated in "The Dynamical Quest":

> Where do orbits go, and what do they do/see when they get there?

The dynamical quest emphasizes the topological (i.e., qualitative) characterization of the long-term behavior of a dynamical system.

4.3   Asymptotic Behavior and Recurrence

In this section we present some more mathematical vocabulary that helps to refine our notions of invariant sets, limit sets (attractors and repellers), asymptotic behavior, and recurrence. For the most part we state the fundamental definitions in terms of maps. The corresponding definitions for flows are completely analogous and can be found in Wiggins [1].

Recurrence is a key theme in the study of dynamical systems. The simplest notion of recurrence is periodicity. Recall that a point of a map is *periodic* of period n if there exists an integer n such that $f^n(p) = p$ and $f^i(p) \neq p$, $0 < i < n$. This notion of recurrence is too restricted since it fails to account for quasiperiodic motions or strange attractors.

We will therefore explore more general notions of recurrence. At the end we will argue that the "chain recurrent set" is the best definition of recurrence that captures most of the interesting dynamics.

4.3.1 Invariant Sets

Formally, a set S is an *invariant set* of a flow if for any $x_0 \in S$ we have $\phi(x_0, t) \in S$ for all $t \in \mathbf{R}$. S is an invariant set of a map if for any $x_0 \in S$, $f^n(x_0) \in S$ for all n. We also speak of a *positively invariant set* when we restrict the definition to positive times, $t \geq 0$ or $n \geq 0$.

Invariant sets are important because they give us a means of decomposing phase space. If we can find a collection of invariant sets, then we can restrict our attention to the dynamics on each invariant set and then try to sew together a global solution from the invariant pieces. Invariant sets also act as boundaries in phase space, restricting trajectories to a subset of phase space.

4.3.2 Limit Sets: α, ω, and Nonwandering

We begin by introducing the ω-limit set, which starts us down the road toward defining an attractor. Let p be a point of a map $f : M \to M$ of the manifold M. Then the ω-limit set of p is

$$\omega(p) = \{a \in M \mid \text{there exists a sequence } n_i \to \infty \text{ such that } f^{n_i}(p) \to a\}.$$

Conversely, by going backwards in time we get the α-limit set of p,

$$\alpha(p) = \{r \in M \mid \text{there exists a sequence } n_i \to \infty \text{ such that } f^{-n_i}(p) \to r\}.$$

These limit sets are the closure of the ends of the orbits. For example, if p is a periodic point, then $\omega(p) = O^+(p)$. It is not difficult to show that $\omega(p)$ is a closed subset of M and that it is invariant under f, i.e., $f(\omega(p)) = \omega(p)$. Finally, a point $p \in M$ is called *recurrent* if it is part of the ω-limit set, i.e., $p \in \omega(p)$.

The *forward limit set*, L^+, is defined as the union of all ω-limit sets; likewise the *backward limit set*, L^-, is defined as the union of all α-limit sets:

$$L^+(f) = \bigcup_{p \in M} \omega(p) \quad \text{and} \quad L^-(f) = \bigcup_{p \in M} \alpha(p).$$

The forward and backward limit sets are not necessarily closed. This observation motivates yet another useful notion of recurrence, the nonwandering set. A point $p \in M$ wanders if there exist a neighborhood U of p and an $m > 0$ such that $f^n(U) \cap U = \emptyset$ for all $n > m$. A *nonwandering point* is one that does not wander. This brings us to the *nonwandering set*, Ω, which is a closed, invariant (under f) subset of M:

$$\Omega(f) = \{q \in M \mid q \text{ is nonwandering}\}.$$

The dynamical decomposition of a set into its wandering and non-wandering parts separates, in mathematical terms, a dynamical system into its transient behavior—the wandering set—and long-term or asymptotic behavior—the nonwandering set.

The ω-limit set and the nonwandering set do not address the question of the stability of an asymptotic motion. To get to the idea of an attractor, we begin with the idea of an attracting set. A closed invariant set $A \subset M$ is an *attracting set* if there is some neighborhood U of A such that for all $x \in U$ and all $n \geq 0$,

$$f^n(x) \in U \quad \text{and} \quad f^n(x) \to A.$$

Moreover, the *domain* or *basin of attraction* of A is given by

$$\bigcap_{n \leq 0} f^n(U).$$

The attracting set can consist of a collection of different sets that are dynamically disconnected. For instance, a single attracting set could consist of two separate periodic orbits. To overcome this last difficulty, we will say that an *attractor* is an *attracting set that contains a dense orbit*. Conversely, a *repeller* is defined as a repelling set with a dense orbit.

Although this is a reasonable definition mathematically, we will see that this is not the most useful definition of an attractor for physical applications or numerical simulations. In these circumstances it will turn out that a more useful (albeit mathematically naive) definition of an attractor or repeller is the closure of a certain collection of periodic orbits. The proper definition of an attractor is yet another hot spot in the creative tension between the rigor demanded by a mathematician and the utility required by a physicist.

4.3.3   Chain Recurrence

In the previous section we attempted to capture all the recurrent behavior of the mapping $f : M \to M$. Let

$$
\begin{aligned}
Per(f) &= \{\text{periodic points of f}\} \\
L^+(f) &= \bigcup_{p \in M} \omega(p) \\
\Omega(f) &= \{\text{nonwandering points}\}
\end{aligned}
$$

Clearly,

$$
Per(f) \subset L^+(f) \subset \Omega(f).
$$

There is no universally accepted notion of a set that contains all the recurrence, but this set ideally ought to be closed and invariant. $Per(f)$ is too small—periodicity is too limited a kind of recurrence. $L^+(f)$ is not necessarily closed. $\Omega(f)$, while closed and invariant, has the drawback that:

$$
\Omega(f|_\Omega) \neq \Omega.
$$

The mapping $f|_\Omega : \Omega \to \Omega$ makes sense, of course, because of Ω's invariance.

In certain contexts the chain recurrent set, which is bigger than Ω, has all the desirable properties. According to the "chain recurrent point of view," all the interesting dynamics take place in the chain recurrent set [4].

Let $\epsilon > 0$. An *ϵ-pseudo orbit* is a finite sequence such that

$$
d[f(x_i), x_{i+1}] < \epsilon, \quad i = 0, 1, \ldots, n-1,
$$

where $d[\cdot, \cdot]$ is a metric on the manifold M. An ϵ-pseudo orbit can be thought of as a "computer-generated orbit" because of the slight roundoff error the computer makes at each stage of an iteration (see Fig. 4.4(a)).

A point $x \in M$ is *chain recurrent* if, for all $\epsilon > 0$, there exists an ϵ-pseudo orbit $x_0, x_1, \ldots, x_n$ such that $x = x_0 = x_n$ (see Fig. 4.4(b)). The *chain recurrent set* is defined as

$$
R(f) = \{\text{chain recurrent points}\}.
$$

The chain recurrent set $R(f)$ is closed and f-invariant. Moreover, $\Omega \subset R$. For proofs, see reference [1] or [4].

(a)   (b)

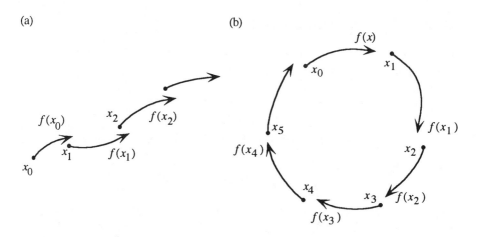

Figure 4.4: (a) An ϵ-pseudo orbit, or computer orbit, of a map; (b) a chain recurrent point.

Figure 4.5: Circle map with two fixed points.

As an example consider the quadratic map, $f_\lambda(x) = \lambda x(1 - x)$, for $\lambda > 2 + \sqrt{5}$ and $M = [-\infty, +\infty]$. It is easy to see from graphical analysis that if $x \notin [0, 1]$ then $\{-\infty\}$ is the attracting limit set. When $x \in [0, 1]$, the limit set Λ is a Cantor set (see section 2.11.1) and the dynamics on Λ are topologically conjugate to a full shift on two symbols, so the periodic orbits are dense in Λ. The chain recurrent set $R(f) = \Lambda \bigcup \{-\infty\}$.

As another example, consider the circle map $f : S^1 \to S^1$ shown in Figure 4.5, in which the only fixed points are x and y. The arrows indicate the direction a point goes in when it is iterated. It is easy to see that

$$\{x, y\} = Per(f) = L^+ = L^- = \Omega.$$

However, $R(f) = S^1$ because an ϵ-pseudo orbit can jump across the fixed points. Chain recurrence is a very weak form of recurrence.

4.4 Expansions and Contractions

In this section we consider how two-dimensional maps and three-dimensional flows transform areas and volumes. Does a map locally expand or contract a region? Does a flow locally expand or contract a volume in phase space? To answer each of these questions we need to calculate the Jacobian of the map and the divergence of the flow [5]. In this section we are concerned with showing how to do these calculations. In section 4.10, where we introduce the tangent map, we provide some geometric insight into these calculations.

4.4.1 Derivative of a Map

To fix notation, let f be a map from $\mathbf{R}^n$ to $\mathbf{R}^m$ specified by m functions,

$$f = (f_1, \ldots, f_m), \tag{4.7}$$

of n variables. Recall that the *derivative*[3] of a map $f : \mathbf{R}^n \to \mathbf{R}^m$ at $\mathbf{x}_0$ is written as $\mathbf{T} = \mathbf{D}f(\mathbf{x}_0)$ and consists of an $m \times n$ matrix called the *matrix of partial derivatives of f at $\mathbf{x}_0$*:

$$\mathbf{D}f(\mathbf{x}_0) = \begin{bmatrix} \frac{\partial f_1}{\partial x_1} & \cdots & \frac{\partial f_1}{\partial x_n} \\ \vdots & & \vdots \\ \frac{\partial f_m}{\partial x_1} & \cdots & \frac{\partial f_m}{\partial x_n} \end{bmatrix}. \tag{4.8}$$

The derivative of f at $\mathbf{x}_0$ represents the *best linear approximation* to f near to $\mathbf{x}_0$.

As an example of calculating a derivative, consider the function from $\mathbf{R}^2$ to $\mathbf{R}^2$ that transforms polar coordinates into Cartesian coordinates:

$$f(r, \theta) = [f_1(r, \theta), f_2(r, \theta)] = (r \cos \theta, r \sin \theta).$$

---

[3]Some books call this the *differential* of f and denote it by $df(\mathbf{x}_0)$.

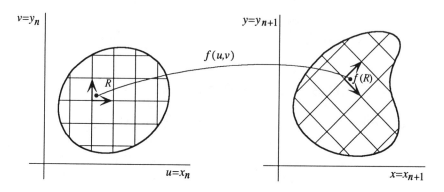

Figure 4.6: Deformation of an infinitesimal region under a map.

The derivative of this particular transformation is

$$\mathbf{D}f(r, \theta) = \begin{bmatrix} \frac{\partial f_1}{\partial r} & \frac{\partial f_1}{\partial \theta} \\ \frac{\partial f_2}{\partial r} & \frac{\partial f_2}{\partial \theta} \end{bmatrix} = \begin{bmatrix} \cos\theta & -r\sin\theta \\ \sin\theta & r\cos\theta \end{bmatrix}.$$

4.4.2 Jacobian of a Map

The derivative contains essential information about the local dynamics of a map. In Figure 4.6 we show how a small rectangular region R of the plane is transformed to $f(R)$ under one iteration of the map $f(u, v) : \mathbf{R}^2 \to \mathbf{R}^2$ where $f_1(u, v) = x(u, v)$ and $f_2(u, v) = y(u, v)$. The *Jacobian* of f, written $\partial(x, y)/\partial(u, v)$, is the determinant of the derivative matrix $\mathbf{D}f(x, y)$ of f:[4]

$$\frac{\partial(x, y)}{\partial(u, v)} = \begin{vmatrix} \frac{\partial x}{\partial u} & \frac{\partial x}{\partial v} \\ \frac{\partial y}{\partial u} & \frac{\partial y}{\partial v} \end{vmatrix} = \frac{\partial x}{\partial u} \cdot \frac{\partial y}{\partial v} - \frac{\partial x}{\partial v} \cdot \frac{\partial y}{\partial u}. \tag{4.9}$$

In the example just considered, $(x, y) = (r\cos\theta, r\sin\theta)$, the Jacobian is

$$\frac{\partial(x, y)}{\partial(r, \theta)} = r(\cos^2\theta + \sin^2\theta) = r.$$

The Jacobian of a map at $\mathbf{x}_0$ determines whether the area about $\mathbf{x}_0$ expands or contracts. If the absolute value of the Jacobian is less than

---

[4]See Marsden and Tromba [5] for the n-dimensional definition.

one, then the map is contracting; if the absolute value of the Jacobian is greater than one, then the map is expanding.

A simple example of a contracting map is provided by the Hénon map for the parameter range $0 \leq \beta < 1$. In this case,

$$f_1(x_n, y_n) = x_{n+1} = \alpha - x_n^2 + \beta y_n,$$
$$f_2(x_n, y_n) = y_{n+1} = x_n,$$

and a quick calculation shows

$$\frac{\partial(x_{n+1}, y_{n+1})}{\partial(x_n, y_n)} = -\beta.$$

The Jacobian is constant for the Hénon map; it does not depend on the initial position (x_0, y_0). When iterating the Hénon map, the area is multiplied each time by β, and after k iterations the size of an initial area a_0 is

$$a = a_0 |\beta^k|.$$

In particular, if $0 \leq \beta < 1$, then the area is contracting.

4.4.3 Divergence of a Vector Field

Recall from a basic course in vector calculus that the divergence of a vector field represents the local rate of expansion or contraction per unit volume [5]. So, to find the local expansion or contraction of a flow we must calculate the divergence of a vector field. The *divergence* of a three-dimensional vector field $\mathbf{F}(x, y, z) = (F_1, F_2, F_3)$ is

$$\text{div } \mathbf{F} = \nabla \cdot \mathbf{F} = \frac{\partial F_1}{\partial x} + \frac{\partial F_2}{\partial y} + \frac{\partial F_3}{\partial z}. \tag{4.10}$$

Let $V(0)$ be the measure of an infinitesimal volume centered at $\mathbf{x}$. Figure 4.7 shows how this volume evolves under the flow; the divergence of the vector field measures the rate at which this initial volume changes,

$$\text{div } \mathbf{F}(\mathbf{x}) = \frac{1}{V(0)} \frac{d}{dt} V(t)|_{t=0}. \tag{4.11}$$

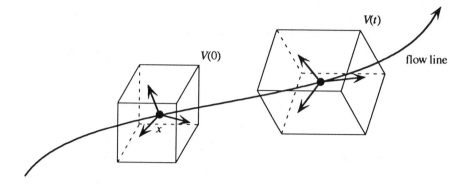

Figure 4.7: Evolution of an infinitesimal volume along a flow line.

For instance, the divergence of the vector field for the Lorenz system is

$$\nabla \cdot \begin{bmatrix} F_1 \\ F_2 \\ F_3 \end{bmatrix} = \nabla \cdot \begin{bmatrix} \sigma(y - x) \\ \rho x - y - xz \\ -\beta z + xy \end{bmatrix} = -(\sigma + 1 + \beta);$$

we find that the flow is globally contracting at a constant rate whenever the sum of σ and β is positive.

4.4.4   Dissipative and Conservative

The local rate of expansion or contraction of a dynamical system can be calculated directly from the vector field or difference equation without explicitly finding any solutions. We say a system is *conservative* if the absolute value of the Jacobian of its map exactly equals one, or if the divergence of its vector field equals zero,

$$\nabla \cdot \mathbf{F} = 0, \qquad (4.12)$$

for all times and all points. A physical system is *dissipative* if it is not conservative.[5] Most of the physical examples studied in this book are dissipative dynamical systems. The phase space of a dissipative dynamical system is continually shrinking onto a smaller region of phase space called the *attracting set*.

---

[5]Note that this definition of dissipative can include expansive systems. These will not arise in the physical examples considered in this book. See Problem 4.12.

4.4.5   Equation of First Variation

Another quantity that can be calculated directly from the vector field is the equation of first variation, which provides an approximation for the evolution of a region about an initial condition $\mathbf{x}$. The flow $\phi(\mathbf{x}, t)$ is a function of both the initial condition and time. We will often be concerned with the stability of an initial point in phase space, and thus we are led to consider the variation about a point $\mathbf{x}$ while holding the time t fixed. Let $\mathbf{D_x}$ denote differentiation with respect to the phase variables while holding t fixed. Then from the differential equation for a flow (eq. (4.3)) we find

$$\mathbf{D_x}\frac{\partial}{\partial t}\phi(\mathbf{x}, t) = \mathbf{D_x}[\mathbf{F}(\phi(\mathbf{x}, t))],$$

which, on applying the chain rule on the right-hand side, yields *the equation of first variation,*

$$\frac{\partial}{\partial t}\mathbf{D_x}\phi(\mathbf{x}, t) = \mathbf{DF}(\phi(\mathbf{x}, t))\mathbf{D_x}\phi(\mathbf{x}, t). \tag{4.13}$$

This is a *linear* differential equation for the operator $\mathbf{D_x}\phi$. $\mathbf{DF}(\phi(\mathbf{x}, t))$ is the derivative of $\mathbf{F}$ at $\phi(\mathbf{x}, t)$. If the vector field $\mathbf{F}$ is n-dimensional, then both $\mathbf{D_x}\mathbf{F}(\phi)$ and $\mathbf{D_x}\phi$ are $n \times n$ matrices.

Turning once again to the vector field

$$\mathbf{F}(x, v) = (v, -x),$$

we find that the equation of first variation for this system is

$$\begin{bmatrix} \dot{\phi}^x_x & \dot{\phi}^x_v \\ \dot{\phi}^v_x & \dot{\phi}^v_v \end{bmatrix} = \begin{bmatrix} 0 & 1 \\ -1 & 0 \end{bmatrix} \begin{bmatrix} \phi^x_x & \phi^x_v \\ \phi^v_x & \phi^v_v \end{bmatrix}.$$

The superscript i to ϕ^i indicates the ith component of flow, and the subscript j to ϕ_j denotes that we are taking the derivative with respect to the jth phase variable. For example, $\phi^v_x = \frac{\partial}{\partial x}\phi^v$. The components of ϕ are the coordinate positions, so they could be rewritten as $[\phi^x(t), \phi^v(t)] = [x(t), v(t)]$. The dot, as always, denotes differentiation with respect to time.

4.5   Fixed Points

An *equilibrium solution* of a vector field $\dot{\mathbf{x}} = \mathbf{f}(\mathbf{x})$ is a point $\bar{\mathbf{x}}$ that does not change with time,

$$\mathbf{f}(\bar{\mathbf{x}}) = 0. \tag{4.14}$$

Equilibria are also known as fixed points, stationary points, or steady-state solutions. A *fixed point* of a map is an orbit which returns to itself after one iteration,

$$\bar{\mathbf{x}} \mapsto \mathbf{f}(\bar{\mathbf{x}}) = \bar{\mathbf{x}}. \tag{4.15}$$

We will tend to use the terminology "fixed point" when referring to a map and "equilibrium" when referring to a flow. The theory for equilibria and fixed points is very similar. Keep in mind, though, that a fixed point of a map could come from a periodic orbit of a flow. This section briefly outlines the theory for flows. The corresponding theory for maps is completely analogous and can be found, for instance, in Rasband [6].

4.5.1   Stability

At least three notions of stability apply to a fixed point: local stability, global stability, and linear stability. Here we will discuss local stability and linear stability. Linear stability often, but not always, implies local stability. The additional ingredient needed is hyperbolicity. This turns out to be quite general: hyperbolicity plus a linearization procedure is usually sufficient to analyze the stability of an attracting set, whether it be a fixed point, periodic orbit, or strange attractor.

The notion of the local stability of an orbit is straightforward. A fixed point is locally *stable* if solutions based near $\bar{\mathbf{x}}$ remain close to $\bar{\mathbf{x}}$ for all future times. Further, if the solution actually approaches the fixed point, i.e., $\mathbf{x}(t) \to \bar{\mathbf{x}}$ as $t \to \infty$, then the orbit is called *asymptotically stable*. Figure 4.8(a) shows a center that is stable, but not asymptotically stable. Centers commonly occur in conservative systems. Figure 4.8(b) shows a sink, an asymptotically stable fixed point that commonly occurs in a dissipative system. A fixed point is *unstable* if it is not stable. A saddle and source are examples of unstable fixed points (see Fig. 0.4).

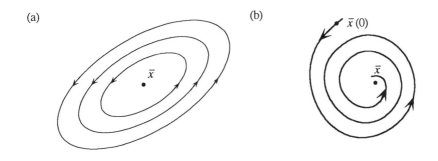

Figure 4.8: (a) A stable fixed point. (b) An asymptotically stable fixed point.

4.5.2 Linearization

To calculate the stability of a fixed point consider a small perturbation, $\mathbf{y}$, about $\bar{\mathbf{x}}$,

$$\mathbf{x} = \bar{\mathbf{x}}(t) + \mathbf{y}. \tag{4.16}$$

The Taylor expansion (substituting eq. (4.16) into eq. (4.2)) about $\bar{\mathbf{x}}$ gives

$$\dot{\mathbf{x}} = \dot{\bar{\mathbf{x}}}(t) + \dot{\mathbf{y}} = \mathbf{f}(\bar{\mathbf{x}}(t)) + \mathbf{Df}(\bar{\mathbf{x}}(t))\mathbf{y} + \text{higher-order terms.} \tag{4.17}$$

It seems reasonable that the motion near the fixed point should be governed by the *linear system*

$$\dot{\mathbf{y}} = \mathbf{Df}(\bar{\mathbf{x}}(t))\mathbf{y}, \tag{4.18}$$

since $\dot{\bar{\mathbf{x}}}(t) = \mathbf{f}(\bar{\mathbf{x}}(t))$. If $\bar{\mathbf{x}}(t) = \bar{\mathbf{x}}$ is an equilibrium point, then $\mathbf{Df}(\bar{\mathbf{x}})$ is a matrix with constant entries. We can immediately write down the solution to this linear system as

$$\mathbf{y}(t) = \exp[\mathbf{Df}(\bar{\mathbf{x}})t]\mathbf{y}_0, \tag{4.19}$$

where $\exp[\mathbf{Df}(\bar{\mathbf{x}})]$ is the evolution operator for a linear system. If we let $A = \mathbf{Df}(\bar{\mathbf{x}})$ denote the constant $n \times n$ matrix, then the linear evolution operator takes the form

$$\exp(At) = \mathbf{id} + At + \frac{1}{2!}A^2t^2 + \frac{1}{3!}A^3t^3 + \cdots \tag{4.20}$$

where **id** denotes the $n \times n$ identity matrix.

The asymptotic stability of a fixed point can be determined by the eigenvalues of the linearized vector field **Df** at $\bar{\mathbf{x}}$. In particular, we have the following test for asymptotic stability: an equilibrium solution of a nonlinear vector field is asymptotically stable if all the eigenvalues of the linearized vector field $\mathbf{Df}(\bar{\mathbf{x}})$ have negative real parts.

If the real part of at least one eigenvalue exactly equals zero (and all the others are strictly less than zero) then the system is still *linearly stable*, but the original nonlinear system may or may not be stable.

4.5.3   Hyperbolic Fixed Points: Saddles, Sources, and Sinks

Let $\mathbf{x} = \bar{\mathbf{x}}$ be an equilibrium point of a vector field. Then $\bar{\mathbf{x}}$ is called a *hyperbolic* fixed point if *none* of the real parts of the eigenvalues of $\mathbf{Df}(\bar{\mathbf{x}})$ is equal to zero. The test for asymptotic stability of the previous section can be restated as: a hyperbolic fixed point is stable if the real parts of all its eigenvalues are negative. A fixed point of a map is hyperbolic if none of the moduli of the eigenvalues equals one.

The motion near a hyperbolic fixed point can be analyzed and brought into a standard form by a linear transformation to the eigenvectors of $\mathbf{Df}(\bar{\mathbf{x}})$. Additional analysis, including higher-order terms, is usually needed to analyze the motion near a nonhyperbolic fixed point.

At last, we can precisely define the terms saddle, sink, source, and center. A hyperbolic equilibrium solution is a *saddle* if the real part of at least one eigenvalue of the linearized vector field is less than zero and if the real part of at least one eigenvalue is greater than zero. Similarly, a *saddle point* for a map is a hyperbolic point if at least one of the eigenvalues of the associated linear map has a modulus greater than one, and if one of the eigenvalues has modulus less than one.

A hyperbolic point of a flow is a *stable node* or *sink* if all the eigenvalues have real parts less than zero. Similarly, if all the moduli are less than one then the hyperbolic point of a map is a *sink*.

A hyperbolic point is an *unstable node* or *source* if the real parts of all the eigenvalues are greater than zero. The moduli of a *source* of a map are all greater than one.

A *center* is a nonhyperbolic fixed point for which all the eigenvalues

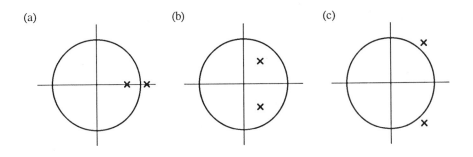

Figure 4.9: Complex eigenvalues for a two-dimensional map with a hyperbolic fixed point: (a) saddle, (b) sink, and (c) source.

are purely imaginary and nonzero (modulus one for maps). For a picture of the elementary equilibrium points in three-dimensional space see Figure 3.10 of Thompson and Stewart [7]. The corresponding stability information for a hyperbolic fixed point of a two-dimensional map is summarized in Figure 4.9.

4.6 Invariant Manifolds

According to our discussion of invariant sets in section 4.3.1, we would like to analyze a dynamical system by breaking it into its dynamically invariant parts. This is particularly easy to accomplish with linear systems because we can write down a general solution for the flow operator as e^{tA} (see section 4.5.2). The eigenspaces of a linear flow or map (i.e., the spaces formed by the eigenvectors of A) are invariant subspaces of the dynamical system. Moreover, the dynamics on each subspace are determined by the eigenvalues of that subspace. If the original manifold is $\mathbf{R}^n$, then each invariant subspace is also a Euclidean manifold which is a subset of $\mathbf{R}^n$. It is sensible to classify each of these invariant submanifolds according to the real parts of its eigenvalues, λ_i:

E^s is the subspace spanned by the eigenvectors of A with $\mathrm{Re}(\lambda_i) < 0$,

E^c is the subspace spanned by the eigenvectors of A with $\mathrm{Re}(\lambda_i) = 0$,

E^u is the subspace spanned by the eigenvectors of A with $\mathrm{Re}(\lambda_i) > 0$.

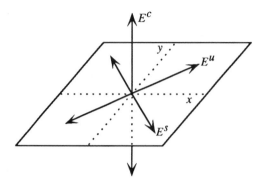

Figure 4.10: Invariant manifolds for a linear flow with eigenvalues $\lambda_i = -1\ (E^s), 0\ (E^c), 2\ (E^u)$.

E^s is called the stable space of dimension n_s, E^c is called the center space of dimension n_c, and E^u is called the unstable space of dimension n_u. If the original linear manifold is of dimension n, then the sum of the dimensions of the invariant subspaces must equal n: $n_u + n_c + n_s = n$. This definition also works for maps when the conditions on λ_i are replaced by *modulus less than one* (E^s), *modulus equal to one* (E^c), and *modulus greater than one* (E^u).

For example, consider the matrix

$$A = \begin{pmatrix} 1 & 2 & 0 \\ 1 & 0 & 0 \\ 0 & 0 & 0 \end{pmatrix}$$

This matrix has eigenvalues $\lambda = -1, 0, 2$ and eigenvectors $(1, -1, 0)$, $(0, 0, 1)$, $(2, 1, 0)$, and the flow on the invariant manifolds is illustrated in Figure 4.10.

4.6.1   Center Manifold Theorem

It is important to keep in mind that we always speak of invariant manifolds based at a point. This point is usually a fixed point $\bar{x}$ of a flow or a periodic point of a map. In the linear setting, the invariant manifold is just a linear vector space. In the nonlinear setting we can also define invariant manifolds that are not linear subspaces but are still manifolds. That is, locally they look like a copy of $\mathbf{R}^n$. These invariant manifolds are a direct generalization of the invariant subspaces of the linear problem. They are the most important geometric structure used in the analysis of a nonlinear dynamical system.

The way to generalize the notion of an invariant manifold from the linear to the nonlinear setting is straightforward. In both the linear and nonlinear settings, the stable manifold is the collection of all orbits that approach a point $\mathbf{x}$. Similarly, the unstable manifold is the collection of all orbits that depart from $\mathbf{x}$. The fact that this notion of an invariant manifold for a nonlinear system is well defined is guaranteed by the *center manifold theorem* [1]:

> **Center Manifold Theorem for Flows.** Let $\mathbf{f}_\mu(\mathbf{x})$ be a smooth vector field on $\mathbf{R}^n$ with $\mathbf{f}_\mu(\bar{\mathbf{x}}) = 0$ and $A = \mathbf{Df}(\bar{\mathbf{x}})$. The spectrum (set of eigenvalues) $\{\lambda_i\}$ of A divides into three sets σ_s, σ_c, and σ_u, where
>
> $$\lambda_i \in \begin{cases} \sigma_s, & \mathrm{Re}(\lambda_i) < 0, \\ \sigma_c, & \mathrm{Re}(\lambda_i) = 0, \\ \sigma_u, & \mathrm{Re}(\lambda_i) > 0. \end{cases}$$
>
> Let E^s, E^c, and E^u be the generalized eigenspaces of σ_s, σ_c, and σ_u. There exist smooth stable and unstable manifolds, called W^s and W^u, tangent to E^s and E^u at $\bar{\mathbf{x}}$, and a center manifold W^c tangent to E^c at $\bar{\mathbf{x}}$. The manifolds W^s, W^c, and W^u are invariant for the flow. The stable manifold W^s and the unstable manifold W^u are unique. The center manifold W^c need not be unique.

W^s is called the *stable manifold,* W^c is called the *center manifold,* and W^u is called the *unstable manifold.* A corresponding theorem for maps also holds and can be found in reference [1] or [6]. Numerical methods for the construction of the unstable and stable manifolds are described in reference [8].

Always keep in mind that flows and maps differ: a trajectory of a flow is a curve in $\mathbf{R}^n$ while the orbit of a map is a discrete sequence of points. The invariant manifolds of a flow are composed from a union of solution curves; the invariant manifolds of a map consist of a union of a

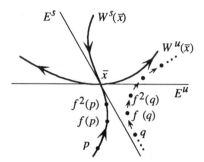

Figure 4.11: Invariant manifolds of a saddle for a two-dimensional map.

discrete collection of points (Fig. 4.11). The distinction is crucial when we come to analyze the global behavior of a dynamical system. Once again, we reiterate that the unstable and stable invariant manifolds are not a single solution, but rather a collection of solutions sharing a common asymptotic past or future.

An example (from Guckenheimer and Holmes [1]) where the invariant manifolds can be explicitly calculated is the planar vector field

$$\dot{x} = x, \quad \dot{y} = -y + x^2.$$

This system has a hyperbolic fixed point at the origin where the linearized vector field is

$$\dot{x} = x, \quad \dot{y} = -y.$$

The stable manifold of the linearized system is just the y-axis, and the unstable manifold of the linearized system is the x-axis (Fig. 4.12(a)). Returning to the nonlinear system, we can solve this system by eliminating time:

$$\frac{\dot{y}}{\dot{x}} = \frac{dy}{dx} = \frac{-y}{x} + x \quad \text{or} \quad y(x) = \frac{x^2}{3} + \frac{c}{x},$$

where c is a constant of integration. It is now easy to see (Prob. 4.20) that (Fig. 4.12(b))

$$W^u(0,0) = \{(x,y) \mid y = \frac{x^2}{3}\} \quad \text{and} \quad W^s(0,0) = \{(x,y) \mid x = 0\}.$$

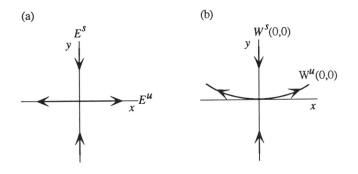

Figure 4.12: (a) Invariant manifolds (at the origin) for the linear approximation. (b) Invariant manifolds for the original nonlinear system.

4.6.2 Homoclinic and Heteroclinic Points

We informally define the unstable manifold and the stable manifold for a hyperbolic fixed point $\bar{x}$ of a map f by

$$W^s(\bar{x}) = \{x \mid \lim_{n \to \infty} f^n(x) = \bar{x}\}$$

and

$$W^u(\bar{x}) = \{x \mid \lim_{n \to \infty} f^{-n}(x) = \bar{x}\}.$$

These manifolds are tangent to the eigenvectors of f at $\bar{x}$.

We are led to study a two-dimensional map f by considering the Poincaré map of a three-dimensional flow in the vicinity of a periodic orbit. This situation is illustrated in Figure 4.13. The map f is *orientation preserving*[6] because it comes from a smooth flow. The periodic orbit of the flow gives rise to the fixed point $\bar{x}$ of the map. The fixed point $\bar{x}$ has a one-dimensional stable manifold $W^s(\bar{x})$ and a one-dimensional unstable manifold $W^u(\bar{x})$. Poincaré was led to his discovery of chaotic behavior and homoclinic tangles (see section 3.6 and Appendix H) by considering the interaction between the stable and unstable manifold of $\bar{x}$. One possible interaction is shown in Figure 4.13(a) where the unstable manifold exactly matches the stable manifold, $W^s(\bar{x}) = W^u(\bar{x})$.

---

[6]Informally, a map of a surface is orientation preserving if the normal vector to the surface is not flipped under the map.

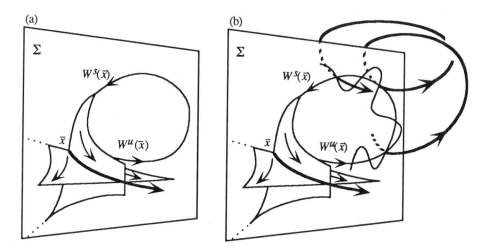

Figure 4.13: (a) Poincaré map in the vicinity of a periodic orbit, $W^s(\bar{x}) = W^u(\bar{x})$. (b) The map shown with a transversal intersection at a homoclinic point.

However, such a smooth match is exceptional. The more common possibility is for a *transversal intersection* between the stable and unstable manifold (Fig. 4.13(b)). The location of the transversal intersection is called a *homoclinic point* when both the unstable and stable manifold emanate from the same periodic orbit (Fig. 4.14(a)). The intersection point is called a *heteroclinic point* when the manifolds emanate from different periodic orbits. A heteroclinic point is shown in Figure 4.14(b)

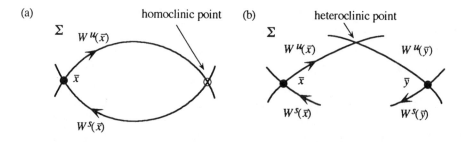

Figure 4.14: (a) A homoclinic point. (b) A heteroclinic point.

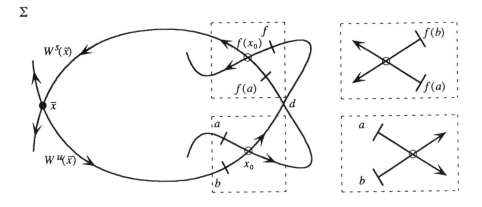

Figure 4.15: The interaction of the stable manifold $W^s(\bar{x})$ and the unstable manifold $W^u(\bar{x})$ with a homoclinic point x_0. The homoclinic point x_0 gets mapped to the homoclinic point $f(x_0)$. The orientation of the map is determined by considering where points a and b in the vicinity of x_0 are mapped.

where the unstable manifold emanating from $\bar{x}$ intersects the stable manifold of a different fixed point $\bar{y}$.

The existence of a single homoclinic or heteroclinic point forces the existence of an infinity of such points. Moreover, it also gives rise to a homoclinic (heteroclinic) tangle. This tangle is the geometric source of chaotic motions.

To see why this is so, consider Figure 4.15. A homoclinic point is indicated at x_0. This homoclinic point is part of both the stable manifold and the unstable manifold,

$$x_0 \in W^s(\bar{x}) \quad \text{and} \quad x_0 \in W^u(\bar{x}).$$

Also shown is a point a that lies on the stable manifold behind x_0 (the direction is determined by the arrow on the manifold), i.e., $a < x_0$ on W^s. Similarly, the point b lies on the unstable manifold with $b < x_0$ on W^u. Now, we must try to find the location of the next iterate of $f(x_0)$ subject to the following conditions:

1. The map f is orientation preserving.

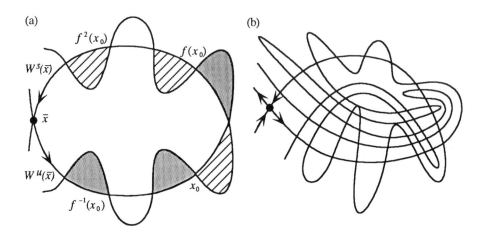

Figure 4.16: (a) Images and preimages of a homoclinic point. (b) A homoclinic tangle resulting from a single homoclinic point.

2. $f(x_0) \in W^s(\bar{x})$ and $f(x_0) \in W^u(\bar{x})$ (all the iterates of a homoclinic point are also homoclinic points).

3. $f(a) < f(x_0)$ on W^s and $f(b) < f(x_0)$ on W^u .

A picture consistent with these assumptions is shown in Figure 4.15. The point $f(x_0)$ must lie at a new homoclinic point (that is, at a new intersection point) ahead of x_0. The first candidate for the location of $f(x_0)$ is the next intersection point, indicated at d. However, $f(x_0)$ could not be located here because that would imply that the map f is orientation reversing (see Prob. 4.21). The next possible location, which does satisfy all the above conditions, is indicated by $f(x_0)$. More complicated constructions could be envisioned that are consistent with the above conditions, but the solution shown in Figure 4.15 is the simplest.

Now $f(x_0)$ is itself a homoclinic point. And the same argument applies again: the point $f^2(x_0)$ must lie closer to $\bar{x}$ and ahead of $f(x_0)$ (Fig. 4.16(a)). In this way a single homoclinic orbit must generate an infinite number of homoclinic orbits. This sequence of homoclinic points asymptotically approaches $\bar{x}$.

Since f arises from a flow, it is a diffeomorphism and thus invertible. Therefore, exactly the same argument applies to the preimages of $\bar{x}_0$. That is, $f^{-n}(x_0)$ approaches $\bar{x}$ via the unstable manifold. The end result of this construction is the violent oscillation of W^s and W^u in the region of $\bar{x}$. These oscillations form the homoclinic tangle indicated schematically in Figure 4.16(b).

The situation is even more complicated than it initially appears. The homoclinic points are not periodic orbits, but Birkhoff and Smith showed that each homoclinic point is an accumulation point for an infinite family of periodic orbits [9]. Thus, each homoclinic tangle has an infinite number of homoclinic points, and in the vicinity of each homoclinic point there exists an infinite number of periodic points. Clearly, one major goal of dynamical systems theory, and nonlinear dynamics, is the development of techniques to dissect and classify these homoclinic tangles. In section 4.8 we will show how the orbit structure of a homoclinic tangle is organized by using a horseshoe map. In Chapter 5 we will continue this topological approach by showing how knot theory can be used to unravel a homoclinic tangle.

4.7 Example: Laser Equations

We now consider a detailed example to help reinforce the barrage of mathematical definitions and concepts in the previous sections. Our example is taken from nonlinear optics and is known as the laser rate equation [10],

$$\mathbf{F} = \begin{pmatrix} F_1 \\ F_2 \end{pmatrix} = \begin{pmatrix} \dot{u} \\ \dot{z} \end{pmatrix} = \begin{pmatrix} zu \\ (1 - \alpha_1 z) - (1 + \alpha_2 z)u \end{pmatrix}.$$

In this model u is the laser intensity and z is the population inversion. The parameters α_1 and α_2 are damping constants. When certain lasers are turned on they tend to settle down to a constant intensity light output (constant u) after a series of damped oscillations (ringing) around the stable steady state solution. This behavior is predicted by the laser rate equation.

To calculate the stability of the steady states we need to know the derivative of $\mathbf{F}$:

$$\mathbf{DF} = \begin{pmatrix} \frac{\partial F_1}{\partial u} & \frac{\partial F_1}{\partial z} \\ \frac{\partial F_2}{\partial u} & \frac{\partial F_2}{\partial z} \end{pmatrix} = \begin{pmatrix} z & u \\ -(1 + \alpha_2 z) & -(\alpha_1 + \alpha_2 u) \end{pmatrix}.$$

4.7.1   Steady States

The steady states are found by setting $\mathbf{F} = 0$:

$$zu = 0,$$
$$(1 - \alpha_1 z) - (1 + \alpha_2 z)u = 0.$$

These equations have two equilibrium solutions. The first, which we label a, occurs at

$$u = 0 \quad \Rightarrow \quad z = \frac{1}{\alpha_1}, \quad a = \left(0, \frac{1}{\alpha_1}\right).$$

The second, which we label b, occurs at

$$z = 0 \quad \Rightarrow \quad u = 1, \quad b = (1, 0).$$

The location in the phase plane of these equilibrium points is shown in Figure 4.17.

The motion in the vicinity of each equilibrium point is analyzed by finding the eigenvalues λ and eigenvectors $\mathbf{v}$,

$$\lambda \mathbf{v} = A \cdot \mathbf{v}, \tag{4.21}$$

of the derivative matrix of $\mathbf{F}$, $A = \mathbf{DF}$, at each fixed point of the flow. At the point a we find

$$\mathbf{DF}|_a = \begin{pmatrix} \frac{1}{\alpha_1} & 0 \\ -(1 + \frac{\alpha_2}{\alpha_1}) & -\alpha_1 \end{pmatrix}.$$

4.7.2   Eigenvalues of a 2×2 Matrix

To calculate the eigenvalues of $\mathbf{DF}$, we recall that the general solution for the eigenvalues of any 2×2 real matrix,

$$A = \begin{pmatrix} a_{11} & a_{12} \\ a_{21} & a_{22} \end{pmatrix}, \tag{4.22}$$

are given by

$$\lambda_+ = \frac{1}{2}[\text{tr}(A) + \sqrt{\Delta}], \quad \lambda_- = \frac{1}{2}[\text{tr}(A) - \sqrt{\Delta}], \qquad (4.23)$$

where

$$\text{tr}(A) = a_{11} + a_{22}, \qquad (4.24)$$
$$\det(A) = a_{11}a_{22} - a_{12}a_{21}, \qquad (4.25)$$
$$\Delta(A) = [\text{tr}(A)]^2 - 4\det(A). \qquad (4.26)$$

Applying these formulas to $\mathbf{DF}|_a$ we find

$$\text{tr}(\mathbf{DF}|_a) = 1/\alpha_1 - \alpha_1,$$
$$\det(\mathbf{DF}|_a) = -1,$$
$$\Delta(\mathbf{DF}|_a) = (1/\alpha_1 - \alpha_1)^2 + 4 = (\alpha_1 + 1/\alpha_1)^2,$$

and the eigenvalues for the fixed point a are

$$\lambda_+ = \frac{1}{\alpha_1} \quad \text{and} \quad \lambda_- = -\alpha_1.$$

The eigenvalue λ_+ is positive, and indicates an unstable direction; the eigenvalue λ_- is negative and indicates a stable direction. The fixed point a is a hyperbolic saddle.

4.7.3   Eigenvectors

The stable and unstable directions, and hence the stable space $E^s(a)$ and the unstable space $E^u(a)$, are determined by the eigenvectors of $\mathbf{DF}|_a$. The stable direction is calculated from

$$\frac{1}{\alpha_1}\begin{pmatrix} \epsilon \\ \eta \end{pmatrix} = \begin{pmatrix} \frac{1}{\alpha_1} & 0 \\ -(1 + \frac{\alpha_2}{\alpha_1}) & -\alpha_1 \end{pmatrix} \cdot \begin{pmatrix} \epsilon \\ \eta \end{pmatrix}.$$

Solving this system of simultaneous equations for ϵ and η gives the unnormalized eigenvector for the unstable space as

$$\mathbf{v}(\lambda_+) = \begin{pmatrix} \epsilon \\ -m\epsilon \end{pmatrix}, \quad m = \frac{\alpha_1 + \alpha_2}{\alpha_1^2 + 1}.$$

Similarly, the stable space is found from the eigenvalue equation for λ_-:

$$-\alpha_1 \begin{pmatrix} \epsilon \\ \eta \end{pmatrix} = \begin{pmatrix} \frac{1}{\alpha_1} & 0 \\ -(1 + \frac{\alpha_2}{\alpha_1}) & -\alpha_1 \end{pmatrix} \cdot \begin{pmatrix} \epsilon \\ \eta \end{pmatrix}.$$

Solving this set of simultaneous equations shows that the stable space is just the z-axis,

$$\mathbf{v}(\lambda_-) = \begin{pmatrix} 0 \\ \eta \end{pmatrix}.$$

In fact, the z-axis is invariant for the whole flow (i.e., if $z = 0$ then $\dot{u} = 0$ for all t) so the z-axis is the global stable manifold at a, $E^s(a) = W^s(a)$.

4.7.4   Stable Focus

To analyze the dynamics in the vicinity of the fixed point b we need to find the eigenvalues and eigenvectors of

$$\mathbf{DF}|_b = \begin{pmatrix} 0 & 1 \\ -1 & -(\alpha_1 + \alpha_2) \end{pmatrix}.$$

The eigenvalues of $\mathbf{DF}|_b$ are

$$\lambda_+ = -\alpha + i\omega \quad \text{and} \quad \lambda_- = -\alpha - i\omega,$$

where

$$\alpha = \frac{\alpha_1 + \alpha_2}{2} \quad \text{and} \quad \omega = \sqrt{1 - \alpha^2}.$$

The fixed point b is a stable focus since the real parts of the eigenvalues are negative. This focus represents the constant intensity output of a laser, and the oscillation about this steady state is the ringing a laser initially experiences when it is turned on.

The global phase portrait, pieced together from local information about the fixed points, is pictured in Figure 4.17.

4.8   Smale Horseshoe

In section 4.6.2 we stressed the importance of analyzing the orbit structure arising within a homoclinic tangle. From a topological and physical point of view, analyzing the orbit structure primarily means answering two questions:

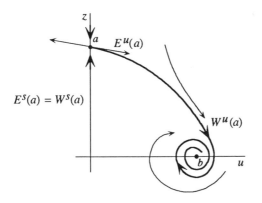

Figure 4.17: Phase portrait for the laser rate equation.

1. What are the relative locations of the periodic orbits?

2. How are the stable and unstable manifolds interwoven within a homoclinic tangle?

By studying the horseshoe example we will see that these questions are intimately connected.

In sections 2.11 and 2.12 we answered the first question for the one-dimensional quadratic map by using symbolic dynamics. For the special case of a chaotic hyperbolic invariant set (to be discussed in section 4.9), Smale found an answer to both of the above questions for maps of any dimension. Again, the solution involves the use of symbolic dynamics.

The prototypical example of a chaotic hyperbolic invariant set is the Smale horseshoe [11]. A detailed knowledge of this example is essential for understanding chaos. The Smale horseshoe (like the quadratic map for $\lambda > 4$) is an example of a chaotic repeller. It is not an attractor. Physical applications properly focus on attractors since these are directly observable. It is, therefore, sometimes believed that the chaotic horseshoe has little use in physical applications. In Chapter 5 we will show that such a belief could not be further from the truth. Remnants of a horseshoe (sometimes called the proto-horseshoe [11]) are buried within a chaotic attractor. The horseshoe (or some other variant of a hyperbolic invariant set) acts as the skeleton on which chaotic and periodic orbits are organized. To quote Holmes, "Horseshoes in a sense provide the 'backbone' for the attractors [12]." Therefore, horseshoes are essential to both the mathematical and physical analysis of a chaotic system.

4.8.1 From Tangles to Horseshoes

The horseshoe map is motivated by studying the dynamics of a map in the vicinity of a periodic orbit with a homoclinic point. Such a system gives rise to a homoclinic tangle. Consider a small box ($ABCD$) in the vicinity of a periodic orbit as seen from the surface of section. This situation is illustrated in Figure 4.18(a). The box is chosen so that the side AD is part of the unstable manifold, and the sides AB and DC are part of the stable manifold. We now ask how this box evolves under forward and backward iterations. The unstable and stable manifolds of the periodic orbit are invariant. Therefore, when the box is iterated, any point of the box that lies on an invariant manifold must always remain on this invariant manifold.

If we iterate points in the box forward, then we generally end up (after a finite number of iterations) with the "horseshoe shape" ($C'D'A'B'$) shown in Figure 4.18(b). The initial segment AD, which lies on the unstable manifold, gets mapped to the segment $A'D'$, which is also part of the unstable manifold. Similarly, if we iterate the box backward we find a backward horseshoe perpendicular to the forward horseshoe (Fig. 4.18(c)). The box of initial points gets compressed along the unstable manifold W^u and stretched along the stable manifold W^s. After a finite number of iterations, the forward image of the box will intersect the backward image of the box. Further iteration produces more intersections.

Each new region of intersection contains a periodic orbit (see Fig. 4.20) as well as segments of the unstable and stable manifolds. That is, the horseshoe can be viewed as generating the homoclinic tangle. Smale realized that this type of horseshoe structure occurs quite generally in a chaotic system. Therefore, he decided to isolate this horseshoe map from the rest of the problem [11].

A schematic for this isolated horseshoe map is presented in Figure

(a)

(b)

(c)

(d)

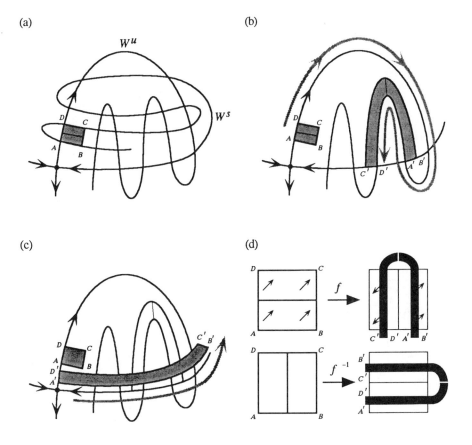

Figure 4.18: Formation of a horseshoe inside a homoclinic tangle.

4.18(d). Like the quadratic map, it consists of a stretch and a fold. The horseshoe map can be thought of as a "thickened" quadratic map. Unlike the quadratic map, though, the horseshoe map is invertible. The future and past of all points are well defined. However, the itinerary of points that get mapped out of the box are ignored. We are only concerned with points that remain in the box under all future and past iterations. These points form the invariant set.

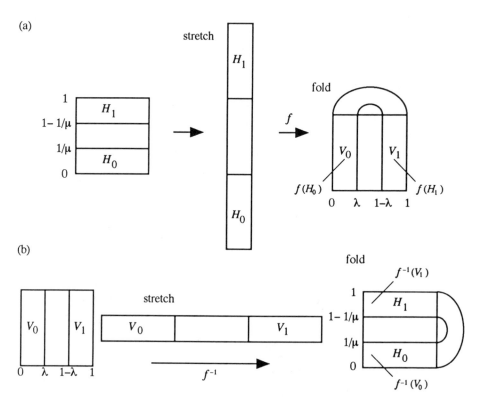

Figure 4.19: (a) Forward iteration of the horseshoe map. (b) Backward iteration of the horseshoe map.

4.8.2   Horseshoe Map

A mathematical discussion of the horseshoe map is provided by Devaney [13] or Wiggins [14]. Here, we will present a more descriptive account of the horseshoe that closely follows Wiggins's discussion.

The forward iteration of the horseshoe map is shown in Figure 4.19(a). The horseshoe is a mapping of the unit square D,

$$f : D \to \mathbf{R}^2, \quad D = \{(x, y) \in \mathbf{R}^2 | 0 \le x \le 1, 0 \le y \le 1\},$$

which contracts the horizontal directions, expands in the vertical direction, and then folds. The mapping is only defined on the unit square. Points that leave the square are ignored. Let the horizontal strip H_0 be all points on the unit square with $0 \le y \le 1/\mu$, and let horizontal strip H_1 be all points with $1 - 1/\mu \leq y \leq 1$. Then a *linear horseshoe map* is defined by the transformation

$$f(H_0): \begin{pmatrix} x \\ y \end{pmatrix} \mapsto \begin{pmatrix} \lambda & 0 \\ 0 & \mu \end{pmatrix} \begin{pmatrix} x \\ y \end{pmatrix}, \qquad (4.27)$$

$$f(H_1): \begin{pmatrix} x \\ y \end{pmatrix} \mapsto \begin{pmatrix} -\lambda & 0 \\ 0 & -\mu \end{pmatrix} \begin{pmatrix} x \\ y \end{pmatrix} + \begin{pmatrix} 1 \\ \mu \end{pmatrix}, \qquad (4.28)$$

where $0 < \lambda < 1/2$ and $\mu > 2$. The horseshoe map takes the horizontal strip H_0 to the vertical strip $V_0 = \{(x, y) | 0 \leq x \leq \lambda\}$, and H_1 to the vertical strip $V_1 = \{(x, y) | 1 - \lambda \leq x \leq 1\}$:

$$f(H_0) = V_0 \quad \text{and} \quad f(H_1) = V_1. \qquad (4.29)$$

The strip H_1 is also rotated by $180°$. The inverse of the horseshoe map f^{-1} is shown in Figure 4.19(b). The inverse map takes the vertical rectangles V_0 and V_1 to the horizontal rectangles H_0 and H_1.

The invariant set Λ of the horseshoe map is the collection of all points that remain in D under all iterations of f,

$$\Lambda = \cdots f^{-2}(D) \bigcap f^{-1}(D) \bigcap D \bigcap f(D) \bigcap f^2(D) \cdots = \bigcap_{n=-\infty}^{\infty} f^n(D).$$

This invariant set consists of a certain infinite intersection of horizontal and vertical rectangles. To keep track of the iterates of the horseshoe map (the rectangles), we will need the symbols $s_i \in \{0, 1\}$ with $i = 0, \pm 1, \pm 2, \ldots$. The symbolic encoding of the rectangles works much the same way as the symbolic encoding of the quadratic map (see section 2.12.2).

The first forward iteration of the horseshoe map produces two vertical rectangles called V_0 and V_1. V_0 is the vertical rectangle on the left and V_1 is the vertical rectangle on the right. The next step is to apply the horseshoe map again, thereby producing $f^2(D)$. As shown in Figure 4.20(a), V_0 and V_1 produce four vertical rectangles labeled (from left to right) V_{00}, V_{01}, V_{11}, and V_{10}. Applying the map yet again produces eight vertical strips labeled V_{000}, V_{001}, V_{011}, V_{010}, V_{110}, V_{111}, V_{101}, and V_{100}. In general, the nth iteration produces 2^n rectangles. The labeling for the vertical strips is recursively defined as follows: if

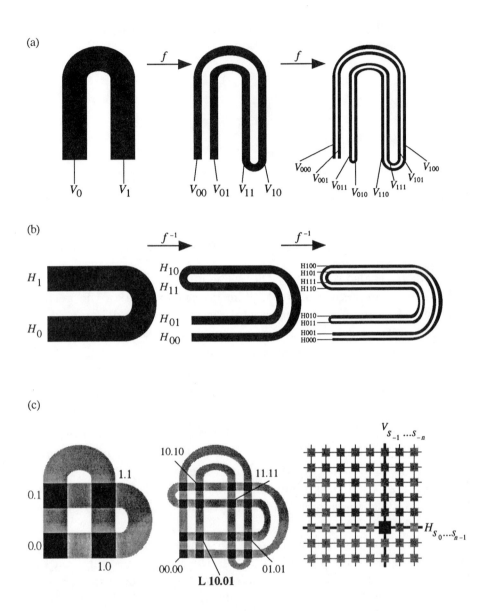

Figure 4.20: (a) Forward iteration of the horseshoe map and symbolic names. (b) Backward iteration. (c) Symbolic encoding of the invariant points constructed from the forward and backward iterations.

the current strip is left of the center, then a 0 is added to the front of the previous label of the rectangle; if it falls to the right, a 1 is added. So, for instance, the rectangle labeled V_1 starts on the right. The rectangle labeled V_{01} originates from strip V_1, but it currently lies on the left. Lastly, the strip V_{101}, starts on the right, then goes to the left, and then returns to the right again. To each vertical strip we associate a *symbolic itinerary*,

$$V_{s_{-1}s_{-2}s_{-3}...s_{-i}...s_{-n}},$$

which gives the approximate orbit (left or right) of a vertical strip after n iterations. The minus sign in the symbolic label indicates that the symbol s_{-i} arises from considering the ith preimage of the particular vertical strip under f. Also note that the vertical strips get progressively thinner, so that after n iterations, each strip has a width of λ^n.

The backward iterates produce 2^n horizontal strips at the nth iteration. The height of each of these horizontal strips is $1/\mu^n$. From the two horizontal strips H_0 and H_1, the inverse map f^{-1} produces four horizontal rectangles labeled (from bottom to top) as H_{00}, H_{01}, H_{11}, and H_{10} (Fig. 4.20(b)). This in turn produces eight horizontal rectangles, H_{000}, H_{001}, H_{011}, H_{010}, H_{110}, H_{111}, H_{101}, and H_{100}. Each horizontal strip can be uniquely labeled with a sequence of 0's and 1's,

$$H_{s_0 s_1 s_2...s_i...s_{n-1}},$$

where the symbol s_0 indicates the current approximate location (bottom or top) of the horizontal rectangle. The fact that the labeling scheme is unique follows from the definition of f and the observation that all of the horizontal rectangles are disjoint. Unlike the vertical strips, the indexing for the horizontal strips starts at 0 and is positive. The need for this indexing convention will become apparent when we specify the labeling of points in the invariant set.

Now, the invariant set Λ of the horseshoe map f is given by the infinite intersection of all the horizontal and vertical strips. The invariant set is a fractal, in fact, it is a product of two Cantor sets. The map f generates a Cantor set in the horizontal direction, and the inverse map f^{-1} generates a Cantor set in the vertical direction. The invariant set is, in a sense, the product of these two Cantor sets.

4.8.3　Symbolic Dynamics

We can identify points in the invariant set according to the following scheme. After one forward iteration and one backward iteration, the invariant set is located within the four shaded rectangular regions shown in Figure 4.20(c). After two forward iterations and two backward iterations, the invariant set is a subset of the 16 shaded regions. The shaded regions are the intersection of the horizontal and vertical strips. To each shaded region we associate a *bi-infinite* symbol sequence,

$$\cdots s_{-n} \cdots s_{-3} s_{-2} s_{-1} . s_0 s_1 s_2 \cdots s_n \cdots,$$

constructed from the label of the vertical and horizontal strips forming a point in the invariant set. The right-hand side of the symbolic name, $s_0 s_1 s_2 \cdots s_n \cdots$, is the label from the horizontal strip $H_{s_0 s_1 s_2 \cdots s_n \cdots}$. The *left-hand side* of the symbolic name, $\cdots s_{-n} \cdots s_{-3} s_{-2} s_{-1}$, is the label from the *vertical strip written backwards*, $V_{s_{-1} s_{-2} s_{-3} \cdots s_{-i} \cdots s_{-n}}$. For instance, the shaded region labeled L in Figure 4.20(c) has a symbolic name "10.01." The ".01" to the right of the dot indicates horizontal strip H_{01}. The "10." ("01" backwards) to the left of the dot indicates that the shaded region comes from the vertical strip V_{01}.

We hone in closer and closer to the invariant set by iterating the horseshoe map both forward and backward. Moreover, the above labeling scheme generates a symbolic name, or symbolic coordinate, for each point of the invariant set. This symbolic name contains information about the dynamics of the invariant point.

To see how this works more formally, let us call Σ the symbol space of all bi-infinite sequences of 0's and 1's. A metric on Σ between the two sequences

$$s = \left(\cdots s_{-n} \cdots s_{-1} . s_0 s_1 \cdots s_n \cdots \right)$$
$$\bar{s} = \left(\cdots \bar{s}_{-n} \cdots \bar{s}_{-1} . \bar{s}_0 \bar{s}_1 \cdots \bar{s}_n \cdots \right)$$

is defined by

$$d[s, \bar{s}] = \sum_{i=-\infty}^{\infty} \frac{\delta_i}{2^{|i|}} \quad \text{where } \delta_i = \begin{cases} 0 & \text{if } s_i = \bar{s}_i, \\ 1 & \text{if } s_i \neq \bar{s}_i. \end{cases} \tag{4.30}$$

Next we define a *shift map* σ on Σ by

$$\sigma(s) = \left(\cdots s_{-n} \cdots s_{-1} s_0 . s_1 s_2 \cdots s_n \cdots \right), \quad \text{i.e., } \sigma(s)_i = s_{i+1}. \tag{4.31}$$

The shift map is continuous and it has two fixed points consisting of a string of all 0's or all 1's. A period n orbit of σ is written as $\overline{s_{-n} \cdots s_{-1}.s_0 s_1 \cdots s_{n-1}}$, where the overbar indicates that the symbolic sequence repeats forever. A few of the periodic orbits and their "shift equivalent" representations are listed below,

Period 1 : $\overline{0.0}, \overline{1.1}$;

Period 2 : $\overline{01.01} \longrightarrow \overline{10.10}$;

Period 3 : $\overline{001.001} \longrightarrow \overline{010.010} \longrightarrow \overline{100.100}$;

$\overline{110.110} \longrightarrow \overline{101.101} \longrightarrow \overline{110.110}$;

and so on. In addition to periodic orbits of arbitrarily high period, the shift map also possesses an uncountable infinity of nonperiodic orbits as well as a dense orbit. See Devaney [13] or Wiggins [14] for the details.

In section 2.11 we showed that the shift map on the space of one-sided symbol sequences is topologically semiconjugate to the quadratic map. A similar results holds for the shift map on the space of bi-infinite sequences and the horseshoe map; namely, there exists a homeomorphism $\phi : \Lambda \to \Sigma$ connecting the dynamics of f on Λ and σ on Σ such that $\phi \circ f = \sigma \circ \phi$:

$$\begin{array}{ccc} \Lambda & \xrightarrow{f} & \Lambda \\ \phi \downarrow & & \downarrow \phi \\ \Sigma & \xrightarrow{\sigma} & \Sigma \end{array}$$

The correspondence between the shift map on Σ and the horseshoe map on the invariant set Λ is pretty easy to see (again, for the mathematical details see Wiggins [14]). The invariant set consists of an infinite intersection of horizontal and vertical strips. These intersection points are labeled by their symbolic itinerary, and the horseshoe map carries one point of the invariant set to another precisely by a shift map. Consider, for instance, the period two orbit

$$\overline{01.01} \xrightarrow{\sigma} \overline{10.10}.$$

The shift map sends $\overline{01.10}$ to $\overline{10.10}$ and back again. This corresponds to an orbit of the horseshoe map that bounces back and forth between the points labeled 01.01 and 10.10 in Λ (see Fig. 4.21). The topological conjugacy between σ and f allows us to immediately conclude that, like the shift map, the horseshoe map has

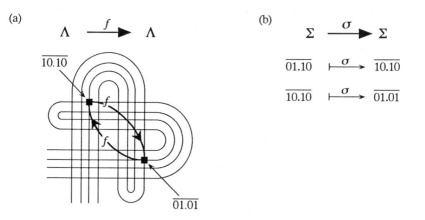

Figure 4.21: Equivalence between the dynamics of the horseshoe map on the unit square and the shift map on the bi-infinite symbol space.

1. a countable infinity of periodic orbits (and all the periodic orbits are hyperbolic saddles);

2. an uncountable infinity of nonperiodic orbits;

3. a dense orbit.

The shift map (and hence the horseshoe map) exhibits sensitive dependence on initial conditions (see section 4.10). As stated above, it possesses a dense orbit. These two properties are generally taken as defining a chaotic set.

4.8.4   From Horseshoes to Tangles

Forward and backward iterations of the horseshoe map generate the locations of periodic points to a higher and higher precision. That is, by iterating the horseshoe map, we can specify the location of a periodic orbit within a homoclinic tangle (of the horseshoe) to any degree of accuracy. For instance, after one iteration, we know the approximate location of a period two orbit. It lies somewhere within the shaded regions labeled 0.1 and 1.0 in Figure 4.20(c). After two iterations, we

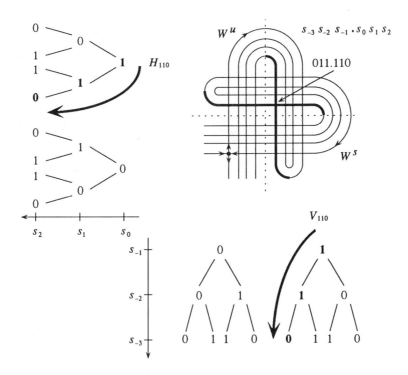

Figure 4.22: Homoclinic (horseshoe) tangle and the labeling scheme for horizontal and vertical branches from a pair of alternating binary trees.

know its position even better. It lies somewhere within the shaded regions labeled 10.10 and 01.01 (Fig. 4.20(c)).

The forward iterates of the horseshoe map produce a "snake" that approaches the unstable manifold W^u of the periodic point. The backward iterates produce another snake that approaches the stable manifold W^s of the periodic point at the origin. Thus, iterating a horseshoe generates a tangle. The relative locations of horizontal and vertical branches of this tangle are the same as those that occur in a homoclinic tangle with a horseshoe arising in a particular flow. This is illustrated in Figure 4.22. We can name the branches of the tangle with the same labeling scheme we used for the horseshoe. For a horseshoe, the labeling scheme is easy to see once we notice that both the horizontal and

Figure 4.23: Examples of horseshoe-like maps that generate hyperbolic invariant sets.

vertical branches are labeled according to the alternating binary tree introduced in section 2.12.2.

The labeling of the horizontal branches is determined by the symbols $s_0 s_1 s_2 \dots$. For instance, the horizontal label for branch H_{110} can be determined by reading down the alternating binary tree as illustrated in Figure 4.22. A second alternating binary tree is used to determine the labeling for the vertical branches. The labeling for the branch V_{110} is indicated in Figure 4.22. The labeling scheme for the horizontal and vertical branches at first appears complicated. However, the branch names are easy to write down once we realize that they can be read directly from the alternating binary tree.

4.9   Hyperbolicity

The horseshoe map is just one of an infinity of possible return maps (chaotic forms) that can be successfully analyzed using symbolic dynamics. A few other possibilities are shown in Figure 4.23. Each different return map generates a different homoclinic tangle, but all these tangles can be dissected using symbolic dynamics. All these maps are similar to the horseshoe because they are topologically conjugate to an appropriate symbol space with a shift map. All these maps possess an invariant Cantor set Λ. These invariant sets all possess a special property that ensures their successful analysis using symbolic dynamics, namely, hyperbolicity.

Recall our definition of a hyperbolic point. A fixed point of a map is hyperbolic if none of the moduli of its eigenvalues exactly equals one. The notion of a hyperbolic invariant set is a generalization of a hyperbolic fixed point. Informally, to define a hyperbolic set we extend this property of "no eigenvalues on the unit circle" to each point of the invariant set. In other words, there is no center manifold for any point of the invariant set. Technically, a set Λ arising in a diffeomorphism $f : \mathbf{R}^2 \to \mathbf{R}^2$ is a hyperbolic set if

1. there exists a pair of tangent lines $E^s(x)$ and $E^u(x)$ for each $x \in \Lambda$ which are preserved by $Df(x)$;

2. $E^s(x)$ and $E^u(x)$ vary smoothly with x;

3. there is a constant $\lambda > 1$ such that $\|Df(x)w\| \geq \lambda \|w\|$ for all $w \in E^u(x)$ and $\|Df^{-1}(x)w\| \geq \lambda \|w\|$ for all $w \in E^s(x)$.

A more complete mathematical discussion of hyperbolicity can be found in Devaney [13] or Wiggins [14]. A general mathematical theory for the symbolic analysis of chaotic hyperbolic invariant sets is described by Devaney [13] and goes under the rubric of "subshifts and transition matrices."

Like the horseshoe, the chaotic hyperbolic invariant sets encountered in the mathematics literature are often chaotic repellers. In physical applications, on the other hand, we are more commonly faced with the analysis of nonhyperbolic chaotic attractors. The extension of symbolic analysis from the (mathematical) hyperbolic regime to the (more physical) nonhyperbolic regime is still an active research question [15].

4.10 Lyapunov Characteristic Exponent

In section 2.10 we informally introduced the Lyapunov exponent as a simple measure of sensitive dependence on initial conditions, i.e., chaotic behavior. The notion of a Lyapunov exponent is a generalization of the idea of an eigenvalue as a measure of the stability of a fixed point or a characteristic exponent [1] as the measure of the stability of a periodic orbit. For a chaotic trajectory it is not sensible to

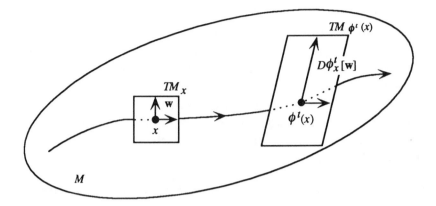

Figure 4.24: Evolution of vectors in the tangent manifold under a flow.

examine the instantaneous eigenvalue of a trajectory. The next best quantity, therefore, is an eigenvalue averaged over the whole trajectory. The idea of measuring the average stability of a trajectory leads us to the formal notion of a Lyapunov exponent. The Lyapunov exponent is best defined by looking at the evolution (under a flow) of the tangent manifold. That is, "sensitive dependence on initial conditions" is most clearly stated as an observation about the evolution of vectors in the tangent manifold rather than the evolution of trajectories in the flow of the original manifold M.

The *tangent manifold* at a point x, written as TM_x, is the collection of all tangent vectors of the manifold M at the point x. The tangent manifold is a linear vector space. The collection of all tangent manifolds is called the *tangent bundle*. For instance, for a surface embedded in $\mathbf{R}^3$ the tangent manifold at each point of the manifold is a tangent plane. More generally, if the original manifold is of dimension n, then the tangent manifold is a linear vector space also of dimension n. For further background material on manifolds, tangent manifolds, and flows, see Arnold [1].

The integral curves of a flow on a manifold provide a smooth *foliation* of that manifold, in the following manner. A point $x \in M$ goes to the point $\phi_t(x) \in M$ under the flow (see Fig. 4.24). Now we make a key observation: the tangent vectors $\mathbf{w} \in TM_x$ are also carried by the flow (this is called a "Lie dragging") so that we can set up a unique correspondence between the tangent vectors in TM_x and the tangent vectors in $TM_{\phi_t(x)}$. Namely, for each $\mathbf{w} \in TM_x$, there exists a unique vector $D\phi_t[\mathbf{w}] \in TM_{\phi^t(x)}$. The *Lyapunov characteristic exponent* of a flow is defined as

$$\lambda(x, \mathbf{w}) = \lim_{t \to \infty} \frac{1}{t} \ln \left(\frac{\|D\phi_t(x)[\mathbf{w}]\|}{\|\mathbf{w}\|} \right). \tag{4.32}$$

That is, the Lyapunov characteristic exponent measures the average growth rate of vectors in the tangent manifold. The corresponding Lyapunov characteristic exponent of a map is [6]

$$\lambda(x, \mathbf{w}) = \lim_{n \to \infty} \frac{1}{n} \ln \|(Df(x))^n[\mathbf{w}]\|, \tag{4.33}$$

where

$$(Df(x))^n = Df(f^{n-1}(x)) \circ \cdots \circ Df(x)[\mathbf{w}]. \tag{4.34}$$

A flow is said to have *sensitive dependence on initial conditions* if the Lyapunov characteristic exponent is positive. From a physical point of view, the Lyapunov exponent is a very useful indicator distinguishing a chaotic from a nonchaotic trajectory [16].

References and Notes

[1] Good general references for the material in this chapter are listed here. For ordinary differential equations, see V. I. Arnold, *Ordinary differential equations* (MIT Press: Cambridge, MA, 1973). Suspensions are discussed in Z. Nitecki, *Differentiable Dynamics* (MIT Press: Cambridge, MA, 1971), and S. Wiggins, *Introduction to applied nonlinear dynamical systems and chaos* (Springer-Verlag: New York, 1990). The standard reference for applied dynamical systems theory is J. Guckenheimer and P. Holmes, *Nonlinear oscillations, dynamical systems, and bifurcations of vector fields*, second printing (Springer-Verlag: New York, 1986).

[2] IHJM stands for Ikeda, Hammel, Jones, and Moloney. See S. K. Hammel, C. K. R. T. Jones, and J. V. Moloney, Global dynamical behavior of the optical field in a ring cavity, J. Opt. Soc. Am. B **2**, 552–564 (1985).

[3] The quotes are from Paul Blanchard. Some of the introductory material is based on notes from Paul Blanchard's 1984 Dynamical Systems Course,

MA771. We are indebted to Paul and to Dick Hall for explaining to us the "chain recurrent point of view."

[4]   C. Conley, Isolated Invariant Sets and the Morse Index, CBMS Conference Series **38** (1978); J. M. Franks, *Homology and Dynamical Systems*, Conference Board of the Mathematical Sciences Regional Conference Series in Mathematics Number 49 (American Mathematical Society, Providence, 1980); R. Easton, Isolating blocks and epsilon chains for maps, Physica **39D**, 95–110 (1989).

[5]   J. Marsden and A. Tromba, *Vector Calculus*, third ed. (W. H. Freeman: New York, 1988).

[6]   S. Rasband, *Chaotic dynamics of nonlinear systems* (John Wiley: New York, 1990). Chapter 7 discusses fixed point theory for maps.

[7]   J. Thompson and H. Stewart, *Nonlinear dynamics and chaos* (John Wiley: New York, 1986).

[8]   T. Parker, and L. Chua, *Practical numerical algorithms for chaotic systems* (Springer-Verlag: New York, 1989).

[9]   A nice description of the homoclinic periodic orbit theorem of Birkhoff and Smith is presented by E. V. Eschenazi, Multistability and Basins of Attraction in Driven Dynamical Systems, Ph.D. Thesis, Drexel University (1988); also see section 6.1 of R. Abraham and C. Shaw, *Dynamics—The geometry of behavior. Part three: Global behavior* (Aerial Press: Santa Cruz, CA, 1984).

[10] H. G. Solari and R. Gilmore, Relative rotation rates for driven dynamical systems, Phys. Rev. **A 37** (8), 3096–3109 (1988); L. M. Narducci and N. B. Abraham, *Laser physics and laser instabilities* (World Scientific: New Jersey, 1988).

[11] S. Smale, *The mathematics of time: Essays on dynamical systems, economic processes, and related topics* (Springer-Verlag: New York, 1980); J. Yorke and K. Alligood, Period doubling cascades for attractors: A prerequisite for horseshoes, Comm. Math. Phys. **101**, 303 (1985).

[12] P. Holmes, Knotted periodic orbits in the suspensions of Smale's horseshoe: Extended families and bifurcation sequences, Physica D **40**, 42–64 (1989).

[13] R. L. Devaney, *An introduction to chaotic dynamical systems,* second edition (Addison-Wesley: New York, 1989).

[14] S. Wiggins, *Introduction to applied nonlinear dynamical systems and chaos* (Springer-Verlag: New York, 1990).

[15] P. Cvitanović, G. H. Gunaratne, and I. Procaccia, Topological and metric properties of Hénon-type strange attractors, Phys. Rev. **A 38** (3), 1503–1520 (1988).

[16] An elementary introduction to Lyapunov exponents is provided by S. Souza-Machado, R. Rollins, D. Jacobs, and J. Hartman, Studying chaotic systems using microcomputers and Lyapunov exponents, Am. J. Phys. **58** (4), 321–329 (1990). For the state of the art in computing Lyapunov exponents, see P. Bryant, R. Brown, and H. Abarbanel, Lyapunov exponents from observed time series, Phys. Rev. Lett. **65** (13), 1523–1526 (1990).

Problems

Problems for section 4.2.

4.1. Show that the Hénon map with $\beta = 0$ reduces to an equivalent form of the quadratic map.

4.2. Introduce the new variable $\tau = \omega t$ to the forced damped pendulum and rewrite the equations for the vector field so that the flow is $\phi(\tau, \theta, v) : S^1 \times \mathbf{R}^2$.

4.3. Write the IHJM optical map as a map of two real variables $f(x, y) : \mathbf{R}^2 \to \mathbf{R}^2$ where $z = x + iy$. Consider separately the cases where the parameters γ and B are real and complex.

Section 4.3.

4.4. What is the ω-limit set of a point x_1 of a period two orbit of the quadratic map: $O(x_1) = \{x_1, x_2\}$?

4.5. Let $f(z) : S^1 \to S^1$, with $z \mapsto e^{2\pi i \theta} z$. Show that if θ is irrational, then $\omega(p) = S^1$.

4.6. Give an example of a map $f : M \to M$ with $L^+ \neq \Omega$.

4.7. Why is the label Ω a good name for the nonwandering set?

Section 4.4.

4.8. Verify that the Jacobian for the Hénon map is

$$\frac{\partial(x_{n+1}, y_{n+1})}{\partial(x_n, y_n)} = -\beta.$$

4.9. Compute the divergence of the vector field for the damped linear oscillator, $\ddot{x} + \alpha \dot{x} + \omega^2 x = 0$.

4.10. Calculate the divergence of the vector field for the Duffing equation, forced damped pendulum, and modulated laser rate equation. For what parameter values are the first two systems dissipative or conservative?

4.11. Calculate the Jacobian for the quadratic map, the sine circle map, the Baker's map, and the IHJM optical map. For what parameter values is the Baker's map dissipative or conservative?

4.12. The definition of a "dissipative" dynamical system in section 4.4.4 actually includes expansive systems. How would you redefine the term dissipative to handle these cases separately?

4.13. Solve the equation of first variation for the vector field $\mathbf{F}(x, v) = (v, -x)$. That is, find the time evolution of $\phi_x^x(t), \phi_v^x(t), \phi_x^v(t), \phi_v^v(t)$.

Section 4.5.

4.14. Explicitly derive the evolution operator for the vector field $\mathbf{F}(x, v) = (v, -x)$ by solving the differential equation for the harmonic oscillator in dimensionless variables.

4.15. Show that the following 2×2 Jordan matrices have the indicated linear evolution operators:

(a)

$$A = \begin{bmatrix} \lambda_1 & 0 \\ 0 & \lambda_2 \end{bmatrix}, \quad e^{tA} = \begin{bmatrix} e^{\lambda_1 t} & 0 \\ 0 & e^{\lambda_2 t} \end{bmatrix};$$

(b)

$$A = \begin{bmatrix} \alpha & -\omega \\ \omega & \alpha \end{bmatrix}, \quad e^{tA} = e^{\alpha t} \begin{bmatrix} \cos \omega t & -\sin \omega t \\ \sin \omega t & \cos \omega t \end{bmatrix};$$

(c)

$$A = \begin{bmatrix} \lambda & 0 \\ 1 & \lambda \end{bmatrix}, \quad e^{tA} = e^{\lambda t} \begin{bmatrix} 1 & 0 \\ t & 1 \end{bmatrix}.$$

4.16. Find an example of a differential equation with an equilibrium point which is linearly stable but not locally stable. Hint: See Wiggins, reference [1].

4.17. Find the fixed points of the sine circle map and the Baker's map.

4.18. Find all the fixed points of the linear harmonic oscillator $\ddot{x} + x + \alpha \dot{x} = 0$ and evaluate their local stability for $\alpha = 1$ and $\alpha < 1$. When are the fixed points asymptotically stable?

4.19. Find the equilibrium points of the unforced Duffing equation and the damped pendulum, and analyze their linear stability.

Section 4.6.

4.20. Calculate $W^u(0,0)$ and $W^s(0,0)$ for the planar vector field $\dot{x} = x$, $\dot{y} = -y+x^2$.

4.21. Draw a picture from Figure 4.15 to show that if $f(x_0)$ is located at d, then the map is orientation reversing.

Section 4.7.

4.22. Follow section 4.7 to construct the phase portrait for the laser rate equation when (a) $\alpha_1 = \alpha_2 = 0$, and (b) $\alpha_1, \alpha_2 < 0$.

Section 4.8.

4.23. Calculate the inverse of the horseshoe map f^{-1} described in section 4.8.2.

Chapter 5

Knots and Templates

5.1 Introduction

Physicists are confronted with a fundamental challenge when studying a nonlinear system; to wit, how are theory and experiment to be compared for a chaotic system? What properties of a chaotic system should a physical theory seek to explain or predict? For a nonchaotic system, a physical theory can attempt to predict the long-term evolution of an individual trajectory. Chaotic systems, though, exhibit sensitive dependence on initial conditions. Long-term predictability is not an attainable or a sensible goal for a physical theory of chaos. What is to be done?

A consensus is now forming among physicists which says that a physical theory for low-dimensional chaotic systems should consist of two interlocking components:

1. a qualitative encoding of the topological structure of the chaotic attractor (symbolic dynamics, topological invariants), and

2. a quantitative description of the metric structure on the attractor (scaling functions, transfer operators, fractal measures).

A physicist's "dynamical quest" consists of first dissecting the topological form of a strange set, and second, "dressing" this topological form with its metric structure.

In this chapter we introduce one beautiful approach to the first part of a physicist's dynamical quest, that is, unfolding the topology of a chaotic attractor. This strategy takes advantage of geometrical properties of chaotic attractors in phase space.

A low-dimensional chaotic dynamical system with one unstable direction has a rich set of recurrence properties that are determined by the unstable saddle periodic orbits embedded within the strange set. These unstable periodic orbits provide a sort of skeleton on which the strange attractor rests. For flows in three dimensions, these periodic orbits are closed curves, or knots. The knotting and linking of these periodic orbits is a bifurcation invariant, and hence can be used to identify or "fingerprint" a class of strange attractors. Although a chaotic system defies long-term predictability, it may still possess good short-term predictability. This short-term predictability fundamentally distinguishes "low-dimensional" chaos from our notion of a "random" process. Mindlin and co-workers [1,2,3], building on work initiated by Solari and Gilmore [4,5], recently developed this basic set of observations into a coherent physical theory for the topological characterization of chaotic sets arising in three-dimensional flows. The approach advocated by Mindlin, Solari, and Gilmore emphasizes the prominent role topology must play in any physical theory of chaos. In addition, it suggests a useful approach toward developing dynamical models directly from experimental time series.

In recent years the ergodic theory of differentiable dynamical systems has played a prominent role in the description of chaotic physical systems [6]. In particular, algorithms have been developed to compute fractal dimensions [7,8], metric entropies [9], and Lyapunov exponents [10] for a wide variety of experimental systems. It is natural to consider such ergodic measures, especially if the ultimate aim is the characterization of turbulent motions, which are, presumably, of high dimension. However, if the aim is simply to study and classify low-dimensional chaotic sets, then topological methods will certainly play an important role. Topological signatures and ergodic measures usually present different aspects of the same dynamical system, though there are some unifying principles between the two approaches, which can often be found via symbolic dynamics [11]. The metric properties of a dynamical system are invariant under coordinate transformations; however, they are not generally stable under bifurcations that occur during parameter changes. Topological invariants, on the other hand, can be stable under parameter changes and therefore are useful in identifying the same dynamical system at different parameter values.

The aim of this chapter is to develop topological methods suitable for the classification and analysis of low-dimensional nonlinear dynamical systems. The techniques illustrated here are directly applicable to a wide spectrum of experiments including lasers [4], fluid systems such as those giving rise to surface waves [12], the bouncing ball system described in Chapter 1, and the forced string vibrations described in Chapter 3.

The major device in this analysis is the template, or knot holder, of the hyperbolic chaotic limit set. The template is a mathematical construction first introduced by Birman and Williams [13] and further developed by Holmes and Williams [14]. Roughly, a *template* is an expanding map on a branched surface. Templates are useful because periodic orbits from the flow of a chaotic hyperbolic dynamical system can be placed on a template in such a way as to preserve their original topological structure. Thus templates provide a visualizable model for the topological organization of the limit set. Templates can also be described algebraically by finite matrices, and this in turn gives us a kind of homology theory for low-dimensional chaotic limit sets.

As recently described by Mindlin, Solari, Natiello, Gilmore, and Hou [2], templates can be reconstructed from a moderate amount of experimental data. This reconstructed template can then be used both to classify the strange attractor and to make specific predictions about the periodic orbit structure of the underlying flow. Strictly speaking, the template construction only applies to flows in the three-sphere, S^3, although it is hoped that the basic methodology illustrated by the template theory can be used to characterize flows in higher dimensions.

The strategy behind the template theory is the following. For a nonlinear dynamical system there are generally two regimes that are well understood, the regime where a finite number of periodic orbits exist and the hyperbolic regime of fully developed chaos. The essential idea is to reconstruct the form of the fully developed chaotic limit set from a non-fully developed (possibly nonhyperbolic) region in parameter space. Once the hyperbolic limit set is identified, then the topological information gleaned from the hyperbolic limit set can be used to make predictions about the chaotic limit set in other (possibly nonhyperbolic) parameter regimes, since topological invariants such as knot types, linking numbers, and relative rotation rates are robust under parameter changes.

In the next section we follow Auerbach and co-workers to show how periodic orbits are extracted from an experimental time series [15, 16]. These periodic cycles are the primary experimental data that the template theory seeks to organize. Section 5.3 defines the core mathematical ideas we need from knot theory: knots, braids, links, Reidemeister moves, and invariants. In section 5.4 we describe a simple, but physically useful, topological invariant called a relative rotation rate, first introduced by Solari and Gilmore [4]. Section 5.5 discusses templates, their algebraic representation, and their symbolic dynamics. Here, we present a new algebraic description of templates in terms of "framed braids," a representation suggested by Melvin [17]. This section also shows how to calculate relative rotation rates from templates. Section 5.6 provides examples of relative rotation rate calculations from two common templates. In section 5.7 we apply the template theory to the Duffing equation. This final section is directly applicable to experiments with nonlinear string motions, such as those described in Chapter 3 [18].

5.2   Periodic Orbit Extraction

Periodic orbits are available in abundance from a single chaotic time series. To see why this is so consider a recurrent three-dimensional flow in the vicinity of a hyperbolic periodic orbit (Fig. 5.1) [19]. Since the flow is recurrent, we can choose a surface of section in the vicinity of this fixed point. This section gives a compact map of the disk onto itself with at least one fixed point. In the vicinity of this fixed point a chaotic limit set (a horseshoe) containing an infinite number of unstable periodic orbits can exist. A single chaotic trajectory meanders around this chaotic limit set in an ergodic manner, passing arbitrarily close to every point in the set including its starting point and *each periodic point.*

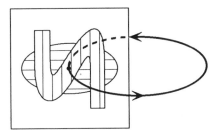

Figure 5.1: A recurrent flow around a hyperbolic fixed point.

(a)   (b)

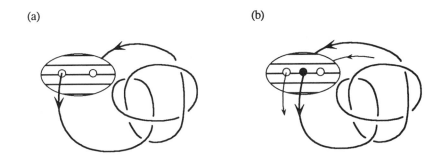

Figure 5.2: (a) A close recurrence of a chaotic trajectory. (b) By gently adjusting a segment of the chaotic trajectory we can locate a nearby periodic orbit. (Adapted from Cvitanović [19].)

Now let us consider a small segment of the chaotic trajectory that returns close to some periodic point. Intuitively, we expect to be able to gently adjust the starting point of the segment so that the segment precisely returns to its initial starting point, thereby creating a periodic orbit (Fig. 5.2).

Based on these simple observations, Auerbach and co-workers [15] showed that the extraction of unstable periodic orbits from a chaotic time series is surprisingly easy. Very recently, Lathrop and Kostelich [16], and Mindlin and co-workers [2], successfully applied this orbit extraction technique to a strange attractor arising in an experimental flow, the Belousov-Zhabotinskii chemical reaction.

As discussed in Appendix H, the idea of using the periodic orbits of a nonlinear system to characterize the chaotic solutions goes back to Poincaré [20]. In a sense, Auerbach and co-workers made the inverse observation: not only can periodic orbits be used to describe a chaotic trajectory, but a chaotic trajectory can also be used to locate periodic orbits.

A real chaotic trajectory of a strange attractor can be viewed as a kind of random walk on the unstable periodic orbits. A segment of a chaotic trajectory approaches an unstable periodic orbit along its stable manifold. This approach can last for several cycles during which time the system has good short-term predictability. Eventually, though, the orbit is ejected along the unstable manifold and proceeds until it is captured by the stable manifold of yet another periodic orbit. A convenient mathematical language to describe this phenomenon is the *shadowing theory* [21] of Conley and Bowen. Informally, we say that a short segment of a chaotic time series shadows an unstable periodic orbit embedded within the strange attractor. This shadowing effect makes the unstable periodic orbits (and hence, the hyperbolic limit set) observable. In this way, horseshoes and other hyperbolic limit sets can be "measured."

This also suggests a simple test to distinguish low-dimensional chaos from noise. Namely, a time series from a chaotic process should have subsegments with strong recurrence properties. In section 5.2.3 we give examples of recurrence plots that show these strong recurrence properties. These recurrence plots give us a quick test for determining if a time series is from a low-dimensional chaotic process.

5.2.1   Algorithm

Let the vector $\mathbf{x}_i$ be a point on the strange attractor where $\mathbf{x}_i$ in our setting is three-dimensional, and the components are either measured directly from experiment (e.g., x, $\dot{x}$, $\ddot{x}$) or created from an embedding $\mathbf{x}_i = (s_i, s_{i+\tau}, s_{i+2\tau})$ of $\{s_i\}_{i=1}^n$, the scalar time series data generated from an experimental measurement of one variable. In the three-dimensional phase space the saddle orbits generally have a one-dimensional repelling direction and a one-dimensional attracting direction. When a trajectory is near a saddle it approximately follows the

motion of the periodic orbit. For a data segment $x_i, x_{i+1}, x_{i+2}, \ldots$ near
a periodic orbit, and an $\epsilon > 0$, we can find a smallest n such that
$\|x_{i+n} - x_i\| < \epsilon$. We will often write $n = kn_0$. In a forced system, n_0 is
simply the number of samples per fundamental forcing period and k is
the integer period. If the system is unforced, then n_0 can be found by
constructing a histogram of the recurrence times as described by Lathrop and Kostelich [16]. Candidates for periodic orbits embedded in
the strange attractor are now given by all x_i which are (k, ϵ) recurrent
points, where $k = n/n_0$ is the period of the orbit.

In practice we simply scan the time series for close returns (strong
recurrence properties),

$$\|x_{i+n} - x_i\| < \epsilon, \tag{5.1}$$

for $k = 1, 2, 3\ldots$ to find the period one, period two, period three orbits,
and so on. When the data are normalized to the maximum of the time
series, then $\epsilon = 0.005$ appears to work well. The recurrence criterion ϵ is
usually relaxed for higher-order orbits and can be made more stringent
for low-order orbits.

In a flow with a moderate number of samples per period, say $n_0 >
32$, the recurrent points tend to be clustered about one another. That
is, if x_i is a recurrent point, then $x_{i+1}, x_{i+2}, \ldots$, are also likely to be
recurrent points for the same periodic orbit, simply because the periodic orbit is being approached from its attracting direction. We call
a cluster of such points a *window*. The saddle orbit is estimated, or
reconstructed, by choosing the orbit with the best recurrence property (minimum ϵ) in a window. Alternatively, the orbit can also be
approximated by averaging over nearby segments, which is the more
appropriate procedure for maps [15].

Data segments of strong recurrence in a chaotic time series are easily
seen in *recurrence plots*, which are obtained by plotting $\|x_{i+n} - x_i\|$ as a
function of i for fixed $n = kn_0$. Different recurrence plots are necessary
for detecting period one $(k = 1)$ orbits, period two $(k = 2)$ orbits, and
so on. Windows in these recurrence plots are clear signatures of near-
periodicity over several complete periods in the corresponding segment
of the time series data. The periodic orbits are reconstructed by choosing the orbit at the bottom of each window. Windows in recurrence
plots also provide a quick test showing that the time series is not being

generated by noise, but may be coming from a low-dimensional chaotic process.

5.2.2   Local Torsion

In addition to reconstructing the periodic orbits, we can also extract from a single chaotic time series the linearized flow in the vicinity of each periodic orbit. In particular, it is possible to calculate the *local torsion*, that is, how much the unstable manifold twists about the fixed point. Let $\mathbf{v}_u$ be the eigenvector corresponding to the largest eigenvalue λ_u about the periodic orbit. That is, $\mathbf{v}_u$ is the local linear approximation of the unstable manifold of the saddle orbit. To find the local torsion, we need to estimate the number of half-twists $\mathbf{v}_u$ makes about the periodic orbit. The operator $S_{T,0}$,

$$S_{T,0} = \exp\left(\int_0^T DF d\eta\right), \tag{5.2}$$

gives the evolution of the variational vector. This variational vector will generate a strip under evolution, and the number of half-turns (the local torsion) is nothing but one-half the linking number (defined in section 5.3.3) of the central line and the end of the strip defined by [22]

$$A = \{\mathbf{x}_0(t) + \eta\mathbf{v}_u(t); 0 \leq t < 2T\}, \tag{5.3}$$

where $\mathbf{x}_0$ is the curve corresponding to the periodic orbit.

The evolution operator can be estimated from a time series by first noting that

$$\exp\left(\int_0^T DF d\eta\right) \approx$$
$$\exp(DF|_T \Delta t) \times \exp(DF|_{T-\Delta T}\Delta t) \times \ldots \times \exp(DF|_0 \Delta t). \tag{5.4}$$

Let $\mathbf{x}_i$ be a recurrent point, and let $\{\mathbf{x}_j\}_{j=1}^n$ be a collection of points in some predefined neighborhood about $\mathbf{x}_i$. Then $DF(\mathbf{x}_i)$ is a 3×3 matrix, and we can use a least-squares procedure on points from the time series in the neighborhood of $\mathbf{x}_i$ to approximate both the Jacobian and the tangent map [10,22]. The local torsion (the number of half-twists about the periodic orbit) is a rather rough number calculated from the product of the Jacobians and hence is insensitive to the numerical details of the approximation.

5.2.3 Example: Duffing Equation

To illustrate periodic orbit extraction we again consider the Duffing oscillator, in the following form:

$$\dot{x}_1 = x_2, \tag{5.5}$$

$$\dot{x}_2 = -\alpha x_2 - x_1 - x_1^3 + f\cos(2\pi x_3 + \phi), \tag{5.6}$$

$$\dot{x}_3 = \omega/2\pi, \tag{5.7}$$

with the control parameters $\alpha = 0.2, f = 27.0$, $\omega = 1.330$, and $\phi = 0.0$ [3]. At these parameter values a strange attractor exists, but the attractor is probably not hyperbolic.

The Duffing equation is numerically integrated for 2^{13} periods with 2^{13} steps per period. Data are sampled and stored every 2^7 steps, so that 2^6 points are sampled per period. A short segment of the chaotic orbit from the strange attractor, projected onto the (x_2, x_1) phase space, is shown in Figure 5.3.

Periodic orbits are reconstructed from the sampled data using a standard Euclidean metric. The program used to extract the periodic orbits of the Duffing oscillator is listed in Appendix F. The distances $d[\mathbf{x}_{i+n} - \mathbf{x}_i]$ are plotted as a function of i for fixed $n = kn_0$, with $n_0 = 2^6$. Samples of these recurrence plots are shown in Figure 5.4 for $k = 2$ and $k = 3$. The windows at the bottom of these plots are clear signatures of near-periodicity in the corresponding segment of the time series data. The segment of length n with the smallest distance (the bottom of the window) was chosen to represent the nearby unstable periodic orbit. Some of the unstable periodic orbits that were reconstructed by this procedure are shown in Figure 5.5. We believe that the orbits shown in Figure 5.5 correctly identify all the period one and some of the period three orbits embedded in the strange attractor at these parameter values. In addition, a period two orbit and higher-order periodic orbits can be extracted (see Table 5.3). However, this may not be all the periodic orbits in the system since some of them may have basins of attraction separate from the basin of attraction for the strange attractor.

The periodic orbits give a spatial outline of the strange attractor. By superimposing the orbits in Figure 5.5 onto one another, we recover the picture of the strange attractor shown in the upper left-hand corner

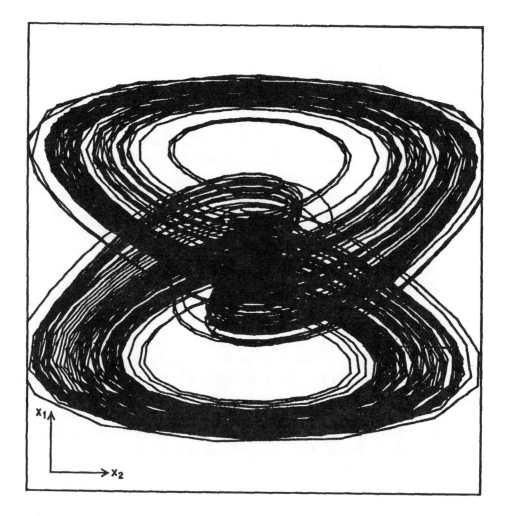

Figure 5.3: Two-dimensional projection of a data segment from the strange attractor of the Duffing oscillator.

$$d[x(i) - x(i+n)]$$

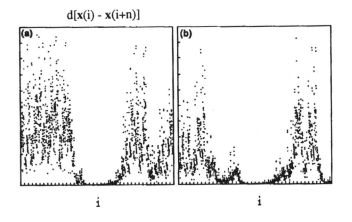

Figure 5.4: Recurrence plots. Distances $d[x_{i+n} - x_i]$ plotted as a function of i for fixed $n = kn_0$, with $n_0 = 2^6$. The windows in these plots show near-periodicity over at least a full period in the corresponding segment of the time series data. The bottom of each window is a good approximation to the nearby unstable periodic orbit: (a) $k = 2$; (b) $k = 3$.

of Figure 5.5. Each periodic orbit indicated in Figure 5.5 is actually a closed curve in three-space (we only show a two-dimensional projection in Fig. 5.5).

Now we make a simple but fundamental observation. Each of these periodic orbits is a knot, and the periodic orbits of the strange attractor are interwoven (tied together) in a very complicated pattern. Our goal in this chapter is to understand the knotting of these periodic orbits, and hence the spatial organization of the strange attractor. We begin by reviewing a little knot theory.

5.3 Knot Theory

Knot theory studies the placement of one-dimensional objects called strings [23,24,25] in a three-dimensional space. A simple and accurate picture of a knot is formed by taking a rope and splicing the ends together to form a closed curve. A mathematician's *knot* is a non-self-intersecting smooth closed curve (a *string*) embedded in three-space. A

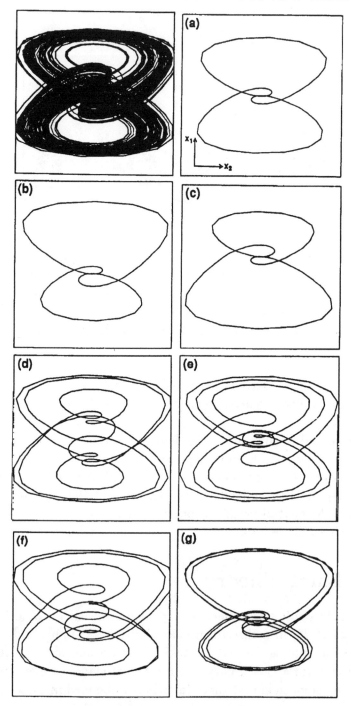

Figure 5.5: Some of the periodic orbits extracted from the chaotic time series data of Figure 5.3: (a) symmetric period one orbit; (b), (c) asymmetric pair of period one orbits; (d), (e) symmetric period three orbits; (f), (g) asymmetric period three orbits.

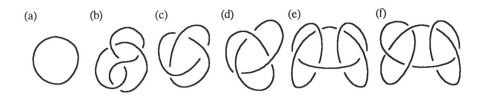

Figure 5.6: Planar diagrams of knots: (a) the trivial or *unknot*; (b) figure-eight knot; (c) left-handed trefoil; (d) right-handed trefoil; (e) square knot; (f) granny knot.

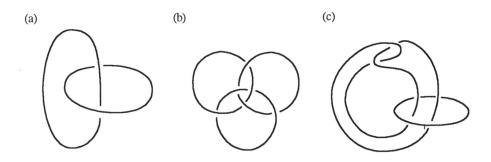

Figure 5.7: Link diagrams: (a) Hopf link; (b) Borromean rings; (c) Whitehead link.

two-dimensional planar diagram of a knot is easy to draw. As illustrated in Figure 5.6, we can project a knot onto a plane using a solid (broken) line to indicate an overcross (undercross). A collection of knots is called a *link* (Fig. 5.7).

The same knot can be placed in space and drawn in planar diagram in an infinite number of different ways. The equivalence of two different presentations of the same knot is usually very difficult to see. Classification of knots and links is a fundamental problem in topology. Given two separate knots or links we would like to determine when two knots are the same or different. Two knots (or links) are said to be *topologically equivalent* if there exists a continuous transformation carrying one knot (or link) into another. That is, we are allowed to deform the knot in any way without ever tearing or cutting the string. For instance, Figure 5.8 shows two topologically equivalent planar di-

Figure 5.8: Equivalent planar diagrams of the trefoil knot.

Figure 5.9: Two knots whose equivalence is hard to demonstrate.

agrams for the trefoil knot and a sequence of "moves" showing their equivalence. The two knots shown in Figure 5.9 are also topologically equivalent. However, proving their equivalence by a sequence of moves is a real challenge.

A periodic orbit of a three-dimensional flow is also a closed noninter-secting curve, hence a knot. A periodic orbit has a natural orientation associated with it: the direction of time. This leads us to study oriented knots. Formally an *oriented knot* is an embedding $S^1 \to \mathbf{R}^3$ where S^1 is oriented. Informally, an oriented knot is just a closed curve with an arrow attached to it telling us the direction along the curve.

The importance of knot theory in the study of three-dimensional flows comes from the following observation. The periodic orbits of a three-dimensional flow form a link. In the chaotic regime, this link is extraordinarily complex, consisting of an infinite number of periodic orbits (knots). As the parameters of the flow are varied the components of this link, the knots, may collapse to points (Hopf bifurcations) or coalesce (saddle-node or pitchfork bifurcations). *But no component of*

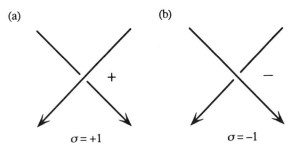

Figure 5.10: Crossing conventions: (a) positive (b) negative.

the link can intersect itself or any other component of the link, because
if it did, then there would be two different solutions based at the same
initial condition, thus violating the uniqueness theorem for differential
equations. The linking of periodic orbits fixes the topological structure
of a three-dimensional flow [14]. Moreover, as we showed in the previous
section, periodic orbits and their linkings are directly available from
experimental data. Thus knot theory is expected to play a key role in
any physical theory of three-dimensional flows.

5.3.1 Crossing Convention

In our study of periodic orbits we will work with oriented knots and
links. To each crossing C in an oriented knot or link we associate a
sign $\sigma(C)$ as shown in Figure 5.10. A positive cross (also known as
right-hand cross, or overcross) is written as $\sigma(C) = +1$. A negative
cross (also known as a left-hand cross, or undercross) is written as
$\sigma(C) = -1$. This definition of crossing is the opposite of the Artin
crossing convention adopted by Solari and Gilmore [4].

5.3.2 Reidemeister Moves

Reidemeister observed that two different planar diagrams of the same
knot represent topologically equivalent knots under a sequence of just
three primary moves, now called Reidemeister moves of type I, II, and
III. These Reidemeister moves, illustrated in Figure 5.11, simplify the
study of knot equivalence by reducing it to a two-dimensional problem.

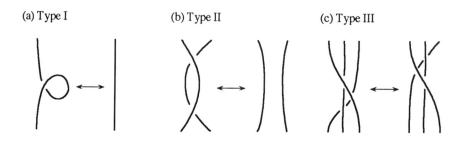

Figure 5.11: Reidemeister moves: (a) untwist; (b) pull apart; (c) middle slide.

The type I move *untwists* a section of a string, the type II move *pulls apart* two strands, and the type III move acts on three strings *sliding* the middle strand between the outer strands. The Reidemeister moves can be applied in an infinite number of combinations. So knot equivalence can still be hard to show using only the Reidemeister moves.

5.3.3 Invariants and Linking Numbers

A more successful strategy for classifying knots and links involves the construction of topological invariants. A *topological invariant* of a knot or link is a quantity that does not change under continuous deformations of the strings. The calculation of topological invariants allows us to bypass directly showing the geometric equivalence of two knots, since distinct knots must be different if they disagree in at least one topological invariant. What we really need, of course, is a *complete set* of calculable topological invariants. This would allow us to definitely say when two knots or links are the same or different. Unfortunately, no complete set of calculable topological invariants is known for knots. However, mathematicians have been successful in developing some very fine topological invariants capable of distinguishing large classes of knots [23].

The linking number is a simple topological invariant defined for a link on two oriented strings α and β. Intuitively, we expect the Hopf link in Figure 5.12(a) to have linking number $+1$ since the two strings are linked once. Similarly, the two strings in Figure 5.12(b) are unlinked

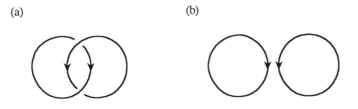

Figure 5.12: Linking numbers: (a) one; (b) zero.

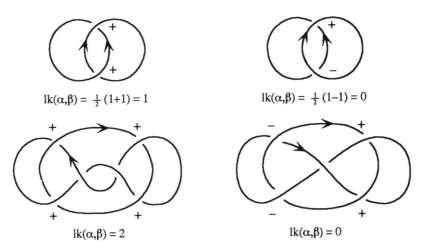

Figure 5.13: Examples of linking number calculations. (Adapted from Kauffman [24].)

and should have linking number 0. The linking number, which agrees with this intuition, is defined by

$$\text{lk}(\alpha, \beta) = \frac{1}{2} \sum_C \sigma(C). \tag{5.8}$$

In words, we just add up the crossing numbers for each cross between the two strings α and β and divide by two. The calculation of linking numbers is illustrated in Figure 5.13. Note that the last example is a planar diagram for the Whitehead link showing that "links can be linked even when their linking number is zero" [24].

The linking number is an integer invariant. More refined algebraic

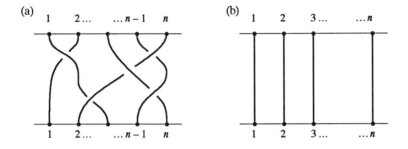

Figure 5.14: (a) A braid on n-strands. (b) A trivial braid.

polynomial invariants can be defined such as the Alexander polynomial and the Jones polynomial [23]. We will not need these more refined invariants for our work here.

5.3.4 Braid Group

Braid theory plays a fundamental role in knot theory since any oriented link can be represented by a closed braid (Alexander's Theorem [24]). The identification between links, braids, and the braid group allows us to pass back and forth between the geometric study of braids (and hence knots and links) and the algebraic study of the braid group. For some problems the original geometric study of braids is useful. For many other problems a purely algebraic approach provides the only intelligible solution.

A geometric braid is constructed between two horizontal level lines with n base points chosen on the upper and lower level lines. From upper base points we draw n strings or *strands* to the n lower base points (Fig. 5.14(a)). Note that the strands have a natural orientation from top to bottom. The trivial braid is formed by taking the ith upper base point directly to the ith lower base point with no crossings between the strands (Fig. 5.14(b)). More typically, some of the strands will intersect. We say that the $i + 1$st strand passes *over* the ith strand if there is a positive crossing between the two strands (see Crossing Convention, section 5.3.1). As illustrated in Figure 5.15(a), an overcrossing (or right-crossing) between the $i + 1$st and ith string is denoted by the

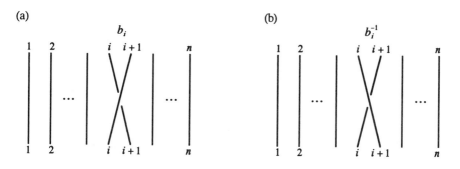

Figure 5.15: Braid operators: (a) b_i, overcross; (b) b_i^{-1}, undercross.

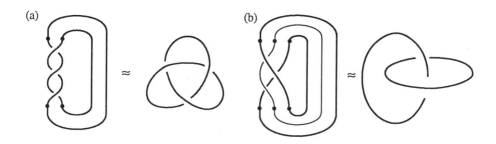

Figure 5.16: (a) Braid of a trefoil knot. (b) Braid of a Hopf link.

symbol b_i. The inverse b_i^{-1} represents an undercross (or left-cross), i.e., the $i + 1$st strand goes under the ith strand (Fig. 5.15(b)).

By connecting opposite ends of the strands we form a *closed braid*. Each closed braid is equivalent to a knot or link, and conversely Alexander's theorem states that any oriented link is represented by a closed braid. Figure 5.16(a) shows a closed braid on two strands that is equivalent to the trefoil knot; Figure 5.16(b) shows a closed braid on three strands that is equivalent to the Hopf link. The representation of a link by a closed braid is not unique. However, only two operations on braids (the Markov moves) are needed to prove the identity between two topologically equivalent braids [24].

A general n-braid can be built up from successive applications of the operators b_i and b_i^{-1}. This construction is illustrated for a braid on four strands in Figure 5.17. The first crossing between the second

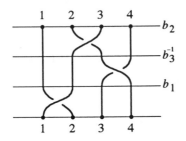

Figure 5.17: Braid on four strands whose braid word is $b_2 b_3^{-1} b_1$.

and third strand is positive, and is represented by the operator b_2. The next crossing is negative, b_3^{-1}, and is between the third and fourth strands. The last positive crossing is represented by the operator b_1. Each geometric diagram for a braid is equivalent to an algebraic *braid word* constructed from the operators used to build the braid. The braid word for our example on four strands is $b_2 b_3^{-1} b_1$.

Two important conventions are followed in constructing a braid word. First, at each level of operation (b_i, b_i^{-1}) only the ith and $i+1$st strands are involved. There are no crossings between any other strands at a given level of operation. Second, it is not the string, but the base point which is numbered. Each string involved in an operation increments or decrements its base point by one. All other strings keep their base points fixed.

The *braid group* on n-strands, B_n, is defined by the operators $\{b_i; i = 1, 2, \ldots n - 1\}$. The identity element of B_n is the trivial n-braid. However, as previously mentioned, the expression of a braid group element (that is, a braid word) is not unique. The topological equivalence between seemingly different braid words is guaranteed by the *braid relations* (Fig. 5.18):

$$b_i b_j = b_j b_i, \quad |i - j| \geq 2, \tag{5.9}$$
$$b_i b_{i+1} b_i = b_{i+1} b_i b_{i+1}. \tag{5.10}$$

The braid relations are taken as the defining relations of the braid group. Each topologically equivalent class of braids represents a collection of words that are different representations for the same braid

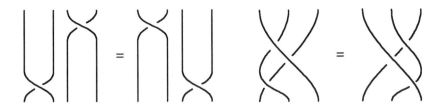

Figure 5.18: Braid relations.

in the braid group. In principle, the braid relations can be used to show the equivalence of any two words in this collection. Finding a practical solution to word equivalence is called the word problem. The word problem is the algebraic analog of the geometric braid equivalence problem.

5.3.5 Framed Braids

In our study of templates we will need to consider braids with "framing." A *framed braid* is a braid with a positive or negative integer associated to each strand. This integer is called the *framing*. We think of the framing as representing an internal structure of the strand. For instance, if each strand is a physical cable then we could subject this cable to a torsional force causing a twist. In this instance the framing could represent the number of half-twists in the cable.

Geometrically, a framed braid can be represented by a *ribbon graph*. Take a braid diagram and replace each strand by a ribbon. To see the framing we twist each ribbon by an integer number of half-turns. A *half-twist* is a rotation of the ribbon through π radians. A positive half-twist is a half-twist with a positive crossing, the rightmost half of the ribbon crosses over the leftmost half of the ribbon. Similarly, a negative half-twist is a negative crossing of the ribbon. Figure 5.19 shows how the framing is pictured as the number and direction of internal ribbon crossings.

This concludes our brief introduction to knot theory. We now turn our attention to discussing how our rudimentary knowledge of knot theory and knot invariants is used to characterize the periodic and chaotic behavior of a three-dimensional flow.

framed braid   ribbon graph

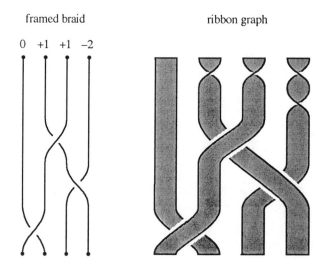

0   +1   +1   −2

Figure 5.19: Geometric representation of a framed braid as a ribbon graph. The integer attached to each strand is the sum of the half-twists in the corresponding branch of the ribbon graph.

5.4   Relative Rotation Rates

Solari and Gilmore [4] introduced the "relative rotation rate" in an attempt to understand the organization of periodic orbits within a flow. The phase of a periodic orbit is defined by the choice of a Poincaré section. Relative rotation rates make use of this choice of phase and are topological invariants that apply specifically to periodic orbits in three-dimensional flows. Our presentation of relative rotation rates closely follows Eschenazi's [25].

As usual, we begin with an example. Figure 5.20 shows a period two orbit, a period three orbit, and their intersections with a surface of section. The relative rotation rate between an orbit pair is calculated beginning with the difference vector formed at the surface of section,

$$\Delta \mathbf{r} = (x_A - x_B, y_A - y_B), \qquad (5.11)$$

where (x_A, y_A) and (x_B, y_B) are the coordinates of the periodic orbits labeled A of period p_A and B of period p_B. In general there will be

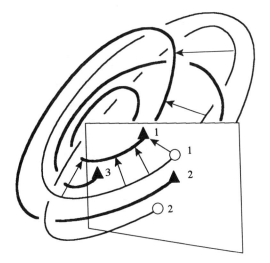

Figure 5.20: Rotations between a period two and period three orbit pair. The rotation of the difference vector between the two orbits is calculated at the surface of section. This vector is followed for $3 \cdot 2 = 6$ full periods, and the number of average rotations of the difference vector is the relative rotation rate between the two orbits. (Adapted from Eschenazi [25].)

$p_A \cdot p_B$ choices of initial conditions from which to form $\Delta \mathbf{r}$ at the surface of section. To calculate the rotation rate, consider the evolution of $\Delta \mathbf{r}$ in $p_A \cdot p_B$ periods as it is carried along by the pair of periodic orbits. The difference vector $\Delta \mathbf{r}$ will make some number of rotations before returning to its initial configuration. This number of rotations divided by $p_A p_B$ is the relative rotation rate. Essentially, the relative rotation rate describes the average number of rotations of the orbit A about the orbit B, or the orbit B about the orbit A. In the example shown in Figure 5.20, the period three orbit rotates around the period two orbit twice in six periods, or one-third average rotations per period.

The general definition proceeds as follows. Let A and B be two orbits of periods p_A and p_B that intersect the surface of section at $(a_1, a_2, ..., a_{p_A})$, and $(b_1, b_2, ..., b_{p_B})$. The *relative rotation rate* $R_{ij}(A, B)$ is

$$R_{ij}(A, B) = \frac{1}{2\pi p_A p_B} \int d[\arctan(\Delta r_2 / \Delta r_1)], \qquad (5.12)$$

or in vector notation,

$$R_{ij}(A, B) = \frac{1}{2\pi p_A p_B} \int \frac{\mathbf{n} \cdot [\Delta \mathbf{r} \times d(\Delta \mathbf{r})]}{\Delta \mathbf{r} \cdot \Delta \mathbf{r}}. \qquad (5.13)$$

The integral extends over $p_A \cdot p_B$ periods, and $\mathbf{n}$ is the unit vector orthogonal to the plane spanned by the vectors $\Delta \mathbf{r}$ and $d\Delta \mathbf{r}$. The indices i and j denote the initial conditions a_i and b_j on the surface of section. In the direction of the flow, a clockwise rotation is positive.[1]

The *self-rotation rate* $R_{ij}(A, A)$ is also well defined by equation (5.13) if we establish the convention that $R_{ii}(A, A) = 0$. The relative rotation rate is clearly symmetric, $R_{ij}(A, B) = R_{ji}(B, A)$. It also commonly occurs that different initial conditions give the same relative rotation rates. Further properties of relative rotation rates, including a discussion of their multiplicity, have been investigated by Solari and Gilmore [4,5].

Given a parameterization for the two periodic orbits, their relative rotation numbers can be calculated directly from equation (5.12) by numerical integration (see Appendix F). There are, however, several

---

[1]As previously mentioned, the crossing convention and this definition of the relative rotation rate are the opposite of those originally adopted by Solari and Gilmore [4].

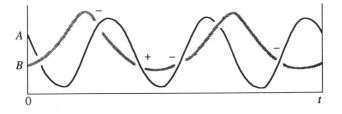

Figure 5.21: The orbit pair of Figure 5.20 arranged as a braid on two strands. The relative rotation rates can be computed by keeping track of all the crossings of the orbit A *over* the orbit B. Each crossing adds or subtracts a half-twist to the rotation rate. The linking number is calculated from the sum of all the crossings of A over B. (Adapted from Eschenazi [25].)

alternative methods for calculating $R_{ij}(A, B)$. For instance, imagine arranging the periodic orbit pair as a braid on two strands. This is illustrated in Figure 5.21 where the orbits A and B are partitioned into segments of length p_A and p_B each starting at $a_i(b_j)$ and ending at $a_{i+1}(b_{j+1})$. We keep track of the crossings between A and B with the counter σ_{ij},

$$\sigma_{ij} = \begin{cases} +1 & \text{if } A_i \text{ crosses over } B_j \text{ from left to right} \\ -1 & \text{if } A_i \text{ crosses over } B_j \text{ from right to left} \\ 0 & \text{if } A_i \text{ does not cross over } B_j. \end{cases} \tag{5.14}$$

Then the relative rotation rates can be computed from the formula

$$R_{ij}(A, B) = \frac{1}{p_A p_B} \sum_n \sigma_{i+n, j+n}, \quad n = 1, 2, 3, ..., p_A p_B. \tag{5.15}$$

Using this same counter, the *linking number* of knot theory is

$$\text{lk}(A, B) = \sum_{i,j} \sigma_{ij}, \quad i = 1, 2, ..., p_A \text{ and } j = 1, 2, ..., p_B. \tag{5.16}$$

The linking number is easily seen to be the sum of the relative rotation rates:

$$\text{lk}(A, B) = \sum_{ij} R_{ij}(A, B). \tag{5.17}$$

An *intertwining matrix* is formed when the relative rotation rates for all pairs of periodic orbits of a return map are collected in a (possibly infinite-dimensional) matrix. Intertwining matrices have been calculated for several types of flows such as the suspension of the Smale horseshoe [4] and the Duffing oscillator [5]. Perhaps the simplest way to calculate intertwining matrices is from a template, a calculation we describe in section 5.5.4.

Intertwining matrices serve at least two important functions. First, they help to predict bifurcation schemes, and second, they are used to identify a return mapping mechanism.

Again, by uniqueness of solutions, two orbits cannot interact through a bifurcation unless all their relative rotation rates are identical with respect to all other existing orbits. With this simple observation in mind, a careful examination of the intertwining matrix often allows us to make specific predictions about the allowed and disallowed bifurcation routes. An intertwining matrix can give rise to bifurcation "selection rules," i.e., it helps us to organize and understand orbit genealogies. A specific example for a laser model is given in reference [4].

Perhaps more importantly, intertwining matrices are used to identify or fingerprint a return mapping mechanism. As described in section 5.2, low-order periodic orbits are easy to extract from both experimental chaotic time series and numerical simulations. The relative rotation rates of the extracted orbits can then be arranged into an intertwining matrix, and compared with known intertwining matrices to identify the type of return map. In essence, intertwining matrices can be used as signatures for horseshoes and other types of hyperbolic limit sets.

If the intertwining comes from the suspension of a map then, as mentioned in section 4.2.3 (see Fig. 4.3), the intertwining matrix with zero global torsion is usually presented as the "standard" matrix. A global torsion of +1 adds a full twist to the suspension of the return map, and this in turn adds additional crossings to each periodic orbit in the suspension. If the global torsion is an integer GT, then this integer is added to each element of the standard intertwining matrix.

Relative rotation rates can be calculated from the symbolic dynamics of the return map [25] or directly from a template if the return map has a hyperbolic regime. We illustrate this latter calculation in section 5.5.4.

5.5 Templates

In section 5.5.1 we provide the mathematical background surrounding template theory. This initial section is mathematically advanced. Section 5.5.2 contains a more pragmatic description of templates and can be read independently of section 5.5.1.

Before we begin our description of templates, we first recall that the dynamics on the attractor can have very complex recurrence properties due to the existence of homoclinic points (see section 4.6.2). Poincaré's original observation about the complexity of systems with transverse homoclinic intersections is stated in more modern terms as [26]

Katok's Theorem. A smooth flow ϕ_t on a three-manifold, with positive topological entropy, has a hyperbolic closed orbit with a transverse homoclinic point.

Templates help to describe the topological organization of chaotic flows in three-manifolds. In our description of templates we will work mainly with forced systems, so the phase space topology is $\mathbf{R}^2 \times S^1$. However, the use of templates works for any three-dimensional flow.

Periodic orbits of a flow in a three-manifold are smooth closed curves and are thus oriented knots. Recall yet again that once a periodic orbit is created (say through a saddle-node or flip bifurcation) its knot type will not change as we move through parameter space. Changing the knot type would imply self-intersection, and that violates uniqueness of the solution. Knot types along with linking numbers and relative rotation rates are topological invariants that can be used to predict bifurcation schemes [14] or to identify the dynamics behind a system [1,2,3,4,5]. The periodic orbits can be projected onto a plane and arranged as a braid. Strands of a braid can pass over or under one another, where our convention for positive and negative crossings is given in section 5.3.1. We next try to organize all the knot information arising from a flow, and this leads us to the notion of a template. Our informal description of templates follows the review article of Holmes [14].

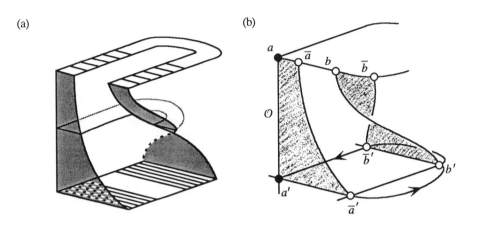

Figure 5.22: (a) Suspension of a Smale horseshoe type return map for a system with a transversal homoclinic intersection. This return map resembles the orientation preserving Hénon map. (b) The dots (both solid and open) will be the boundary points of the branches in a template. The pieces of the unstable manifold W^u on the intervals $(a, \bar{a})$ and $(b, \bar{b})$ generate two ribbons.

5.5.1   Motivation and Geometric Description

Before giving the general definition of a template, we begin by illustrating how templates, or knot holders, can arise from a flow in $\mathbf{R}^3$. In accordance with Katok's theorem, let $\mathcal{O}$ be a closed hyperbolic orbit with a transversal intersection and a return map resembling a Smale horseshoe (Figure 5.22(a)). For a specific physical model, the form of the return map can be obtained either by numerical simulations or by analytical methods as described by Wiggins [27]. The only periodic orbit shown in Figure 5.22(b) is given by a solid dot ($\bullet$). The points of transversal intersection indicated by open dots (o) are not periodic orbits, but—according to the Smale-Birkhoff homoclinic theorem [28]—they are accumulation points for an infinite family of periodic orbits (see section 4.6.2).

The periodic orbits in the suspension of the horseshoe map have a complex knotting and linking structure, which was first explored by Birman and Williams [13] using the template construction.

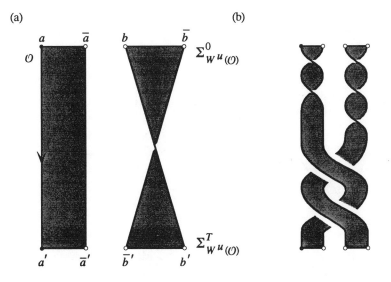

Figure 5.23: Suspension of intervals on $W^u(\mathcal{O})$: (a) global torsion is 0; (b) global torsion is -1.

Let us assume that our example is from a forced system. Then the simplest suspension consistent with the horseshoe map is shown in Figure 5.23(a). However, this is not the only possible suspension. We could put an arbitrary number of full twists around the homoclinic orbit $\mathcal{O}$. The number of twists is called the *global torsion*, and it is a topological invariant of the flow. In Figure 5.23(b) the suspension of the horseshoe with a global torsion of -1 is illustrated by representing the lift[2] of the boundaries on $W^u(\mathcal{O})$ of the horseshoe as a braid of two ribbons.

Note that adding a single twist adds one to the linking number of the boundaries of the horseshoe, and this in turn adds one to the relative rotation rates of all periodic orbits within the horseshoe. That is, a change in the global torsion changes the linking and knot types, but it does so in a systematic way. In the horseshoe example the global torsion is the relative rotation rate of the period one orbits.

To finish the template construction we identify certain orbits in the suspension. Heuristically, we project down along the stable mani-

---

[2]For our purposes, a lift is a suspension consisting of flow with a simple twist.

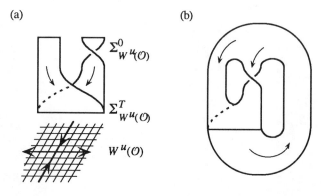

Figure 5.24: Template construction: (a) identify the branch ends at Σ^T, i.e., "collapse onto $W^u(\mathcal{O})$," and (b) identify Σ^0 and Σ^T to get a "braid template."

folds $W^s(\mathcal{O})$ onto the unstable manifold $W^u(\mathcal{O})$, i.e., we "collapse onto $W^u(\mathcal{O})$." For the Smale horseshoe, this means that we first identify the ends of the ribbons (now called *branches*) at Σ^T in Figure 5.24(a), and next identify Σ^0 and Σ^T. The resulting braid template for the horseshoe is shown in Figure 5.24(b). The template itself may now be deformed to several equivalent forms (not necessarily resembling a braid) including the standard horseshoe template illustrated in Figure 5.25 [14].

For this particular example it is easy to see that such a projection is one-to-one on periodic orbits. Each point of the limit set has a distinct symbol sequence and thus lies on a distinct leaf of the stable manifold $W^s(\mathcal{O})$. This projection takes each leaf of $W^s(\mathcal{O})$ onto a distinct point of $W^u(\mathcal{O})$. In particular, for each periodic point of the limit set there is a unique point on $W^u(\mathcal{O})$.

Each periodic orbit of the map corresponds to some knot in the template. Since the collapse onto $W^u(\mathcal{O})$ is one-to-one, we can use the standard symbolic names of the horseshoe map to name each knot in the template (sections 4.8.2–4.8.3). Each knot will generate a symbolic sequence of 0s and 1s indicated in Figure 5.25. Conversely, each periodic symbolic sequence of 0s and 1s (up to cyclic permutation) will generate a unique knot. The three simplest periodic orbits and their symbolic

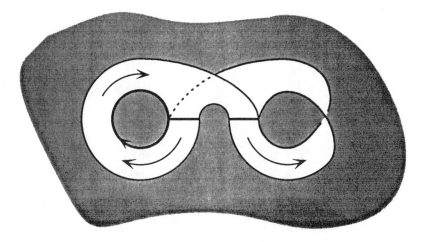

Figure 5.25: Standard Smale horseshoe template. Each periodic (infinite) symbolic string of 0s and 1s generates a knot.

names are illustrated in Figure 5.26.

If a template has more than two branches, then we number the k branches of the template with the numbers $\{0, 1, 2, \ldots, k - 1\}$. In this way we associate a symbol from the set $\{0, 1, 2, \ldots, k - 1\}$ to each branch. A periodic orbit of period n on the template generates a sequence of n symbols from the set $\{0, 1, 2, \ldots, k-1\}$ as it passes through the branches. Conversely, each periodic word (up to cyclic permutations) generates a unique knot on the template. The template itself is not an invariant object. However, from the template one can easily calculate invariants such as knot types and linking numbers. In this sense it is a knot holder.

The branches of a template are joined (or glued) at the *branch lines*. In a *braid template*, all the branches are joined at the same branch line. Figure 5.24(b) is an example of a braid template, and Figure 5.25 is an example of a (nonbraided) template holding the same knots. Forced systems always give rise to braid templates. We will work mostly with *full braid templates*, i.e., templates which describe a full shift on k symbols. Franks and Williams have shown that any embedded template can be arranged, via isotopy, as a braid template [26].

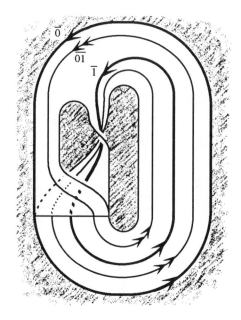

Figure 5.26: Some periodic orbits held by the horseshoe template. Note that the orbits $\overline{0}$ and $\overline{1}$ are unlinked, but the orbits $\overline{1}$ and $\overline{01}$ are linked once.

The template construction works for any hyperbolic flow in a three-manifold. To accommodate the unforced situation we need the following more general definition of a template.

Definition. A *template* is a branched surface T and a semiflow $\bar{\phi}_t$ on T such that the branched surface consists of the joining charts and the splitting charts shown in Figure 5.27.

The semiflow fails to be a flow because inverse orbits are not unique at the branch lines. In general the semiflow is an expanding map so that some sections of the semiflow may also spill over at the branch lines. The properties that the template $(T, \bar{\phi}_t)$ are required to satisfy are described by the following theorem [13].

Birman and Williams Template Theorem (1983). Given a flow ϕ_t on a three-manifold M^3 having a hyperbolic chain recurrent set

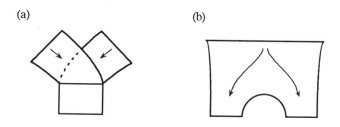

Figure 5.27: Template building charts: (a) joining chart and (b) splitting chart.

> there is a template $(T, \bar{\phi}_t)$, $T \subset M^3$, such that the periodic orbits under ϕ_t correspond (with perhaps a few specified exceptions) one-to-one to those under $\bar{\phi}_t$. On any finite subset of the periodic orbits the correspondence can be taken via isotopy.

The correspondence is achieved by collapsing onto the (strong) stable manifold. Technically, we establish the equivalence relation between elements in the neighborhood, N, of the chain recurrent set, as follows: $x_1 \sim x_2$ if $\|\phi_t(x_1) - \phi_t(x_2)\| \to 0$ as $t \to \infty$. In other words, x_1 and x_2 are equivalent if they lie in the same connected component of some local stable manifold of a point $x \in N$.

Orbits with the same asymptotic future are identified regardless of their past. By throwing out the history of a symbolic sequence, we can hope to establish an ordering relation on the remaining symbols and thus develop a symbolic dynamics and kneading theory for templates similar to that for one-dimensional maps. The symbolic dynamics of orbits on templates, as well as their kneading and bifurcation theory, is discussed in more detail in the excellent review article by Holmes [14].

The "few specified exceptions" will become clear when we consider specific examples. In some instances it is necessary to create a few period one orbits in $\bar{\phi}_t$ that do not actually exist in the original flow ϕ_t. These virtual orbits can sometimes be identified with points at infinity in the chain recurrent set.

Some examples of two-branch templates that have arisen in physical problems are shown in Figure 5.28: the Smale horseshoe with global torsion 0 and +1, the Lorenz template, and the Pirogon. The Lorenz template is the first knot holder originally studied by Birman and Williams [13]. It describes some of the knotting of orbits in the Lorenz equation for thermal convection. The horseshoe template describes some of the knots in the modulated laser rate equations mentioned in section 4.1 [4].

5.5.2   Algebraic Description

In addition to the geometric view of a template, it is useful to develop an algebraic description. Braid templates are described by three pieces of algebraic data. The first is a braid word describing the crossing structure of the k branches of the template. The second is the framing describing the twisting in each branch, that is, the *local* or *branch torsion* internal to each branch. The third piece of data is the "layering information" or *insertion array*, which determines the order in which branches are glued at the branch line. We now develop some conventions for drawing a geometric template. In the process we will see that the first piece of data, the braid word, actually contains the last piece of data, the insertion array. Thus we conclude that a template is just an instance of a framed braid, a braid word with framing (see sections 5.3.3–5.3.5).

Drawing Conventions

A graph of a template consists of two parts: a ribbon graph and a layering (or insertion) graph (Fig. 5.29). In the upper section of a template we draw the ribbon graph, which shows the intertwining of the branches as well as the internal twisting (local torsion) within each branch. The lower section of the template shows the layering information, that is, the order in which branches are glued at the branch line. By convention, we usually confine the expanding part of the semiflow on the template to the layering graph: the branches of the layering graph get wider before they are glued at the branch line. We also often draw the local torsion as a series of half-twists at the top of the ribbon graph.

In setting up the symbolic dynamics on the templates (that is, in naming the knots) we follow two important conventions. First, at the

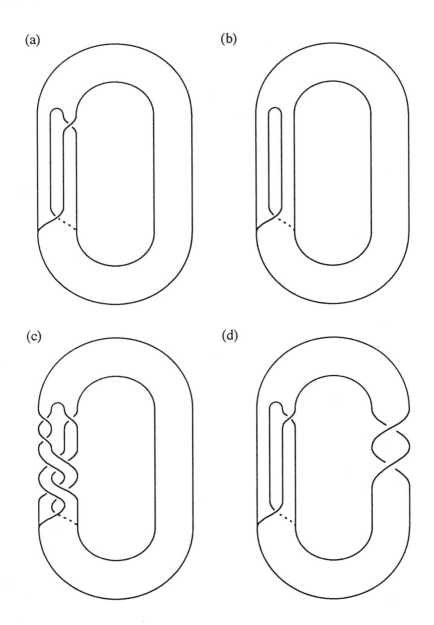

Figure 5.28: Common two-branch templates: (a) Smale horseshoe with global torsion 0; (b) Lorenz flow, showing equivalence to Lorenz mask; (c) Pirogon; (d) Smale horseshoe with global torsion +1. (Adapted from Mindlin et al. [1].)

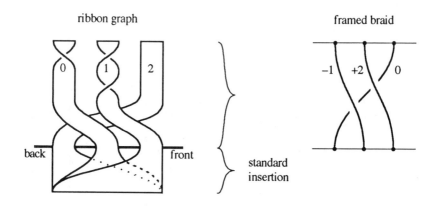

Figure 5.29: Representation of a template as a framed braid using standard insertion.

top of the ribbon graph we label each of the k branches from *left to right* with a number from the labeling set $0, 1, 2, \ldots, i, \ldots, k-1$. Second, from now on we will always arrange the layering graph so that the branches of the template are ordered *back to front* from *left to right*. This second convention is called the *standard insertion.*

The standard insertion convention follows from the following observation. Any layering graph can always be isotoped to the standard form by a sequence of branch moves that are like type II Reidemeister moves. This is illustrated in Figure 5.30, where we show a layering graph in nonstandard form and a simple branch exchange that puts it into standard form.

The adoption of the standard insertion convention allows us to dispense with the need to draw the layering graph. The insertion information is now implicitly contained in the lower ordering (left to right, back to front) of the template branches. We see that the template is well represented by a ribbon graph or a framed braid. We will often continue to draw the layering graph. However, if it is not drawn then we are following the standard insertion convention. A template with standard insertion is a framed braid.

These conventions and the framed braid representation of a template are illustrated in Figure 5.31 for a series of two-branch templates. We show the template, its version following standard insertion, and the

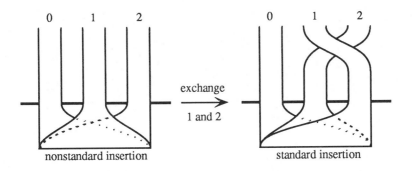

Figure 5.30: The layering graph can always be moved to standard form, back to front.

ribbon graph (framed braid) without the layering graph from which we can write a braid word with framing. We also show the "braid linking matrix" for the template, which is introduced in the next section.

Braid Linking Matrix

A nice characterization of some of the linking data of the knots held by a template is given by the braid linking matrix. In particular, in the second half of section 5.5.4 we show how to calculate the relative rotation rates for all pairs of periodic orbits from the braid linking matrix. The *braid linking matrix* is a square symmetric $k \times k$ matrix defined by[3]

$$
B = \begin{cases}
b_{ii} & : \text{ the sum of half-twists in the } i\text{th branch,} \\
b_{ij} & : \text{ the sum of the crossings between the} \\
& \quad i\text{th and the } j\text{th branches of the ribbon graph} \\
& \quad \text{with standard insertion.}
\end{cases} \tag{5.18}
$$

The ith diagonal element of B is the local torsion of the ith branch. The off-diagonal elements of B are twice the linking numbers of the ribbon graph for the ith and jth branches. The braid linking matrix

---

[3]The braid linking matrix is equivalent to the orbit matrix and insertion array previously introduced by Mindlin et al. [1]. See Melvin and Tufillaro for a proof [17].

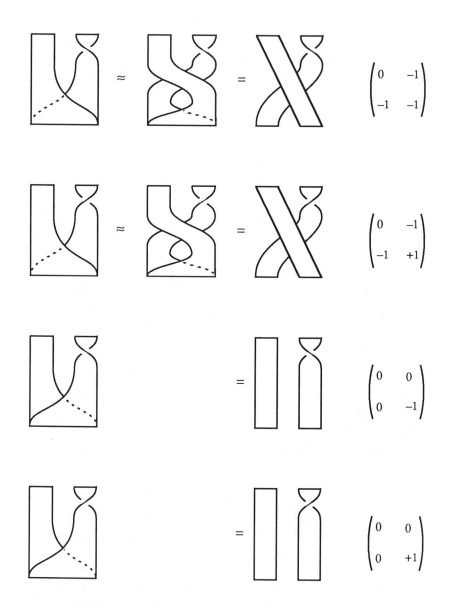

Figure 5.31: Examples of two-branched templates, their corresponding ribbon graphs (framed braids) with standard insertion, and their braid linking matrices.

describes the linking of the branches within a template and is closely related to the linking of the period one orbits in the underlying flow [17]. For the example shown in Figure 5.29, the braid linking matrix is

$$B = \begin{pmatrix} -1 & 0 & -1 \\ 0 & 2 & -1 \\ -1 & -1 & 0 \end{pmatrix}.$$

The braid linking matrix also allows us to compute how the strands of the framed braid are permuted. At the top of the ribbon graph the branches of the template are ordered $0, 1, \ldots, i, \ldots, k - 1$. At the bottom of the ribbon graph each strand occupies some possibly new position. The new ordering, or *permutation* σ_B, of the strands is given by

$$\sigma_B(i) = i \quad - \# \text{ odd entries } b_{ij} \text{ with } j < i \\ + \# \text{ odd entries } b_{ij} \text{ with } j > i. \tag{5.19}$$

Informally, to calculate the permutation on the ith strand, we examine the ith row of the braid linking matrix, adding the number of odd entries to the right of the ith diagonal element to i, and subtracting the number of odd entries to the left.

For example, for the template shown in Figure 5.29 we find that $\sigma_B(0) = 0 + 1 = 1$, $\sigma_B(1) = 1 + 1 - 0 = 2$, and $\sigma_B(2) = 2 - 2 = 0$. The permutation is $\sigma_B = (012)$. That is, the first strand goes to the second position, and the third strand goes to the first position. The second strand goes to the front third position.

5.5.3 Location of Knots

Given a knot on a template, the symbolic name of the knot is determined by recording the branches over which the knot passes. Given a template and a symbolic name, how do we draw the correct knot on a template?

There are two methods for finding the location of a knot on a template with k branches. The first is global and consists of finding the locations of all knots up to a length (period) n by constructing the appropriate k-ary tree with n levels. The second method, known as "kneading theory," is local. Kneading theory is the more efficient method of solution when we are dealing with just a few knots.

Trees

A branch of a template is called *orientation preserving* if the local torsion (the number of half-twists) is an even integer. Similarly, a branch is called *orientation reversing* if the local torsion is an odd integer. A convenient way to find the relative location of knots on a k-branch template is by constructing a k-ary tree which encodes the ordering of points on the orientation preserving and reversing branches of the template.

The *ordering tree* is defined recursively as follows. At the first level $(n = 1)$ we write the symbolic names for the branches from left to right as $0, 1, 2, \ldots, k-1$. The second level $(n = 2)$ is constructed by recording the symbolic names at the first level of the tree according to the ordering rule: if the ith branch of the template is orientation preserving then we write the branch names in forward order $(0, 1, 2, \ldots, k - 1)$; if the ith branch is orientation reversing then we write the symbolic names at the first level in reverse order $(k - 1, k - 2, \ldots, 2, 1, 0)$. The $n + 1$st level is constructed from the nth level by the same ordering rule: if the ith symbol (branch) at the nth level is orientation preserving then we record the ordering of the symbols found at the nth level; if the ith symbol labels an orientation reversing branch then we reverse the ordering of the symbols found at the nth level.

This rule is easier to use than to state. The ordering rule is illustrated in Figure 5.32 for a three-branch template. Branch 0 is orientation reversing; branches 1 and 2 are both orientation preserving. Thus, we reverse the order of any branch at the $n + 1$st level whose nth level is labeled with 0. In this example we find

$$
\begin{array}{ccc}
0 & 1 & 2 \\
(2,1,0) & (0,1,2) & (0,1,2) \\
((2,1,0),(2,1,0),(0,1,2)) & ((2,1,0),(0,1,2),(0,1,2)) & ((2,1,0),(0,1,2),(0,1,2)) \\
\vdots & \vdots & \vdots
\end{array}
$$

and so on.

To find the ordering of the knots on the template we *read down* the k-ary tree recording the branch names through which we pass (see Fig. 5.32). The ordering at the nth level of the tree is the correct ordering

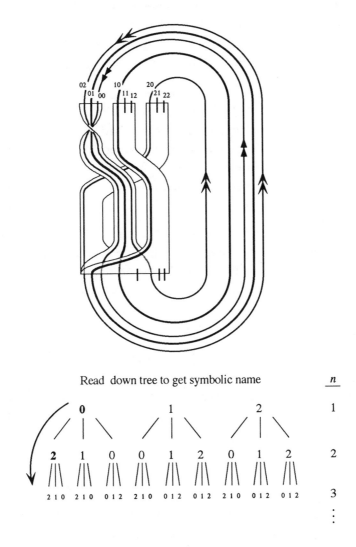

Figure 5.32: Example: location of period two knots $(n = 2)$.

for all the knots of period n on the template. Returning to our example, we find that the ordering up to period two is $02 \prec 01 \prec 00 \prec 10 \prec 11 \prec 12 \prec 20 \prec 21 \prec 22$. The symbol $\prec$ is read "precedes" and indicates the ordering relation found from the ordering tree (the order induced by the template).

To draw the knots with the desired symbolic name on a template, we use the ordering found at the bottom of the k-ary tree. Lastly, we draw connecting line segments between "shift equivalent" periodic orbits as illustrated in Figure 5.32. For instance, the period two orbit $\overline{02}$ is composed of two shift equivalent string segments 02 and 20, which belong to the branches 0 and 2 respectively.

Kneading Theory

The limited version of kneading theory [14] needed here is a simple rule which allows us to determine the relative ordering of two or more orbits on a template. From an examination of the ordering tree, we see that the ordering relation between two itineraries $\mathbf{s} = \{s_0, s_1, \ldots, s_i, \ldots, s_n\}$ and $\mathbf{s}'$ is given by $\mathbf{s} \prec \mathbf{s}'$ if $s_i = s_i'$ for $0 \leq i < n$, and $s_n < s_n'$ when the number of symbols in $\{s_0, \ldots, s_{n-1}\}$ representing orientation reversing branches is even, or $s_n > s_n'$ when the number of symbols representing orientation reversing branches is odd.

As an example, consider the orbits $\overline{012}$ and $\overline{011}$ on the template shown in Figure 5.32. We first construct all cyclic permutations of these orbits,

$$
\begin{array}{ll}
012 & 011 \\
201 & 101 \\
120 & 110.
\end{array}
$$

Next we sort these permutations in ascending order,

$$011 \quad 012 \quad 101 \quad 110 \quad 120 \quad 201.$$

Last, we note that the only orientation reversing branch is 0, so we need to reverse the ordering of the points 012 and 011, yielding

$$012 \prec 011 \prec 101 \prec 110 \prec 120 \prec 201,$$

which agrees with the ordering shown on the ordering tree in Figure 5.32.

5.5.4 Calculation of Relative Rotation Rates

Relative rotation rates can be calculated from the symbolic dynamics of the return map or directly from the template. We now illustrate this latter calculation for the zero global torsion lift of the horseshoe. We then describe a general algorithm for the calculation of relative rotation rates from the symbolics.

Horseshoe Example

Consider two periodic orbits A and B of periods p_A and p_B. At the surface of section, a periodic orbit is labeled by the set of initial conditions $(x_1, x_2, ..., x_n)$, each x_i corresponding to some cyclic permutation of the symbolic name for the orbit. That is, it amounts to a choice of "phase" for the periodic orbit. For instance, the period three orbit "011" on the standard horseshoe template gives rise to three symbolic names (011, 110, 101). When calculating relative rotation rates it is important to keep track of this phase since different permutations can give rise to different relative rotation rates.

To calculate the relative rotation rate between two periodic orbits we first represent each orbit by some symbolic name (choice of phase). Next, we form the composite template of length $p_A \cdot p_B$. This is illustrated for the period three orbit 110 and the period one orbit 000 in Figure 5.33(a). The two periodic orbits can now be extracted from the composite template and presented as two strands of a pure braid of length $p_A \cdot p_B$ with the correct crossing data (Figure 5.33(b)). The self-rotation rate is calculated in a similar way. The case of the period two orbit in the horseshoe template is illustrated in Figure 5.33(c,d).

General Algorithm

Although we have illustrated this process geometrically, it is completely algorithmic and algebraic. For a general braid template, the relative ordering of the orbits at each branch line is determined from the symbolic names and kneading theory. Given the ordering at the branch lines, and the form of the template, all the crossings between orbits are determined, and hence so are the relative rotation rates. Gilmore developed a computer program [29] that generates the full spectrum

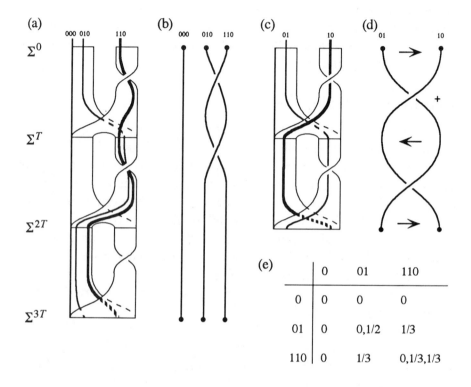

Figure 5.33: Relative rotation rates from the standard horseshoe template: (a) composite template for the orbits 110 and 000; (b) the periodic orbits represented as pure braids; (c) composite template for calculating self-rotation rate of 01; (d) pure braid of 01 and 10; (e) the intertwining matrix for the orbits 0, 01, and 110.

of relative rotation rates when supplied with only the periodic orbit matrix and insertion array (for a definition of periodic orbit matrix, also known as the template matrix, see ref. [1]), i.e., purely algebraic data. Here we describe an alternative algorithm based on the framed braid presentation and the braid linking matrix. This algorithm is implemented as a *Mathematica* package, listed in Appendix G.

To calculate the relative rotation rate between two orbits we need to keep track of three pieces of crossing data: (1) crossings between two knots within the same branch (recorded by the branch torsion); (2) crossings between two knots on separate branches (recorded by the branch crossings); and (3) any additional crossings occurring at the insertion layer (calculated from kneading theory). One way to organize this crossing information is illustrated in Figure 5.34, which shows the braid linking matrix for a three-branch template and two words for which we wish to calculate the relative rotation rate.

Formulas can be written down describing the relative rotation rate calculation [17], but we will instead try to describe in words the "relative rotation rate arithmetic" that is illustrated in Figure 5.34. To calculate the relative rotation rate between two orbits we use the following sequence of steps:

1. Write the braid linking matrix for the template with standard insertion (rearrange branches as necessary until you reach back to front form).

2. In the row labeled α write the word w_1 until the length of the row equals the least common multiple (LCM) between the lengths of w_1 and w_2; do the same with word w_2 in row β.

3. Above these rows create a new row (called the zeroth level) formed by the braid linking matrix elements $b_{\alpha_i \beta_i}$, where i indexes the rows.

4. Find all identical *blocks* of symbols, that is, all places where the symbolics in both words are identical (these are boxed in Figure 5.34). Wrap around from the end of the rows to the beginning of the rows if appropriate.

5. The remaining groups of symbols are called *unblocked* regions; for the unblocked regions write the zeroth-level value mod 2 (if even record a zero, if odd record a one) directly below the word rows at the first level.

6. For the <u>blocked</u> regions sum the zeroth-level values above the block (i.e., add up all the entries at the zeroth level that lie directly above a block) and write this sum mod 2 at the second level.

7. For the <u>unblocked</u> regions look for a sign change (orientation reversing branches) from one pair of symbols to the next (i.e., $\alpha_i < \beta_i$ but $\alpha_{i+u} > \beta_{i+u}$, or $\alpha_i > \beta_i$ but $\alpha_{i+u} < \beta_{i+u}$) and write a 1 at the second level if there is a sign change, or write a 0 if there is no change of sign. The counter u gives the integer distance to the next unblocked region. Wrap around from the end of the rows to the beginning of the rows if necessary.

8. *Group* all terms in the first and second levels as indicated in Figure 5.34. Add all terms in each group mod 2 and write at the third level.

9. Sum all the terms at the zeroth level, and write the sum to the right of the zeroth row.

10. Sum all the terms at the third level, and write the sum to the right of the third row.

11. To calculate the relative rotation rate, add the sums of the zeroth level and the third level, and divide by 2× the *LCM* found in step 2.

The rules look complicated, but they can be mastered in just a few minutes, after which time the calculation of relative rotation rates becomes just an exercise in the rotation rate arithmetic.

5.6 Intertwining Matrices

With the rules learned in the previous section we can now calculate relative rotation rates and intertwining matrices directly from templates. For reference we present a few of these intertwining matrices below.

5.6.1 Horseshoe

The intertwining matrix for the zero global torsion lift of the Smale horseshoe is presented in Table 5.1.

5.6.2 Lorenz

The intertwining matrix for the relative rotation rates from the Lorenz template is presented in Table 5.2. Adding a global torsion of one (a full twist) adds two to the braid linking matrix, and it adds one to each relative rotation rate, i.e., each entry of the intertwining matrix.

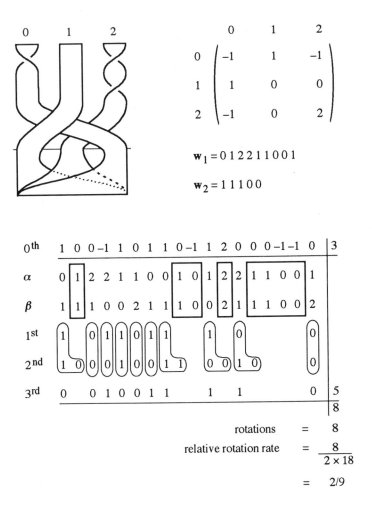

Figure 5.34: Example of the relative rotation rate arithmetic.

	$\overline{0}$	$\overline{1}$	$\overline{01}$	$\overline{001}$	$\overline{011}$	$\overline{0001}$	$\overline{0011}$	$\overline{0111}$
$\overline{0}$	0							
$\overline{1}$	0	0						
$\overline{01}$	0	$\frac{1}{2}$	$0,\frac{1}{2}$					
$\overline{001}$	0	$\frac{1}{3}$	$\frac{1}{3}$	$0,\frac{1}{3},\frac{1}{3}$				
$\overline{011}$	0	$\frac{1}{3}$	$\frac{1}{3}$	$\frac{1}{3}$	$0,\frac{1}{3},\frac{1}{3}$			
$\overline{0001}$	0	$\frac{1}{4}$	$\frac{1}{4}$	$\frac{1}{4}$	$\frac{1}{4}$	$0,\frac{1}{4},\frac{1}{4},\frac{1}{4}$		
$\overline{0011}$	0	$\frac{1}{4}$	$\frac{1}{4}$	$\frac{1}{4}$	$\frac{1}{4}$	$\frac{1}{4}$	$0,\frac{1}{4},\frac{1}{4},\frac{1}{4}$	
$\overline{0111}$	0	$\frac{1}{2}$	$\frac{1}{4}$	$\frac{1}{3}$	$\frac{1}{3}$	$\frac{1}{4}$	$\frac{1}{4}$	$0,\frac{1}{2},\frac{1}{4},\frac{1}{2}$

Table 5.1: Horseshoe intertwining matrix.

	$\overline{0}$	$\overline{1}$	$\overline{01}$	$\overline{001}$	$\overline{011}$	$\overline{0001}$	$\overline{0011}$	$\overline{0111}$
$\overline{0}$	0							
$\overline{1}$	0	0						
$\overline{01}$	0	0	$0,\frac{1}{2}$					
$\overline{001}$	0	0	$\frac{1}{6}$	$0,\frac{1}{3},\frac{1}{3}$				
$\overline{011}$	0	0	$\frac{1}{6}$	$0,0,\frac{1}{3}$	$0,\frac{1}{3},\frac{1}{3}$			
$\overline{0001}$	0	0	0	$\frac{1}{6}$	$\frac{1}{12}$	$0,\frac{1}{4},\frac{1}{4},\frac{1}{4}$		
$\overline{0011}$	0	0	$\frac{1}{4}$	$\frac{1}{6}$	$\frac{1}{6}$	$0,0,\frac{1}{4},\frac{1}{4}$	$0,\frac{1}{4},\frac{1}{4},\frac{1}{4}$	
$\overline{0111}$	0	0	0	$\frac{1}{12}$	$\frac{1}{6}$	$0,0,0,\frac{1}{4}$	$0,0,\frac{1}{4},\frac{1}{4}$	$0,\frac{1}{4},\frac{1}{4},\frac{1}{4}$

Table 5.2: Lorenz intertwining matrix.

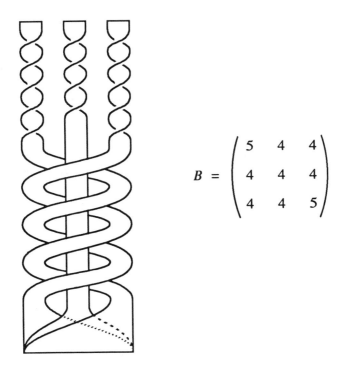

$$B = \begin{pmatrix} 5 & 4 & 4 \\ 4 & 4 & 4 \\ 4 & 4 & 5 \end{pmatrix}$$

Figure 5.35: Template for the Duffing oscillator for the parameter regime explored in section 5.2.3.

5.7 Duffing Template

In this final section we will apply the periodic orbit extraction technique and the template theory to a chaotic time series from the Duffing oscillator. For the parameter regime discussed in section 5.2.3, Gilmore and Solari [4,29] argued on theoretical grounds that the template for the Duffing oscillator is that shown in Figure 5.35. The resulting intertwining matrix up to period three is presented in Table 5.3. All the relative rotation rates calculated from the periodic orbits extracted from the chaotic time series agree (see Appendix G) with those found in Table 5.3, which were calculated from the braid linking matrix for the template shown in Figure 5.35.

However, not all orbits (up to period three) were found in the chaotic

	$\overline{0}$	$\overline{1}$	$\overline{2}$	$\overline{01}$	$\overline{02}$	$\overline{12}$	$\overline{001}$	$\overline{002}$	$\overline{011}$	$\overline{012}$	$\overline{021}$	$\overline{022}$	$\overline{112}$	$\overline{122}$
$\overline{0}$	0													
$\overline{1}$	2	0												
$\overline{2}$	2	2	0											
$\overline{01}$	$\frac{5}{2}$	2	2	$0,\frac{5}{2}$										
$\overline{02}$	$\frac{5}{2}$	$\frac{5}{2}$	$\frac{5}{2}$	$\frac{5}{2}$	$0,\frac{5}{2}$									
$\overline{12}$	2	2	$\frac{5}{2}$	2	$\frac{5}{2}$	$0,\frac{5}{2}$								
$\overline{001}$	$\frac{7}{3}$	2	2	$\frac{7}{3}$	$\frac{5}{2}$	2	$0,\frac{7}{3},\frac{7}{3}$							
$\overline{002}$	$\frac{7}{3}$	$\frac{7}{3}$	$\frac{7}{3}$	$\frac{7}{3}$	$\frac{5}{2}$	$\frac{7}{3}$	$\frac{7}{3}$	$0,\frac{7}{3},\frac{7}{3}$						
$\overline{011}$	$\frac{7}{3}$	2	2	$\frac{7}{3}$	$\frac{5}{2}$	2	$\frac{7}{3}$	$\frac{7}{3}$	$0,\frac{7}{3},\frac{7}{3}$					
$\overline{012}$	$\frac{7}{3}$	$\frac{7}{3}$	$\frac{7}{3}$	$\frac{7}{3}$	$\frac{5}{2}$	$\frac{7}{3}$	$\frac{7}{3}$	$\frac{7}{3}$	$\frac{7}{3}$	$0,\frac{7}{3},\frac{7}{3}$				
$\overline{021}$	$\frac{7}{3}$	$\frac{7}{3}$	$\frac{7}{3}$	$\frac{7}{3}$	$\frac{5}{2}$	$\frac{7}{3}$	$\frac{7}{3}$	$\frac{7}{3}$	$\frac{7}{3}$	$\frac{7}{3}$	$0,\frac{7}{3},\frac{7}{3}$			
$\overline{022}$	$\frac{7}{3}$	$\frac{7}{3}$	$\frac{7}{3}$	$\frac{7}{3}$	$\frac{5}{2}$	$\frac{7}{3}$	$\frac{7}{3}$	$\frac{7}{3}$	$\frac{7}{3}$	$\frac{7}{3}$	$\frac{7}{3}$	$0,\frac{7}{3},\frac{7}{3}$		
$\overline{112}$	2	2	$\frac{7}{3}$	2	$\frac{5}{2}$	$\frac{7}{3}$	2	$\frac{7}{3}$	2	$\frac{7}{3}$	$\frac{7}{3}$	$\frac{7}{3}$	$0,\frac{7}{3},\frac{7}{3}$	
$\overline{122}$	2	2	$\frac{7}{3}$	2	$\frac{5}{2}$	$\frac{7}{3}$	2	$\frac{7}{3}$	2	$\frac{7}{3}$	$\frac{7}{3}$	$\frac{7}{3}$	$\frac{7}{3}$	$0,\frac{7}{3},\frac{7}{3}$

Table 5.3: Duffing intertwining matrix, calculated from the template for the Duffing oscillator. All the periodic orbits were extracted from a single chaotic time series except for those in italics, the so-called "pruned" orbits.

time series. In particular, the orbits $\overline{02}$, $\overline{002}$, and $\overline{022}$ did not appear to be present. Such orbits are said to be "pruned."

The template theory helps to organize the periodic orbit structure in the Duffing oscillator and other low-dimensional chaotic processes. Mindlin and co-workers [1,2] have carried the template theory further than our discussion here. In particular, they show how to extract not only periodic orbits, but also templates from a chaotic time series. Thus, the template theory is a very promising first step in the development of topological models of low-dimensional chaotic processes.

References and Notes

[1]   G. B. Mindlin, X.-J. Hou, H. G. Solari, R. Gilmore, and N. B. Tufillaro, Classification of strange attractors by integers, Phys. Rev. Lett. **64** (20), 2350–2353 (1990).

[2]   G. B. Mindlin, H. G. Solari, M. A. Natiello, R. Gilmore, and X.-J. Hou, Topological analysis of chaotic time series data from the Belousov-Zhabotinskii reaction, J. Nonlinear Sci. **1** (2), 147–173 (1991).

[3]   N. B. Tufillaro, H. Solari, and R. Gilmore, Relative rotation rates: Fingerprints for strange attractors, Phys. Rev. A **41** (10), 5717–5720 (1990).

[4]   H. G. Solari and R. Gilmore, Relative rotation rates for driven dynamical systems, Phys. Rev. A **37** (8), 3096–3109 (1988).

[5]   H. G. Solari and R. Gilmore, Organization of periodic orbits in the driven Duffing oscillator, Phys. Rev. A **38** (3), 1566–1572 (1988).

[6]   J.-P. Eckmann and D. Ruelle, Ergodic theory of chaos and strange attractors, Rev. Mod. Phys. **57** (3), 617–656 (1985).

[7]   P. Grassberger and I. Procaccia, Measuring the strangeness of strange attractors, Physica **9D**, 189–202 (1983).

[8]   T. C. Halsey, M. H. Jensen, L. P. Kadanoff, I. Procaccia, and B. I. Shraiman, Fractal measures and their singularities: The characterization of strange sets, Phys. Rev. A **33**, 1141 (1986); erratum, Phys. Rev. A **34**, 1601 (1986).

[9]   I. Procaccia, The static and dynamic invariants that characterize chaos and the relations between them in theory and experiments, Physica Scripta **T9**, 40 (1985).

[10] J.-P. Eckmann, S. O. Kamphorst, D. Ruelle, and S. Ciliberto, Lyapunov exponents from time series, Phys. Rev. A **34**, 497 (1986).

[11] J. M. Franks, *Homology and dynamical systems*, Conference Board of the Mathematical Sciences Regional Conference Series in Mathematics Number 49 (American Mathematical Society: Providence, 1980).

[12] F. Simonelli and J. P. Gollub, Surface wave mode interactions: Effects of symmetry and degeneracy, J. Fluid Mech. **199**, 471–494 (1989).

[13] J. Birman and R. Williams, Knotted periodic orbits in dynamical systems I: Lorenz's equations, Topology **22**, 47–82 (1983); J. Birman and R. Williams, Knotted periodic orbits in dynamical systems II: Knot holders for fibered knots, Contemporary Mathematics **20**, 1–60 (1983).

[14] P. Holmes, Knots and orbit genealogies in nonlinear oscillators, in *New directions in dynamical systems*, edited by T. Bedford and J. Swift, London Mathematical Society Lecture Notes 127 (Cambridge University Press: Cambridge, 1988), pp. 150–191; P. J. Holmes and R. F. Williams, Knotted periodic orbits in suspensions of Smale's horseshoe: Torus knots and bifurcation sequences, Archives of Rational Mechanics Annals **90**, 115–194 (1985).

[15] D. Auerbach, P. Cvitanović, J.-P. Eckmann, G. Gunaratne, and I. Procaccia, Exploring chaotic motion through periodic orbits, Phys. Rev. Lett. **58** (23), 2387–2389 (1987).

[16] D. P. Lathrop and E. J. Kostelich, Characterization of an experimental strange attractor by periodic orbits, Phys. Rev. A **40** (7), 4028–4031 (1989).

[17] P. Melvin and N. B. Tufillaro, Templates and framed braids, Phys. Rev. A **44** (6), R3419–3422 (1991).

[18] N. B. Tufillaro, Chaotic themes from strings, Ph.D. Thesis, Bryn Mawr College (1990).

[19] P. Cvitanović, Chaos for cyclists, in *Noise and chaos in nonlinear dynamical systems*, edited by F. Moss (Cambridge University Press: Cambridge, 1988).

[20] H. Poincaré, *Les méthodes nouvelles de la mécanique céleste*, Vol. 1–3, (Gauthier-Villars: Paris, 1899); reprinted by Dover, 1957. English translation: New methods of celestial mechanics (NASA Technical Translations, 1967). See Volume I, Section 36.

[21] R. Easton, Isolating blocks and epsilon chains for maps, Physica **39 D**, 95–110 (1989).

[22] D. L. González, M. O. Magnasco, G. B. Mindlin, H. Larrondo and L. Romanelli, A universal departure from the classical period doubling spectrum, Physica **39 D**, 111–123 (1989).

[23] V. F. R. Jones, Knot theory and statistical mechanics, Sci. Am. (November 1990), 98–103.

[24] M. Wadati, T. Deguchi, and Y. Akutsu, Exactly solvable models and knot theory, Phys. Reps. **180** (4&5), 247–332 (1989); L. H. Kauffman, *On knots* (Princeton University Press: Princeton, NJ, 1987).

[25] E. V. Eschenazi, Multistability and basins of attraction in driven dynamical systems, Ph.D. Thesis, Drexel University (1988).

[26] J. Franks and R. F. Williams, Entropy and knots, Trans. Am. Math Soc. **291**, 241–253 (1985).

[27] S. Wiggins, *Global bifurcations and chaos: analytical methods* (Springer-Verlag: New York, 1988).

[28] J. Guckenheimer and P. Holmes, *Nonlinear oscillations, dynamical systems and bifurcation of vector fields* (Springer-Verlag: New York, 1983).

[29] R. Gilmore, Relative rotation rates from templates, a Fortran program, private communication (1989). Address: Dept. of Physics, Drexel University, Philadelphia, PA 19104-9984.

Problems

Problems for section 5.3.

5.1. Calculate the linking numbers for the links shown in Figures 5.7(a) and (c). Choose different orientations for the knots and recalculate.

5.2. Calculate the linking numbers between the three orbits shown in Figure 5.26.

5.3. Calculate the linking numbers between the orbits 00, 01, and 02 in Figure 5.32.

5.4. Write the braid words for the braids shown in Figure 5.16.

5.5. Verify that the braid group (section 5.3.4) is, in fact, a group.

5.6. Find an equivalent braid to the braid shown in Figure 5.17 and write down its corresponding braid word. Use the braid relations to show the equivalence of the two braids.

5.7. Write down the braid words for the braids shown in Figure 5.18. Use the braid relations to demonstrate the equivalence of the braids as shown.

5.8. Write down the braid word for the braid in Figure 5.19.

Section 5.4.

5.9. Calculate the relative rotation rates from both the A orbit and the B orbit in Figure 5.20. That is, verify the equivalence of $R(A, B)$ and $R(B, A)$ in this instance. Use the geometric method illustrated in Figure 5.21 (which is taken from Figure 5.20). Attempt the same calculation directly from Figure 5.20.

5.10. Show that the sum of the relative rotation rates is the linking number (see reference [4] for more details).

5.11. Show the equivalence of equations (5.12) and (5.13).

Section 5.5.

5.12. Draw three different three-branch templates and sketch their associated return maps. Assume a linear expansive flow on each branch. Construct their braid linking matrices.

5.13. Show that the Lorenz template arises from the suspension of a discontinuous map. Consider the evolution of a line segment connecting the two branches at the middle.

5.14. Draw the template in Figure 5.32 as a ribbon graph and as a framed braid. What is its braid linking matrix?

5.15. Verify that the strands of a braid are permuted according to equation (5.19).

5.16. Verify the relative rotation rates in Figure 5.33(e) by the relative rotation rate arithmetic described in section 5.5.4. Add the orbit $\overline{010}$ to the table.

Section 5.6.

5.17. Calculate the intertwining matrix for the orbits shown in Figure 5.32 up to period two.

Section 5.7.

5.18. Verify the $\overline{012}$ row in Table 5.3 by the relative rotation arithmetic.

Appendix A: Bouncing Ball Code

The dynamical state of the bouncing ball system is specified by three variables: the ball's height y, its velocity v, and the time t. The time also specifies the table's vertical position $s(t)$. Let t_k be the time of the kth table-ball collision, and v_k the ball's velocity immediately after the kth collision. The system evolves according to the *velocity* and *phase* equations:

$$v_k - \dot{s}(t_k) = -\alpha\{v_{k-1} - g[t_k - t_{k-1}] - \dot{s}(t_k)\},$$

$$s(t_k) = s(t_{k-1}) + v_{k-1}[t_k - t_{k-1}] - \frac{1}{2}g[t_k - t_{k-1}]^2,$$

which are collectively called the *impact map*. The velocity equation states that the relative ball-table speed just after the kth collision is a fraction α of its value just before the kth collision. The phase equation determines t_k (given t_{k-1}) by equating the table position and the ball position at time t_k; t_k is the smallest strictly positive solution of the phase equation. The simulation of the impact map on a microcomputer presents no real difficulties. The only somewhat subtle point arises in finding an effective algorithm for solving the phase equation, which is an implicit algebraic equation in t_k.

We choose to solve for t_k by the *bisection method* [1] because of its great stability and ease of coding. Other zero-finding algorithms, such as Newton's method, are not recommended because of their sensitivity to initial starting values. All bisection methods must be supplied with a natural step size for the method at hand. The step interval must be large enough to work quickly, yet small enough so as not to include more than one zero on the interval. Our solution to the problem is documented in the function *findstep()* of the C program below. In essence, our step-finding method works as follows.

If the relative impact velocity is large then t_k and t_{k-1} will not be close, so the step size need only be some suitable fraction of the forcing period. On the other hand, if t_k and t_{k-1} are close then the step size needs to be some fraction of the interval. We approximate the interval between t_k and t_{k-1} by noting that the relative velocity between the ball and the table always starts out positive and must be zero before they collide again. Using the fact that the time between collisions is small, it is easy to show that

$$\tau_k \approx \frac{A\omega \cos(\delta_{k-1}) - v_{k-1}}{A\omega^2 \sin(\delta_{k-1}) - g},$$

where δ_{k-1} is the phase of the previous impact and τ_k is the time it takes the relative velocity to reach zero. τ_k provides the correct order of magnitude for the step size. This algorithm is coded in the following C program.

---

[1] For a discussion of the bisection method for finding the real zeros of a continuous function, see R. W. Hamming, *Introduction to applied numerical analysis* (McGraw-Hill: New York, 1971), p. 36.

```
/* bb.c bouncing ball program
   copyright 1985 by Nicholas B. Tufillaro
   Date written:  25 July 1985
   Date last modified:  5 August 1987
   This program simulates the dynamics of a bouncing ball subject to
   repeated collisions with an oscillating wall.
   The INPUT is:
           delta v0 Amplitude Frequency damping cycles
   where

   delta: the initial position of ball between 0 and 1, this is the
          phase of forcing frequency that you start at (phase mod 2*PI)
   v0:    the initial velocity of the ball, this must be greater than
          the initial velocity of the wall.
   A:     Amplitude of the oscillating wall
   Freq:  Frequency of the oscillating wall in hertz
   damp:  (0-1) the impact parameter describing the energy loss per
          collision.  No energy loss (conservative case) when d = 1,
          maximum dissipation occurs at d = 0.
   cycles:the total length the simulation should be run in terms of
          the number of forcing oscillations.
          Units:  CGS assumed
          Compile with:  cc bb.c -lm -O -o bb
   Bugs:
*/


#include <stdio.h>
#include <math.h>

/* CONSTANTS (CGS Units) */

#define STEPSPERCYCLE         (256)
#define TOLERANCE             (1e-12)
#define MAXITERATIONS         (1024)
#define PI                    (3.14159265358979323846)
#define G                     (981) /* earth's gravitational constant */
#define STUCK                 (-1)
#define EIGHTH                (0.125)

/* Macros */

#define max(A, B) ((A) > (B) ? (A) : (B))
#define min(A, B) ((A) < (B) ? (A) : (B))
```

```
/* Comments of variables
   t is time since last impact
   ti is time of last impact
   tau is ti + t, time since start of simulation
   xi is position of ball at last impact
   vi is velocity of ball at last impact
   w is velocity of wall
   v is velocity of ball
*/

/* Global Variables */

double delta;      /* initial phase (position) of ball */
double v0;         /* initial ball velocity */
double A;          /* table amplitude */
double freq;       /* frequency of forcing */
double damp;       /* impact parameter */
double omega;      /* angular frequency 2*PI*f = 2*PI/T */
double T;          /* period of forcing */
double cycles;     /* length of simulation */

/* Functions */

double s(), w(), x(), v(), d(), acc();
double find_step();
double checkstep();
double root();
double fmod(), asin();

main()
{
   int stuckcount;
   double t, ti, tau, dt, tstop, xi, vi, tj, xj, vj;
   double t_alpha, t_beta, t_ph; /* variables for sticking case */

   /* read input parameters */
   scanf("%lf%lf%lf%lf%lf%lf", &delta, &v0, &A, &freq, &damp, &cycles);
   T = 1.0/freq;
   tstop = T*cycles;
   omega = 2*PI*freq;
   delta = delta/freq;
   if( v0 < w(0.0) ) {
      printf("Error:  Initial velocity less than wall velocity \n");
      printf("Wall velocity:  %g \n", w(0.0));
```

```
    exit();
}

t = 0; ti = 0; tau = ti + t; xi = s(tau); vi = v0;
dt = find_step(ti, xi, vi);
t = dt; tau = ti + t; stuckcount = 0;

while( tau <= tstop) {
    t += dt; tau = ti + t;
    if( impact( tau, t, xi, vi)) {
        t = root(t-dt, t, t, xi, vi, ti);
        tj = ti + t;
        xj = s(tj);
        vj = (1+damp)*w(tj) - damp*v(t,vi);
        t = 0; xi = xj; vi = vj; ti = tj;
        dt = find_step(ti, xi, vi);
        if (dt == STUCK ) { /* sticking solution */
            if (A*omega*omega < G) {
                printf("Stuck forever with table\n");
                exit();
            }
            if(fabs(G/(A*omega*omega)) < 1.0)
                t_alpha = (1.0/omega)*asin(G/(A*omega*omega));
            else
                t_alpha = T*0.25;
            t_beta = 0.5*T-t_alpha;
            dt = (t_beta-t_alpha)/STEPSPERCYCLE;
            t_ph = fmod(ti+delta,T);
            if(!((t_ph > t_alpha) && (t_ph < t_beta))) {
                stuckcount +=1;
                if (t_alpha < t_ph)
                    tj = T + ti + t_alpha - t_ph;
                else
                    tj = ti + t_alpha - t_ph;
                xj = s(tj); vj = w(tj);
                t = 0; xi = xj; vi = vj; ti = tj;
            }
            if(stuckcount == 2) {
                printf("%g %g\n", fmod(tj/T,1.0), v(t,vj));
                printf("Ball Stuck Twice\n");
                exit();
            }
        }
    }
    if(tau > tstop/10.0)
```

```
                printf("%g %g\n", fmod(tj/T,1.0), v(t,vj));
            dt = checkstep(dt,ti,xi,vi);
            t = dt; tau = ti + dt;
        }
    }
}

double s(tau)              /* wall position */
double tau;
{
   return(A*(sin(omega*(tau+delta))+1));
}

double w(tau)              /* wall velocity */
double tau;
{
   return(A*omega*cos(omega*(tau+delta)));
}

double acc(tau)            /* table acceleration */
double tau;
{
   return(-A*omega*omega*sin(omega*(tau+delta)));
}

double x(t,xi,vi)          /* ball position */
double t, xi, vi;
{
   return(xi+vi*t-0.5*G*t*t);
}

double v(t,vi)             /* ball velocity */
double t, vi;
{
   return(vi-G*t);
}

double d(tau,t,xi,vi)   /* distance between ball and wall */
double tau, t, xi, vi;
{
   return(x(t,xi,vi)-s(tau));
}

int impact(tau,t,xi,vi) /* find when ball is below wall */
```

```
double tau,t,xi,vi;
{
   return(d(tau,t,xi,vi) <= 0.0);
}


/* pick a good step for finding zero */
double find_step(ti,xi,vi)
double ti,xi,vi;
{
   double t_max, tstep;

   if(vi-w(ti) <= 0.0) { /* should be alpha = 0, or sticking */
      return(STUCK);
   }
   if(-acc(ti) - G != 0.0)
      t_max = fabs((w(ti)-vi)/(-acc(ti)-G));
   else
      t_max = T/STEPSPERCYCLE;
   if(t_max < T*TOLERANCE) {
      return(STUCK);
   }
      tstep = min(EIGHTH*t_max,T/STEPSPERCYCLE);
   return(tstep);
}


double root(a,b,t,xi,vi,ti)
double a,b;
double t, xi, vi, ti;
{
   double m;
   int count;
   count = 0;
   while(1) {
      count += 1;
      if(count > MAXITERATIONS) {
         printf("ERROR: infinite loop in root\n");
         exit();
      }
      if(d(ti+a,a,xi,vi)*d(ti+b,b,xi,vi) > 0.0) {
         printf("root finding error:  no zero on interval\n");
         exit();
      }
      m = (a+b)/2.0;
      if((d(ti+m,m,xi,vi) == 0.0 ) || (b-a) < T*TOLERANCE)
```

```
                  return(m);
          else if ((d(ti+a,a,xi,vi)*d(ti+m,m,xi,vi)) < 0.0 )
              b = m;
          else
              a = m;
      }
}

double checkstep(dt,ti,xi,vi)
double dt, ti, xi, vi;
{
    int count;

    for(count=0; d(ti+dt,dt,xi,vi) < 0.0; ++count) {
        dt = EIGHTH*dt;
        if (count > 10) {
            printf("Error:  Can't calculate dt\n");
            exit();
        }
    }
    return(dt);
}
```

Appendix B: Exact Solutions for a Cubic Oscillator

The free conservative cubic oscillator, equation (3.19), arose in the single-mode model of string vibrations, and admits exact solutions in two special circumstances. The first is the case of circular motion at a constant radius R. In Figure 3.3 we could imagine circular orbits arising when the restoring force on the string just balances the centrifugal force. Plugging the ansatz

$$\mathbf{r} = (R\cos\omega t, \ R\sin\omega t) \tag{B.1}$$

into equation (3.6) we see that it is indeed a solution provided the frequency is adjusted to

$$\omega_c^2 = \omega_0^2(1 + KR^2) . \tag{B.2}$$

The second solution appears when we consider planar motion. If all the motion is confined to the x–z plane, then the system is a single degree of freedom oscillator, whose equation of motion in the dimensionless form obtained from equation (3.11) is

$$x'' + x + \beta x^3 = 0. \tag{B.3}$$

The exact solution to equation (B.3) is [1]

$$\tau = \frac{1}{(1+4\beta E)^{1/4}} \left[K\left(\frac{a^2}{a^2+b^2}\right) - F\left(\arccos\frac{x}{a}, \frac{a^2}{a^2+b^2}\right)\right], \tag{B.4}$$

where $F(\theta, \phi)$ is an elliptic integral of the first kind and $K(\phi) = F(\pi/2, \phi)$ [2]; E is the energy constant

$$E = \frac{1}{2}x'^2 + \frac{1}{2}x^2 + \frac{\beta}{4}x^4 \tag{B.5}$$

and

$$b^2, a^2 = \frac{1}{\beta}(\sqrt{1+4\beta E} \pm 1) . \tag{B.6}$$

As in circular motion, the frequency in planar motion is again shifted to a new value given by

$$\gamma_p = \frac{\pi}{2} \frac{(1+4\beta E)^{1/4}}{I[a^2/(a^2+b^2)]} . \tag{B.7}$$

The exact solutions for circular and planar motion are useful benchmarks for testing limiting cases of more general, but not necessarily exact, results.

An exact solution for the more general case of an anharmonic oscillator,

$$x'' + a_1 + a_2 x + a_3 x^2 + a_4 x^3 = 0$$

is provided by Reynolds [3].

References and Notes

[1] K. Banerjee, J. Bhattacharjee, and H. Mani, Classical anharmonic oscillators: Rescaling the perturbation series, Phys. Rev. A **30** (2), 1118–1119 (1984).

[2] For an elementary discussion of elliptic functions, see Appendix F of E. A. Jackson, *Perspectives of nonlinear dynamics*, Vol. 1 (Cambridge University Press: New York, 1989), pp. 404–408.

[3] M. J. Reynolds, An exact solution in nonlinear oscillators, J. Phys. A: Math. Gen. **22**, L723–L726 (1989). Also, see N. Euler, W.-H. Steeb, and K. Cyrus, On exact solutions for damped anharmonic oscillators, J. Phys. A: Math. Gen. **22**, L195–L199 (1989).

Appendix C: Ode Overview

Ode renders a numerical solution to the initial value problem for many families of first-order differential equations. It is a programming language that resembles the mathematical language so that the problem posed by a system is easy to state, thereby making a numerical solution readily available. *Ode* solves higher-order systems since a simple procedure converts an *n*th-order equation into *n* first-order equations. Three distinct numerical methods are implemented at present: Runge-Kutta-Fehlberg (default), Adams-Moulton, and Euler. The Adams-Moulton and Runge-Kutta routines are available with adaptive step size [1]. The *Ode User's Manual* provides both a tutorial on applying *Ode* and a discussion of its design and implementation [2].

The user need only be familiar with the fundamentals of the UNIX operating system to access and run this numerical software. *Ode* provides:

- A simple problem-oriented user interface,

- A table-driven grammar, simplifying extensions and changes to the language,

- A structure designed to ease the introduction of new numerical methods, and

- Remarkable execution speed and capacity for large problems, for an "interpretive" system.

Ode currently runs on a wide range of microcomputers and mainframes that support the UNIX operating system. *Ode* was developed at Reed College in the summer of 1981 under a UNIX operating system and is public domain software. The program is currently in use at numerous educational and industrial sites (Reed College, Tektronix Inc., U.C. Berkeley, Bell Labs, etc.), and the program and documentation are available from some electronic networks, such as Internet.

Ode solves the initial value problem for a family of first-order differential equations when provided with an explicit expression for each equation. *Ode* parses a set of equations, initial conditions, and control parameters, and then provides an efficient numerical solution. *Ode* makes the initial value problem easy to express; for example, the *Ode* program

```
# an ode to Euler
y  = 1
y' = y
print y from 1
step 0, 1
```

prints 2.718282.

A UNIX Shell can be used as a control language for *Ode*. Indeed, this allows *Ode* to be used in combination with other graphical or analytical tools commonly available with the UNIX operating system. For instance, the following *shell script* [3] could be used to generate a bifurcation diagram for the Duffing equation:

```
:   shell script using Ode to construct a bifurcation diagram
:   for the Duffing oscillator.
:   Scan in F: 50 ---> 60.
:   Create a file "bif.data" with the initial conditions
:   before running this shell script.

for I in 0 1 2 3 4 5 6 7 8 9
do
for J in 0 1 2 3 4 5 6 7 8 9
do
for K in 0 2 4 6 8
do
tail -1 bif.data > lastline.tmp
xo=`awk '{print $1}' lastline.tmp`
vo=`awk '{print $2}' lastline.tmp`
ode <<marker >>bif.data

alpha   = 0.0037
beta    = 86.2
gamma   = 0.99
F       = 5$I.$J$K
x0      = $xo
v0      = $vo
theta0  = 0

x'      = v
v'      = -(alpha*v + x + beta*x^3) + F*cos(theta)
theta'  = gamma

F = 5$I.$J$K; t = 0; theta = 0;
print x, v, F every 64 from (2*PI*200)/(0.99)
step 0, (2*PI*400)/(0.99), (2*PI)/(64*0.99)

marker
done
done
done
```

In addition to *Ode*, there exist many other packages that provide numerical solutions to ordinary differential equations. However, if you wish to write your own routines see Chapter 15 of *Numerical Recipes* [4].

References and Notes

[1] K. Atkinson, *An introduction to numerical analysis* (John Wiley: New York, 1978). Pages 380–384 contain a review of the literature on the numerical solution of ordinary differential equations.

[2] N. B. Tufillaro and G. A. Ross, *Ode*—A program for the numerical solution of ordinary differential equations, Bell Laboratories Technical Memorandum, TM: 83-52321-39 (1983), based on the *Ode User's Manual* (Reed College Academic Computer Center, 1981). At last report the latest version of the source code and documentation were located on the REED VAX. Contact the Reed College Academic Computer Center at 1-503-771-1112 for more details.

[3] For a tutorial on the UNIX shell, see S. R. Bourne, *The UNIX system* (Addison-Wesley: Reading, MA, 1983).

[4] W. Press, B. Flannery, S. Teukolsky, and W. Vetterling, *Numerical recipes in C* (Cambridge University Press: New York, 1988).

Appendix D: Discrete Fourier Transform

The following C routine calculates a discrete Fourier transform and power spectrum for a time series. It is only meant as an illustrative example and will not be very useful for large data sets, for which a fast Fourier transform is recommended.

```
/* Discrete Fourier Transform and Power Spectrum
   Calculates Power Spectrum from a Time Series
   Copyright 1985 Nicholas B. Tufillaro
*/

#include <stdio.h>
#include <math.h>

#define PI (3.1415926536)
#define SIZE 512

double ts[SIZE], A[SIZE], B[SIZE], P[SIZE];

main()
{
    int i, k, p, N, L;
    double avg, y, sum, psmax;

    /* read in and scale data points */
    i = 0;
    while(scanf("%lf", &y) != EOF) {
        ts[i] = y/1000.0;
        i += 1;
    }
    /* get rid of last point and make sure #
       of data points is even */
    if((i%2) == 0)
        i -= 2;
    else
        i -= 1;
    L = i; N = L/2;
    /* subtract out dc component from time series */
    for(i = 0, avg = 0; i < L; ++i) {
        avg += ts[i];
    }
    avg = avg/L;
    /* now subtract out the mean value from the time series */
    for(i = 0; i < L; ++i) {
```

```
        ts[i] = ts[i] - avg;
   }
   /* o.k. guys, ready to do Fourier transform */
   /* first do cosine series */
   for(k = 0; k <= N; ++k) {
      for(p = 0, sum = 0; p < 2*N; ++p) {
         sum += ts[p]*cos(PI*k*p/N);
      }
      A[k] = sum/N;
   }
   /* now do sine series */
   for(k = 0; k < N; ++k) {
      for(p = 0, sum = 0; p < 2*N; ++p) {
         sum += ts[p]*sin(PI*k*p/N);
      }
      B[k] = sum/N;
   }
   /* lastly, calculate the power spectrum */
   for(i = 0; i <= N; ++i) {
      P[i] = sqrt(A[i]*A[i]+B[i]*B[i]);
   }
   /* find the maximum of the power spectrum to normalize */
   for(i = 0, psmax = 0; i <= N; ++i) {
      if(P[i] > psmax)
         psmax = P[i];
   }
   for(i = 0; i <= N; ++i) {
      P[i] = P[i]/psmax;
   }
   /* o.k., print out the results: k, P(k) */
   for(k = 0; k <= N; ++k) {
      printf("%d %g\n", k, P[k]);
   }
}
```

Appendix E: Hénon's Trick

The numerical calculation of a Poincaré map from a cross section at first appears to be a rather tedious problem. Consider, for example, calculating the Poincaré map for an arbitrary three-dimensional flow

$$\frac{dx}{dt} = f(x, y, z), \quad \frac{dy}{dt} = g(x, y, z), \quad \frac{dz}{dt} = h(x, y, z),$$

with a planar cross section

$$\Sigma = (x, y, z = 0).$$

A simpleminded approach to this problem would involve numerically integrating the equations of motion until the cross section is pierced by the trajectory. An intersection of the trajectory with the cross section is determined by testing for a change in sign of the variable in question. Once an intersection is found, a bisection algorithm could be used to hone-in on the surface of section to any desired degree of accuracy.

Hénon, however, suggested a very clever procedure that allows one to find the intersection point of the trajectory and the cross section in one step [1]. Suppose that between the nth and the $n + 1$st steps we find a change of sign in the z coordinate:

$$(t_n, \ x_n, \ y_n, \ z_n < 0) \quad \text{and} \quad (t_n + \Delta t, \ x_{n+1}, \ y_{n+1}, \ z_{n+1} > 0).$$

To find the exact value in t at which the z coordinate equals zero we can change t from the independent to a dependent variable, and change z from a dependent variable to the independent variable. This is accomplished by dividing the first two equations by dt/dz and inverting the last equation:

$$\frac{dx}{dz} = \frac{f(x, y, z)}{h(x, y, z)}, \quad \frac{dy}{dz} = \frac{g(x, y, z)}{h(x, y, z)}, \quad \frac{dt}{dz} = \frac{1}{h(x, y, z)}.$$

This new system can be numerically integrated forward one step, $\Delta z = -z_n$, with the initial values x_n, y_n, t_n. A simple numerical integration method, such as a Runge-Kutta procedure, is ideally suited for this single integration step.

As an application of Hénon's method consider the swinging Atwood's machine (SAM) defined by the conservative equations of motion [2]:

$$\dot{v} = \frac{1}{1 + \mu} \left(ru^2 + \cos\theta - \mu \right),$$

$$\dot{u} = -\frac{1}{r} \left(2vu + \sin\theta \right),$$

$$\dot{r} = v,$$

$$\dot{\theta} = u.$$

A cross section for this system is defined by $(r, \dot{r}, \theta = 0)$. So we need to change the independent variable from t to θ. When θ changes sign, we can find the Poincaré map by numerically integrating the equations:

$$\frac{dv}{d\theta} = \frac{1}{u(1+\mu)} \left(ru^2 + \cos\theta - \mu \right),$$

$$\frac{du}{d\theta} = -\frac{1}{ur} \left(2vu + \sin\theta \right),$$

$$\frac{dr}{d\theta} = \frac{v}{u}$$

$$\frac{dt}{d\theta} = \frac{1}{u}.$$

Pictures of the resulting Poincaré map are found in reference [2].

References and Notes

[1] M. Hénon, On the numerical computation of Poincaré maps, Physica **5D**, 412–414 (1982). Also see Hao B.-L., *Elementary symbolic dynamics* (World Scientific: New Jersey, 1989), pp. 260–262.

[2] N. B. Tufillaro, Motions of a swinging Atwood's machine, J. Physique **46**, 1495–1500 (1985).

Appendix F: Periodic Orbit Extraction Code

The first C routine below was used to extract the periodic orbits from a chaotic time series arising from the Duffing oscillator, as described in section 5.2. This code is specific to the Duffing oscillator and is provided because it illustrates the coding techniques needed for periodic orbit extraction. The second routine takes a pair of extracted periodic orbits and calculates their relative rotation rates according to equation (5.12). The input to the relative rotation rate program is a pair of periodic orbits plus a first line containing some header information about the input file.

```c
/* fp.c
   Find all Periodic orbits of period P and tolerance TOL.
   Copyright 1989 by Nicholas B. Tufillaro
                        Department of Physics
                        Bryn Mawr College, Bryn Mawr, PA 19010-2899 USA
*/

#include <stdio.h>
#include <math.h>

/* This Array Maximum must be greater then 2*P*STEPS + 1 */
#define AMAX 1000
#define DIST(X1,X2,Y1,Y2)
        ((float)sqrt((double)((X2-X1)*(X2-X1)+(Y2-Y1)*(Y2-Y1))))

main(ac, av)
char **av;
{
    int m, n, cnt, P, STEPS, LEN, TWOLEN, SHORTERPERIODICORBIT;
    float d, ds, phe[AMAX], x[AMAX], y[AMAX], TOL, SHRTOL;
    float oldphe;

    /* process command line arguments: P TOL */
    --ac; P = atoi(*++av); --ac; TOL = (float)atof(*++av);
    /* set "global" variables */
    STEPS = 64; LEN = P*STEPS; TWOLEN = 2*LEN;
    SHRTOL = 2*TOL;
    /* initialization, get first LEN points */
    for(n = 0; n < LEN; ++n) {
        if(scanf("%f %f %f", &phe[n], &x[n], &y[n]) == EOF) {
            printf("Not enough orbits for computations.\n");
            exit();
        }
    }
}
```

```
/* find periodic orbits */
for(;;) {
   for(n = LEN; n < TWOLEN; ++n) {
      if(scanf("%f %f %f", &phe[n], &x[n], &y[n]) == EOF) {
         exit();
      }
   }
   for(n = 0; n < LEN; ++n) {
      d = DIST(x[n], x[n+LEN], y[n], y[n+LEN]);
      if(d < TOL) {
         /* Identify shorter periodic orbits, if any.
            This is a kludge.
         */
         for(m = 1, SHORTERPERIODICORBIT = 0; m < P; ++m) {
            ds = DIST(x[n], x[n+m*STEPS], y[n], y[n+m*STEPS]);
            if(ds < SHRTOL) {
               SHORTERPERIODICORBIT = 1;
            }
         }
         /* End of kludge. */
         if(!SHORTERPERIODICORBIT) {
         if(fabs(phe[n]-oldphe) > 1.0) {
            oldphe = phe[n];
            printf("\n%f %f %f\n\n", -1.0, (float) P, (float) cnt);
            for(m = n; m <= n+LEN; ++m) {
               printf("%f %f %f\n", phe[m], x[m], y[m]);
            }
            cnt += 1;
         }
         }
      }
   }
   for(n = 0; n < LEN; ++n) {
      phe[n] = phe[n+LEN]; x[n] = x[n+LEN]; y[n] = y[n+LEN];
   }
}
}
```

```
/* rrr.c
    calculate Relative Rotation Rates of two periodic orbits
    of periods PA, PB.
    Copyright 1989 by Nicholas B. Tufillaro
    Department of Physics, Bryn Mawr College
    Bryn Mawr, PA 19010-2899 USA

    INPUT:
        Data file with phase, x, and y listed for points in two
        periodic orbits. For the first input line of each
        periodic orbit, use -1.0 for phase and give the period in x.
        For this first point y is ignored.
*/

#include <stdio.h>
#include <math.h>

#define PI 3.14159265
#define ARG(X,Y) (float)(atan2((double)Y,(double)X))

main()
{
    int m, n, M, N, I[10], J[10];
    float phe, x, y, PA, PB, A[3][1000], B[3][1000];
    int i, j, q, Q;
    float rx[10000], ry[10000], RR[10][10];

    if(scanf("%f %f %f", &phe, &x, &y)==EOF) {
        printf("Error: empty input file\n");
        exit();
    }
    if(phe != -1.0) {
        printf("Error: Input file does not begin with -1.0\n");
        exit();
    }
    if(phe == -1.0) {
        PA = x;
        for(n = 0, m = 0; ; ++n) {
            if(scanf("%f%f%f", &phe, &x, &y)==EOF) {
                printf("Error: not enough orbits\n");
                exit();
            }
            if(phe == -1.0) {
                break;
```

```
      }
      if(fmod(phe,1.0) == 0.0) {
            I[m] = n;
            m +=1;
      }
      A[0][n] = phe; A[1][n] = y; A[2][n] = x;
   }
   PB = x;
   M = n;
   for(n = 0, m = 0; scanf("%f%f%f", &phe, &x, &y) != EOF; ++n) {
      if(phe == -1.0) {
         printf("Error: Too many orbits\n");
         exit();
      }
      if(fmod(phe,1.0) == 0.0) {
         J[m] = n;
         m += 1;
      }
      B[0][n] = phe; B[1][n] = y; B[2][n] = x;
   }
   N = n;
}
Q = (int)(PA)*(N-1) + 1;
for(i = 0; i < PA; ++i) {
   for(j = 0; j < PB; ++j) {
      for(m = I[i], n = J[j], q = 0; q < Q; ++q, ++m, ++n) {
         if(m == 0)  m = M-1;
         if(n == 0)  n = N-1;
         if(m == M)  m = 1;
         if(n == N)  n = 1;
         rx[q] = B[1][n]-A[1][m]; ry[q] = B[2][n]-A[2][m];
      }
      for(q = 0; q < Q - 1; ++q) {
         if(ARG(rx[q],ry[q]) > PI/2 && ARG(rx[q],ry[q]) < PI
            && ARG(rx[q+1],ry[q+1]) > -PI &&
               ARG(rx[q+1],ry[q+1]) < -PI/2)
            RR[i][j] += 2*PI + ARG(rx[q+1],ry[q+1]) -
               ARG(rx[q],ry[q]);
         else if(ARG(rx[q],ry[q]) > -PI &&
            ARG(rx[q],ry[q]) < -PI/2 &&
            ARG(rx[q+1],ry[q+1]) > PI/2 &&
            ARG(rx[q+1],ry[q+1]) < PI)
               RR[i][j] += ARG(rx[q+1],ry[q+1])-
                  ARG(rx[q],ry[q]) - 2*PI;
```

```
                else
                    RR[i][j] += ARG(rx[q+1],ry[q+1])-
                              ARG(rx[q],ry[q]);
            }
        }
    }
    printf("\nROTATION RATES\n\n");
    printf("PA: %g   PB: %g\n\n", PA, PB);
    for(i = 0; i < PA; ++i) {
        for(j = 0; j < PB; ++j) {
            printf("Index: %d, %d.  Rotations: %g   Rel. Rot.: %g\n",
                i+1, j+1, RR[i][j]/(2*PI), RR[i][j]/(2*PI*PA*PB));
        }
    }
    printf("\n");
}
```

Appendix G: Relative Rotation Rate Package

The following *Mathematica* package calculates relative rotation rates and intertwining matrices by the methods described in section 5.5.4.

```
(*
    Relative Rotation Rates --- Mathematica Package
    Date: 10/04/90
    Last Modified:


Authors: Copyright 1990
A. Lorentz, N. Tufillaro and P. Melvin.
Departments of Mathematics and Physics
Bryn Mawr College, Bryn Mawr, PA 19010-2899 USA


Bugs:
    Many of these symbolic computations are exponential time algorithms,
so they are slow for long periodic orbits and templates with many
branches. A "C" version of these routines exists which runs considerably
faster. The algorithm for the calculation of the relative
rotation rate from a single word pair, however, is polynomial time.


About the Package:
This collection of routines automates the process for the symbolic
calculation of relative rotation rates, and the intertwining matrix
for an arbitrary template. The template is represented algebraically
by a framed braid matrix, which is specified by the global variable
"bm" in these routines.  This variable must be defined by the user
when these routines are entered.
    There are three major routines:

        RelRotRate[word pair], AllRelRotRates[word pair], and
        Intertwine[start row, stop row].

The input for the RelRotRate programs is a word pair, which is just
a list of lists. For example, a valid input for these programs is

        { {0,1,0,1,1,0}, {1,1,0} }

where the first word of the word pair is "010110" and the second word
is "110". The input for the Intertwine program is just two integers,
"start row" and "stop row". For instance, Intertwine[2,3] would produce
all relative rotation rates for all words of length between 2 and 3.
    In addition, the routine
```

```
        GetFunCyc[branches, period]
```

generates all fundamental cycles of length "period" for a template with
"b" branches or a symbolic alphabet of "b" letters. This routine can
be useful for the cycle expansion techniques (see, R. Artuso, E. Aurell,
and P. Cvitanovic, Recycling of strange sets I and II, Nonlinearity
Vol 3., Num. 2, May 1990, p. 326.).

References:
[1] G. B. Mindlin, X.-J. Hou, H. G. Solari, R. Gilmore, and N. B.
Tufillaro, Classification of strange attractors by integers, Phys. Rev.
Lett. 64 (20), 2350 (1990).
[2] N. B. Tufillaro, H. G. Solari, and R. Gilmore, Relative rotation rates:
 fingerprints for strange attractors, Phys. Rev. A 41 (10), 5717 (1990).
[3] H. G. Solari and R. Gilmore, Organization of periodic orbits in the
        driven Duffing oscillator, Phys. Rev. A 38 (3), 1566 (1988).
[4] H. G. Solari and R. Gilmore, Relative rotation rates for driven
        dynamical systems, Phys. Rev. A 37 (8), 3096 (1988).
*)

(*
    The template braid matrix is a global variable that should be
    defined by the user before this package is used. Comment out the
    default setting for the braid matrix.
    Examples for the braid matrix are presented below.

    Global Variable Abbreviation.
    bm braid matrix --- algebraic description of template
*)

(* Braid Matrix Example: The Horseshoe Template *)
(*
bm = {
    { 0, 0 },
    { 0, 1 }
};
*)

(* Braid Matrix Example: Second Iterate of The Horseshoe Template *)
bm = {
    { 0, 0, 0, 0},
    { 0, 1, 1, 1},
    { 0, 1, 2, 1},
```
```

```
   { 0, 1, 1, 1}
};

(* Calculate the relative rotation rate for a pair of words.
  Variables local to RelRotRate.
  wordp      wordpair
  lcm        lowest common multiple of word pair lengths
  ewordp     expanded word pair
  top        top list
  bottom     bottom list
  rrr        relative rotation rate of word pair
*)

RelRotRate[wordp_List] :=
Block[
   {
     lcm,
     ewordp, top, bottom,
     rrr,
 },

If[wordp[[1]] == wordp[[2]], Return[0]];        (* Self rotation rate *)

   ewordp = ExpandWordPair[wordp];
   top = GetTop[ewordp];
   bottom = GetBottom[ewordp];
   lcm = Length[top];

   rrr    = (SumAll[top] + SumAll[bottom])/(2 lcm);
   Return[rrr]
]

(* Permute the word pair list to generate
   all possible relative rotation rates *)

AllRelRotRates[wordp_List] :=
Block[
   {i, pa, pb, lcm, wordstep, rrr,
   firstword, secondword, rrrs},

   firstword = wordp[[1]];
   secondword = wordp[[2]];
   pa = Length[firstword];
   pb = Length[secondword];
```

```
   lcm = LCM[pa, pb];
   wordstep = (lcm*lcm)/(pa*pb);
   rrrs = {};
   For[i = 0, i < lcm, i += wordstep,
      rrr = RelRotRate[{firstword, secondword}];
      rrrs = Append[rrrs, {firstword, secondword, rrr}];
      firstword = RotateRight[firstword, wordstep];
   ];
   Return[rrrs];
]

(* Calculate the Intertwining Matrix of all orbits of length "startrow"
   to length "stoprow", startrow <= stoprow. *)

Intertwine[startrow_Integer, stoprow_Integer] :=
Block[
   {b, i, j, k, rowsize, rsum, colsize, csum, mult,
    cycls = {}, rcycls = {}, ccycls = {}, rrr = {}, rrrs = {}},

   b = Length[bm];

   rowsize = 0; rsum = 0;
   For[i = startrow, i <= stoprow, ++i,
      cycls = GetFunCyc[b, i]; rowsize = Length[cycls];
      rsum += rowsize;
      For[j = 1, j <= rowsize, ++j,
         rcycls = Append[rcycls, cycls[[j]]];
      ];
   ];
   colsize = 0; csum = 0;
   For[i = 1, i <= stoprow, ++i,
      cycls = GetFunCyc[b, i]; colsize = Length[cycls];
      csum += colsize;
      For[j = 1, j <= colsize, ++j,
         ccycls = Append[ccycls, cycls[[j]]];
      ];
   ];
   rrr = {}; rrrs = {};
   For[i = 1, i <= rsum, ++i,
      For[j = 1, j <= csum, ++j,
         rrr = AllRelRotRates[{rcycls[[i]], ccycls[[j]]}];
         mult = Length[rrr];
         rrrs = {rrr[[1,1]], rrr[[1,2]]};
         For[k = 1, k <= mult, ++k,
```

```
            rrrs = Append[rrrs, rrr[[k,3]]];
        ];
        Print[rrrs];
        If[rcycls[[i]] == ccycls[[j]], Break[]];
    ];
  ];
]


(* Subroutines for major programs: RelRotRate, AllRelRotRates,
   Intertwine *)

ExpandWordPair[wp_List] :=
Block[

{i, lcm, firstword, secondword},

firstword = wp[[1]]; secondword = wp[[2]];
lcm = LCM[Length[firstword], Length[secondword]];
Return[
   { Flatten[ Table[firstword, {i, lcm/Length[firstword]  }]],
  Flatten[ Table[secondword, {i, lcm/Length[secondword] }]] }
];
]


GetTop[ewp_List] :=
Block[
   {i, lcm},
   lcm = Length[ ewp[[1]] ];
   Return[
      Table[ bm[[ ewp[[1,i]]+1, ewp[[2,i]]+1]], {i, lcm} ]
   ];
]


GetBottom[ewp_List] :=
Block[
   {i=0, s=0, j=0, prevj=0, nextj=0, cnt=0, lcm=0, sgn=0,
   top={}, ewpd={}, b1={}, b2={}, b3={}, bot={}},

   lcm    = Length[ewp[[1]]];
   top    = GetTop[ewp];
   ewpd   = ewp[[1]] - ewp[[2]];

   (* initialize rows b1, b2, b3 *)
   For[i = 1, i < lcm+1, i++,
```

```
      AppendTo[b1,0]; AppendTo[b2,0]; AppendTo[b3,0];
   ];

   (* calculate b1 row *)
   For[i = 1, i < lcm + 1, i++,
      If[ewpd[[i]] != 0, b1[[i]] = Mod[top[[i]],2]];
   ];

   (* calculate b2 row *)
   For[s = 0, ewpd[[s+1]] == 0, ++s];
   For[i = 1, i < lcm + 1, i++,
      j = i + s; If[j > lcm, j -= lcm];
      prevj = j - 1; If[prevj < 1, prevj += lcm];
      If[0 == ewpd[[j]],
         cnt = 0;
         For[k = j, ewpd[[k]] == 0, k++,
            cnt = cnt + top[[k]];
            If[k == lcm, k = 0];
            i++;
         ];
         cnt = Mod[cnt, 2];
         b2[[prevj]] = cnt;
      ];
   ];

   (* calculate b3 row *)
   For[i = 1, i < lcm + 1, i++,
      j = i + s; If[j > lcm, j -= lcm];
      nextj = j + 1; If[nextj > lcm, nextj -= lcm];
      k = nextj;
      While[0 == ewpd[[nextj]],
         ++nextj; ++i;
         If[nextj > lcm, nextj -= lcm];
      ];
      If[Negative[ewpd[[j]]*ewpd[[nextj]]], sgn = 1, sgn = 0];
      b3[[j]] = sgn;
   ];

   bot = Mod[b1 + b2 + b3, 2];
   Return[bot];
]


SumAll[l_List] :=
```

```
Block[{i},
Return[Sum[l[[i]], {i, Length[l]}]]
]
```

(* Generates the fundamental cycles of length "period" for a template
   with "b" branches *)

```
GetFunCyc[b_Integer, period_Integer] :=
Block[
{cycles},
                                 (* get all cycles of length *)
cycles = GetAllCyc[b, period];   (* "period", and b "branches" *)
cycles = DelSubCyc[b, cycles];   (* delete nonfundamental subcycles *)
cycles = DelCycPerm[cycles];     (* delete cyclic permutations
                                    of fundamental cycles *)
Return[cycles];
]
```

(* Subroutines for GetFunCyc *)

```
GetAllCyc[branches_Integer, levels_Integer] :=
Block[
   {n, m, i, j,
    roots, tree, nextlevel, cycle, cycles},

   For[i = 0; roots = {}, i < branches, ++i,
    roots = Append[roots, i];
   ];

   (* creates full n-ary tree recursively *)
   tree = {roots};
   For[n = 1, n < levels, ++n,
     nextlevel = {};
    For[m = 1, m <= branches, ++m,
      nextlevel = Append[nextlevel, Last[tree]];
    ];
    tree = Append[tree, Flatten[nextlevel]];
   ];

   (* reads up each branch of tree to root, from left to right *)
   cycles = {};
   For[i = 1, i <= branches^levels, ++i,
     cycle = {};
       For[j = levels, j > 0, --j,
```

```
            k = Ceiling[i/(branches^(levels-j))];
            cycle = Prepend[cycle, tree[[j]][[k]]];
        ];
          cycles = Append[cycles, cycle];
    ];
    Return[cycles];
]


DelSubCyc[b_Integer, allcycls_List] :=
Block[
    {i, j, k, levels, period, numofdivs, numofsub, copies, wordpos,
     cycles = {}, div = {}, word = {}, droplist = {},
     subcycles = {}, subword = {}, nonfun = {}, funcycls = {}},

    (* Initializations and gets divisor list *)
    cycles = allcycls; levels = Length[ cycles[[1]] ];
    period = levels; div = Divisors[levels];

    (* Creates nonfundamental words from periodic orbits created from
       divisor list *)
    numofdivs = Length[div];
    For[i = 1, i < numofdivs, ++i,      (* go throw divisor list *)
        copies = period/div[[i]];
        subcycles = GetAllCyc[b, div[[i]]];
        numofsub = Length[subcycles];
        (* create subwords of lengths found in divisor list *)
        For[j = 1; subword = {}, j <= numofsub, ++j,
            subword = subcycles[[j]];
            (* expand subwords to length of periodic orbits *)
            For[k = 1; word = {}, k <= copies, ++k,
                word = Flatten[Append[word, subword]];
            ];
            (* find positions of nonfundamental cycles in all cycles *)
            wordpos = Flatten[Position[cycles, word]][[1]];
            droplist = Union[Append[droplist, wordpos]];
        ];
    ];

    (* this is a kludge to delete nonfundamental cycles *)
    For[i = 1; nonfun = {}, i <= Length[droplist], ++i,
        nonfun = Append[nonfun, cycles[[droplist[[i]]]]];
    ];
    funcycls = Complement[cycles, nonfun];
```

```
   Return[funcycls];
]

DelCycPerm[funcycls_List] :=
Block[
   {size, i, j, period,
    cycs, word },
   cycs = funcycls;
   size = Length[cycs]; period = Length[cycs[[1]]];
   For[i = 1, i < size, ++i,
      word = cycs[[i]];
      For[j = 1, j < period, ++j,
         word = RotateLeft[word];
         cycs = Complement[cycs, {word}];
         size = size-1;
      ];
   ];
   Return[cycs];
]
```

Appendix H: Historical Comments

How hard are the problems posed by classical mechanics? The beast in the machine, the beast we now call chaos, was discovered about a century ago by the French scientist and mathematician Henri Poincaré during the course of his investigations on "the three-body problem," the great unsolved problem of classical mechanics. Poincaré, in his magnum opus on the three-body problem, *New Methods of Celestial Mechanics*, writes:[1]

> 397. When we try to represent the figure formed by these two curves and their intersections in a finite number, each of which corresponds to a doubly asymptotic solution, these intersections form a type of trellis, tissue, or grid with infinitely serrated mesh. Neither of the two curves must ever cut across itself again, but it must bend back upon itself in a very complex manner in order to cut across all of the meshes in the grid an infinite number of times.

> The complexity of this figure will be striking, and I shall not even try to draw it. Nothing is more suitable for providing us with an idea of the complex nature of the three-body problem, and of all the problems of dynamics in general, where there is no uniform integral and where the Bohlin series are divergent.

Poincaré is describing his discovery of homoclinic solutions (homoclinic intersections, or homoclinic tangles[2]). The existence of these homoclinic solutions "solved" the three-body problem insofar as it proved that no solution of the type envisioned by Jacobi or Hamilton could exist. Volume III of Poincaré's *New Methods of Celestial Mechanics* (1892–1898), from which the above quote is taken, is the foundational work of modern dynamical systems theory.

Poincaré's great theorem of celestial mechanics, as stated in his prize-winning essay to the King of Sweden, is the following:[3]

> The canonical equations of celestial mechanics do not admit (except for some exceptional cases to be discussed separately) any analytical and uniform integral besides the energy integral.

Simply put, Poincaré's theorem says that there does not exist a solution to the three-body problem of the type assumed by the Hamilton-Jacobi method or any other method seeking an analytic solution to the differential equations of motion.

---

[1] H. Poincaré, *Les méthodes nouvelles de la mécanique céleste*, Vol. 1–3 (Gauthier-Villars: Paris, 1899); reprinted by Dover, 1957. English translation: New methods of celestial mechanics (NASA Technical Translations, 1967).

[2] R. Abraham and C. Shaw, *Dynamics—The geometry of behavior*, Vol. 1–4 (Aerial Press: Santa Cruz, CA, 1988).

[3] H. Poincaré, Sur le problème des trois corps et les équations de la dynamique, Acta Math. **13**, 1–271 (1890).

Poincaré "solved" the three-body problem by showing that no solution exists—at least not of the type assumed by scientists of his era.

But Poincaré did much more. Confronted with his discovery of homoclinic tangles, Poincaré went on to reinvent what is meant by a solution. In the process, Poincaré laid the foundations for several new branches of mathematics including topology, ergodic theory, homology theory, and the qualitative theory of differential equations.[4] Poincaré also pointed out the possible uses of periodic orbits in taming the beast:[5]

> 36. . . . It seems at first that this fact can be of no interest whatever for practice. In fact, there is a zero probability for the initial conditions of the motion to be precisely those corresponding to a periodic solution. However, it can happen that they differ very little from them, and this takes place precisely in the case where the old methods are no longer applicable. We can then advantageously take the periodic solution as first approximation, as intermediate orbit, to use Gyldén's language.
>
> There is even more: here is a fact which I have not been able to demonstrate rigorously, but which seems very probable to me, nevertheless.
>
> Given equations of the form defined in art. 13 and any particular solution of these equations, we can always find a periodic solution (whose period, it is true, is very long), such that the difference between the two solutions is as small as we wish, during as long a time as we wish. In addition, these periodic solutions are so valuable for us because they are, so to say, the only breach by which we may attempt to enter an area heretofore deemed inaccessible.

The periodic orbit theme is pursued in Chapter 5.

Shortly thereafter Hadamard (1898) produced the first example of an abstract system exhibiting chaos—the geodesics on a surface of constant negative curvature.[6] Hadamard's example was later generalized by Anosov and is still one of the best mathematical examples of chaos in its most extreme form.

In America, George David Birkhoff continued in the way of Poincaré. He examined the use of maps[7] instead of flows, and began the process of hunting and naming different critters that he called limit sets of the alpha and omega variety.[8]

---

[4]F. Browder, *The mathematical heritage of Henri Poincaré*, Vol. 1–2, Proc. Sym. in Pure Math. Vol. 39 (American Mathematical Society: Providence, RI, 1983).

[5]R. MacKay and J. Meiss, eds., *Hamiltonian dynamical systems* (Adam Hilger: Philadelphia, 1987).

[6]J. Hadamard, Les surfaces à curbures opposés et leurs lignes géodésiques, Journ. de Math. **4** (5), 27–73 (1898).

[7]G. D. Birkhoff, Surface transformations and their dynamical applications, Acta Math. **43**, 1–119 (1922).

[8]G. D. Birkhoff, *Collected mathematical works*, Vol. 1–3 (Dover: New York,

Still, without the aid of the computer, Birkhoff was often fooled by the beast into thinking that nonintegrability implies complete ergodicity. It took the work of Kolmogorov, Arnold, and Moser in the early 1960s to show that the beast is more subtle in its destructive tendencies, and that it prefers to chew on rational frequencies.[9]

Few physicists were concerned about the beast, or even aware of its existence, during the first half of the twentieth century; however, this is more than understandable when you consider that they had their hands full with quantum mechanics and some less savory inventions. Still, there were some notable exceptions, such as those discussed by Brillouin in his book *Scientific Uncertainty, and Information*.[10]

Near the end of the Second World War, Cartwright and Littlewood[11] and Levinson[12] showed that the beast liked to play the numbers, and could generate solutions as random as a coin toss. At first they did not quite believe their results; however, experiments with a simple circuit (the van der Pol oscillator) forced them to accept the idea that a fully deterministic system can produce random results—what we now think of as the definition of chaos.

In 1960, the young topologist Steven Smale was sitting on the beach in Rio playing with the beast when he first saw that in its heart lay a horseshoe.[13] Smale found that he could name the beast, a hyperbolic limit set, even if he had trouble seeing it since it was very, very thin. Moreover, Smale found a simple way to unravel the horseshoe via symbolic dynamics.[14] Now that the beast was named and dissected, mathematicians had some interesting mathematical objects to play with, and Smale and his friends were off and running.

At about the same time the meteorologist Ed Lorenz, a former student of G. D. Birkhoff, discovered that the beast was not only in the heavens, as evidenced in the three-body problem, but may well be around us all the time in the atmosphere.[15] Whereas Smale showed us how to name the beast, Lorenz allowed us to see the beast with computer simulations and judiciously chosen models. Lorenz pointed out the role return maps can play in understanding real systems, how ubiquitous the beast really is, and how we can become more familiar with the beast through

---

1968).

[9]J. Moser, *Stable and random motions in dynamical systems*, Ann. Math. Studies 77 (Princeton University Press: Princeton, NJ, 1973).

[10]L. Brillouin, *Scientific uncertainty, and information* (Academic Press: New York, 1964).

[11]M. L. Cartwright and N. Littlewood, On nonlinear differential equations of the second order, I, J. Lond. Math. Soc. 20, 180–189 (1945).

[12]N. Levinson, A second-order differential equation with singular solutions, Ann. Math. 50, 127–153 (1949).

[13]S. Smale, *The mathematics of time: Essays on dynamical systems, economic processes, and related topics* (Springer-Verlag: New York, 1980).

[14]R. L. Devaney, *An introduction to chaotic dynamical systems*, second ed. (Addison-Wesley: New York, 1989).

[15]E. N. Lorenz, Deterministic nonperiodic flow, J. Atmos. Sci. 20, 130–141 (1963).

simple models such as the logistic map.[16]

This two-pronged approach of using abstract topological methods on the one hand and insightful computer experiments on the other is central to the methodology employed when studying nonlinear dynamical systems. More than anything else, it probably best defines the "nonlinear dynamical systems method" as it has developed since Lorenz and Smale. In fact, J. von Neumann helped invent the electronic computer mainly to solve, and provide insight into, nonlinear equations. In 1946 in an article called "On the principles of large scale computing machines," he wrote:

> Our present analytic methods seem unsuitable for the solution of the important problems arising in connection with non-linear partial differential equations and, in fact, with virtually all types of non-linear problems in pure mathematics. . . .
>
> . . . really efficient high-speed computing devices may, in the field of non-linear partial differential equations as well as in many other fields which are now difficult or entirely denied of access, provide us with those heuristic hints which are needed in all parts of mathematics for genuine progress.

Thank heavens for the Martians.

---

[16] E. N. Lorenz, The problem of deducing the climate from the governing equations, Tellus **16**, 1–11 (1964).

Appendix I: Projects

In this appendix we provide a brief guide to the literature on some topics that might be suitable for an advanced undergraduate research project.

1. Acoustics

    (a) Review article
        W. Lauterborn and U. Parlitz, Methods of chaos physics and their application to acoustics, J. Acoust. Soc. Am. **84** (6), 1975–1993 (1988).

    (b) Oscillations in gas columns
        T. Yazaki, S. Takashima, and F. Mizutani, Complex quasiperiodic and chaotic states observed in thermally induced oscillations of gas columns, Phys. Rev. Lett. **58** (11), 1108–1111 (1987).

    (c) Wineglass
        A. French, A study of wineglass acoustics, Am. J. Phys. **51** (8), 688–694 (1983).

2. Biology

    (a) Review article
        L. Olsen and H. Degn, Chaos in biological systems, Quart. Rev. Biophys. **18** (2), 165–225 (1985).

    (b) Brain waves
        P. Rapp, T. Bashore, J. Martinerie, A. Albano, I. Zimmerman, and A. Mees, Dynamics of brain electrical activity, Brain Topography **2** (1&2), 99–118 (1989).

    (c) Gene structure
        H. Jeffrey, Chaos game representation of gene structure, Nucleic Acids Res. **18** (8), 2163–2170 (1990).

3. Chemistry

    (a) Chemical clocks
        S. Scott, Clocks and chaos in chemistry, New Scientist (2 December 1989), 53–59; E. Mielczarek, J. Turner, D. Leiter, and L. Davis, Chemical clocks: Experimental and theoretical models of nonlinear behavior, Am. J. Phys. **51** (1), 32–42 (1983).

4. Electronics

    (a) Analog simulation of laser rate equations
        M. James and F. Moss, Analog simulation of a periodically modulated laser model, J. Opt. Soc. Am. B **5** (5), 1121–1127 (1988); for more details about the circuit contact F. Moss, Dept. of Physics, University of St. Louis, St. Louis, MO 63121.

(b) Circuits

J. Lesurf, Chaos on the circuit board, New Scientist (30 June 1990), 63–66. A. Rodriguez-Vazquez, J. Huertas, A. Rueda, B. Perez-Verdu, and L. Chua, Chaos from switched-capacitor circuits: Discrete maps, Proc. IEEE **75** (8), 1090–1106 (1987).

(c) Double scroll

T. Matsumoto, L. Chua, and M. Komuro, The double scroll bifurcations, Circuit theory and applications **14**, 117–146 (1986); T. Matsumoto, L. Chua, and M. Komuro, The double scroll, IEEE Trans. on Circuits and Systems CAS-32 (8), 798–818 (1985); L. Chua, M. Komuro, and T. Matsumoto, The double scroll family, IEEE Trans. on Circuits and Systems CAS-33 (11), 1073–1118 (1986); T. Weldon, An inductorless double scroll circuit, Am. J. Phys. **58** (10), 936–941 (1990).

5. Hydrodynamics

(a) Dripping faucet

R. Cahalan, H. Leidecker, and G. Cahalan, Chaotic rhythms of a dripping faucet, Computing in Physics, 368–383 (Jul/Aug 1990); R. Shaw, *The dripping faucet as a model chaotic system* (Aerial Press, Santa Cruz, CA 1984); H. Yepez, N. Nuniez, A. Salas Brito, C. Vargas, and L. Viente, Chaos in a dripping faucet, Eur. J. Phys. **10**, 99–105 (1989).

(b) Hele-Shaw cell

J. Nye, H. Lean, and A. Wright, Interfaces and falling drops in Hele-Shaw cell, Eur. J. Phys. **5**, 73–80 (1984).

(c) Lorenz loop

S. Dodd, Chaos in a convection loop, Reed College Senior Thesis (1990); M. Gorman, P. Widmann, and K. Robbins, Nonlinear dynamics of a convection loop, Physica **19D**, 255–267 (1986); P. Widmann, M. Gorman, and K. Robbins, Nonlinear dynamics of a convection loop II, Physica **36D**, 157–166 (1989).

(d) Surface waves

R. Apfel, "Whispering" waves in a wineglass, Am. J. Phys. **53** (11), 1070–1073 (1985); S. Douady and S. Fauve, Pattern selection in Faraday instability, Europhys. Lett. **6** (3), 221–226 (1988); J. P. Gollub, Spatiotemporal chaos in interfacial waves (To appear in: *New perspectives in turbulence*, edited by S. Orszag and L. Sirovich, Springer-Verlag); J. Miles and D. Henderson, Parametrically forced surface waves, Annu. Rev. Fluid Mech. **22**, 143–165 (1990); V. Nevolin, Parametric excitation of surface waves, J. Eng. Phys. (USSR) **47**, 1482–1494 (1984).

6. Mechanics

(a) Artistic mobiles
O. Viet, J. Wesfreid, and E. Guyon, Art cinétique et chaos mécanique, Eur. J. Phys. **4**, 72–76 (1983).

(b) Bicycle stability
Y. Hénaff, Dynamical stability of the bicycle, Eur. J. Phys. **8**, 207–210 (1987); J. Papadopoulos, Bicycling handling experiments you can do (preprint, Cornell University, 1987); C. Miller, The physical anatomy of steering stability, Bike Tech (October 1983), 8–11.

(c) Compass in a B-field
H. Meissner and G. Schmidt, A simple experiment for studying the transition from order to chaos, Am. J. Phys. **54** (9), 800–804 (1986); V. Croquette and C. Poitou, Cascade of period doubling bifurcations and large stochasticity in the motions of a compass, J. Physique Lettres **42**, L-537–L-539 (1981).

(d) Compound pendulum
N. Pedersen and O. Soerensen, The compound pendulum in intermediate laboratories and demonstrations, Am. J. Phys. **45** (10), 994–998 (1977).

(e) Coupled pendula
K. Nakajima, T. Yamashita, and Y. Onodera, Mechanical analogue of active Josephson transmission line, J. Appl. Phys. **45** (7), 3141–3145 (1974).

(f) Elastic pendulum
E. Breitenberger and R. Mueller, The elastic pendulum: A nonlinear paradigm, J. Math. Phys. **22** (6), 1196–1210 (1981); M. Olsson, Why does a mass on a spring sometimes misbehave? Am. J. Phys. **44** (12), 1211–1212 (1976); H. Lai, On the recurrence phenomenon of a resonant spring pendulum, Am. J. Phys. **52** (3), 219–223 (1984); L. Falk, Recurrence effects in the parametric spring pendulum, Am. J. Phys. **46** (11), 1120–1123 (1978); M. Rusbridge, Motion of the sprung pendulum, Am. J. Phys. **48** (2), 146–151 (1980); T. Cayton, The laboratory spring-mass oscillator: An example of parametric instability, Am. J. Phys. **47** (8), 723–732 (1977); J. Lipham and V. Pollak, Constructing a "misbehaving" spring, Am. J. Phys. **46** (1), 110–111 (1978).

(g) Forced beam
See Appendix C of F. Moon, *Chaotic vibrations* (John Wiley & Sons, New York, 1987).

(h) Forced pendulum
R. Leven, B. Pompe, C. Wilke, and B. Koch, Experiments on periodic and chaotic motions of a parametrically forced pendulum, Physica **16D**, 371–384 (1985).

(i) Impact oscillator

H. Isomaki, J. Von Boehm, and R. Raty, Devil's attractors and chaos of a driven impact oscillator, Phys. Lett. **107A** (8), 343–346 (1985); A. Brahic, Numerical study of a simple dynamical system, Astron. & Astrophys. **12**, 98–110 (1971); C. N. Bapat and C. Bapat, Impact-pair under periodic excitation, J. Sound Vib. **120** (1), 53–61 (1988).

(j) Impact pendulum

D. Moore and S. Shaw, The experimental response of an impacting pendulum system, Int. J. Nonlinear Mech. **25** (1), 1–16 (1990); S. Shaw and R. Rand, The transition to chaos in a simple mechanical system, Int. J. Nonlinear Mech. **24** (1), 41–56 (1989).

(k) Impact printer

P. Tung and S. Shaw, The dynamics of an impact print hammer, J. of Vibration, Acoustics, Stress, and Reliability in Design **110**, 193–200 (April 1988).

(l) Swinging Atwood's machine

J. Casasayas, A. Nunes, and N. B. Tufillaro, Swinging Atwood's machine: Integrability and dynamics, J. de Physique **51**, 1693–1702 (1990); J. Casasayas, N. B. Tufillaro, and A. Nunes, Infinity manifold of a swinging Atwood's machine, Eur. J. Phys. **10** (10), 173–177 (1989); N. B. Tufillaro, A. Nunes, and J. Casasayas, Unbounded orbits of a swinging Atwood's machine, Am. J. Phys. **56** (12), 1117–1120 (1988); N. B. Tufillaro, Integrable motion of a swinging Atwood's machine, Am. J. Phys. **54** (2), 142–143 (1986); N. B. Tufillaro, Motions of a swinging Atwood's machine, J. de Physique **46**, 1495–1500 (1985); N. B. Tufillaro, Collision orbits of a swinging Atwood's machine, J. de Physique **46**, 2053–2056 (1985); N. B. Tufillaro, T. A. Abbott, and D. J. Griffiths, Swinging Atwood's machine, Am. J. Phys. **52** (10), 895–903 (1984); B. Bruhn, Chaos and order in weakly coupled systems of nonlinear oscillators, Physica Scripta **35**, 7–12 (1987).

(m) Swinging track

J. Long, The nonlinear effects of a ball rolling in a swinging quarter circle track, Reed College Senior Thesis (1990); R. Benenson and B. Marsh, Coupled oscillations of a ball and a curved-track pendulum, Am. J. Phys. **56** (4), 345–348 (1988).

(n) Wedge

N. Whelan, D. Goodings and J. Cannizzo, Two balls in one dimension with gravity, Phys. Rev. A **42** (2), 742–754 (1990); H. Lehtihet and B. Miller, Numerical study of a billiard in a gravitational field, Physica **21D**, 93–104 (1986); A. Matulich and B. Miller, Gravity in one dimension: Stability of a three particle system, Celest. Mech. **39**, 191–198 (1986); B. Miller and K. Ravishankar, Stochastic modeling of a billiard in a gravitational field, J. Stat. Phys. **53** (5/6), 1299–1314 (1988).

7. Optics

    (a) Laser instabilities
    N. B. Abraham, A new focus on laser instabilities and chaos, Laser Focus, May 1983.

    (b) Light absorbing fluid
    G. Indebetouw and T. Zukowski, Nonlinear optical effects in absorbing fluids: Some undergraduate experiments, Eur. J. Phys. **5**, 129–134 (1984).

    (c) Maser equations
    D. Kleinman, The maser rate equations and spiking, Bell System Tech. J. (July 1964), 1505–1532.

    (d) Semiconductor lasers
    H. Winful, Y. Chen, and J. Liu, Frequency locking, quasiperiodicity, and chaos in modulated self-pulsing semiconductor lasers, Appl. Phys. Lett. **48** (10), 616–618 (1986); J. Camparo, The diode laser in atomic physics, Contemp. Phys. **26** (5), 443–477 (1985).

    (e) Diode-pumped YAG laser
    N. B. Abraham, L. Molter, and G. Alman, Experiments with a diode-pumped Nd-YAG laser, private communication. Address: Department of Physics, Bryn Mawr College, Bryn Mawr, PA 19101-2899 USA.

8. Spatial-Temporal Chaos

    (a) Capillary ripples
    N. B. Tufillaro, Order-disorder transition in capillary ripples, Phys. Rev. Lett. **62** (4), 422–425 (1989); H. Riecke, Stable wave-number kinks in parametrically excited standing waves, Europhys. Lett. **11** (3), 213–218 (1990); A. B. Ezerskii, M. I. Rabinovitch, V. P. Reutov, and I. M. Starobinets, Spatiotemporal chaos in the parametric excitations of a capillary ripple, Sov. Phys. JEPT **64** (6), 1228–1236 (1987); W. Eisenmenger, Dynamic properties of the surface tension of water and aqueous solutions of surface active agents with standing capillary waves in the frequency range from 10 kc/s to 1.5 Mc/s, Acustica **9**, 327–341 (1959).

    (b) Cellular automata
    J. P. Reilly and N. B. Tufillaro, Worlds within worlds—An introduction to cellular automata, The Physics Teacher (February 1990), 88–91.

    (c) Coupled lattice maps
    R. Kapral, Pattern formation in two-dimensional arrays of coupled, discrete-time oscillators, Phys. Rev. A **31** (6), 3868–3879 (1985).

    (d) Ferrohydrodynamics
    E. M. Karp, Pattern selection and pattern competition in a ferrohydrodynamic system, Reed College Senior Thesis (1990).

(e) Video feedback

J. Crutchfield, Space-time dynamics in video feedback, Physica 10D, 229–245 (1984); V. Golubev, M. Rabinovitch, V. Talanov, V. Shklover, and V. Yakhno, Critical phenomena in inhomogeneous excitable media, JEPT Lett. **42** (3), 99–102 (1985); G. Hausler, G. Seckmeyer, and T. Weiss, Chaos and cooperation in nonlinear pictorial feedback systems, Appl. Opt. **25** (24), 4656–4672 (1986).

Commonly Used Notation

α	coefficient of restitution
$\alpha(p)$	α-limit set of p
b_i	braid operator
b_{ij}	braid linking matrix item
B	braid linking matrix
B_n	braid group on n strands
β	normalized nonlinear term; fractional binary number
β_i	binary digit
C	complex plane
$C(\epsilon)$	correlation integral
D	trapping region; unit square
$d[s,t]$	metric
$\mathrm{d}f^n(x)$	derivative of nth composite of f
$\mathbf{D}f$	derivative matrix of f
∇	divergence operator
$\nabla \cdot \mathbf{F}, \mathrm{div}\mathbf{F}$	divergence of vector field $\mathbf{F}$
E^s, E^c, E^u	stable space, center space, unstable space
f_λ	quadratic map
$(f^n)'(x)$	derivative of nth composite of f
g	acceleration of gravity
γ	normalized frequency
h	topological entropy
H_k	power spectrum amplitude
H_0, H_1	horizontal strips of horseshoe
I	unit interval
K	coefficient of nonlinear term in Duffing equation
l	equilibrium string length
l_0	relaxed string length
L^+, L^-	forward, backward limit sets
$\mathrm{lk}(\alpha, \beta)$	linking number of the knots α and β
λ	quadratic map parameter; string damping; eigenvalue; Lyapunov characteristic exponent
Λ	invariant set, invariant limit set
M	manifold
M^n	n-dimensional manifold
ν	correlation dimension
$\mathcal{O}$	orbit
ω	angular frequency
$\omega(p)$	ω-limit set of p
Ω	nonwandering set
P	Poincaré map

P_A	period of orbit A
Per(f)	set of periodic points of f
ϕ, ϕ_t	flow
$\mathbf{R}, \mathbf{R}^n$	real numbers; real n-space
$R(f)$	chain recurrent set of f
$R_{ij}(A, B)$	relative rotation rate of the periodic orbits A and B
$\mathbf{s}$	one-sided symbol sequence $\{s_0 s_1 s_2 \ldots\}$ or
	two-sided symbol sequence $\{\ldots s_{-2} s_{-1}.s_0 s_1 s_2 \ldots\}$
S^1	unit circle
σ	shift map; sign of a crossing
$\sigma_s, \sigma_c, \sigma_u$	set of stable, center, unstable eigenvalues
σ_B	permutation of braid strands in a template
Σ	cross section; Poincaré section; symbol space of bi-infinite sequences
Σ_2	sequence space on two symbols
Σ^{θ_0}	global cross section
T	period; tension; branched surface
T^n	n-torus
τ	dimensionless time
θ	phase variable
V	potential energy function
V_0, V_1	vertical strips of horseshoe
W^s, W^c, W^u	stable, center, unstable manifolds
x^*	periodic point
$\dot{x}, dx/dt$	time derivative of x
$\circ$	composite of functions
$\mid \, \mid$	absolute value
$\parallel \, \parallel$	norm
$\cup$	union
$\cap$	intersection
$\in$	a member of
$\prec$	precedes
$\triangleright$	implies

Index

Bouncing Ball User's Guide

TABLE OF CONTENTS

INTRODUCTION

Bouncing Ball is a program written for the Apple® Macintosh™ computer that simulates the dynamics of a ball bouncing on a sinusoidally oscillating table. Bouncing Ball 2.0 requires the use of a Macintosh with at least one megabyte (1024K) of memory, at least the 128K version of ROM (all Macintoshes except the original 128K Macintosh and the original Macintosh 512 have such a version), and the 6.0 System set (System 6.0 and Finder 6.1) or later. If you use Bouncing Ball with MultiFinder™, please read "MultiFinder and Memory Considerations" in the section "Tips and Techniques."

This manual explains how to use Bouncing Ball. Before you start, you should already know how to perform basic operations with your Macintosh. You should know how to use the Finder™ and the mouse, manipulate windows, scroll, pull down menus, and choose commands. If you need information about any of these topics, consult your owner's guide.

This manual has six main sections:

- Getting Started
- Types of Simulation
- Bouncing Ball Windows
- Bouncing Ball Menus
- Tips and Techniques
- Selected Error Messages

Getting Started shows you how to create simulations and describes the sample files that accompany Bouncing Ball.

Types of Simulation describes the three different types of simulations Bouncing Ball can run.

Bouncing Ball Windows describes the windows Bouncing Ball uses to display data from the various simulations.

Bouncing Ball Menus contains a directory of all Bouncing Ball menu commands, arranged in the order they appear on the pull-down menus.

Tips and Techniques describes various ways to tailor your use of Bouncing Ball. It includes a discussion of the memory requirements of Bouncing Ball.

Selected Error Messages lists the error messages that Bouncing Ball generates and describes what to do when they appear.

GETTING STARTED

1. Start the Macintosh by turning it on. If you do not have a hard disk, insert a Macintosh system disk.

2. Insert the Bouncing Ball disk and open it, if it is not already open.

3. Open the Bouncing Ball folder.

4. Open the Bouncing Ball application.

5. After the program starts up, choose **Initial Conditions** from the **Control** menu. Change the initial ball velocity and phase (phase is measured in relation to the table, originally from 0 to 1, 1 corresponding to 2π radians) to some other values, or leave them the same, then click the OK button.

The simulation should now be running. Note that there are four windows currently displayed; additional windows can be activated through the **Windows** menu. The upper left window, *Trajectory*, contains a graph of the motion of the ball and table, the horizontal axis representing time. The lower left window, *Animation*, contains an animation of the ball and table. The upper right window, *Untitled* (when you open a simulation or save one, this window takes on the name of the file; it will often be called the *Impact Data* window), lists the table parameters (amplitude, frequency, and damping) and, for each collision between the ball and table, the sequential number of the collision from the beginning of the simulation, the ball's velocity after impact, the phase (again, from 0 to 1), and the total simulation time elapsed. The lower right window, *Impact Map*, plots, for each collision, the ball's outgoing velocity versus the phase.

In addition to the initial velocity and phase, you can control parameters relating to the table—amplitude, frequency, and damping—by choosing **Table Parameters** from the **Control** menu.

For instance, to illustrate the period-doubling route to chaos in the bouncing ball system, we increase the table amplitude by small increments in successive simulations. First reset the initial conditions to the original values, velocity = 10 cm/s, phase = 0. The default values for the table parameters, which lead to a period one orbit, are 0.01 cm for the amplitude, 60 Hz for the frequency, and 0.5 for the damping coefficient. To generate a period two orbit, increase the table amplitude to 0.011 cm by choosing **Table Parameters** from the **Control** menu and changing the table amplitude from 0.01 to 0.011. Now click OK to run the simulation with this new parameter value. Follow the same procedure to increase the table amplitude to 0.012 to produce a chaotic trajectory.

An alternate way to choose new initial conditions is to click in the body of the impact map; the **Initial Conditions** dialog box will come up with velocity and phase corresponding to the point of the mouse-click. In addition, clicking in certain areas of the various windows calls forth appropriate dialog boxes. These active areas are known as "hot spots." Such hot spots are located along the two axes of the *Impact Map* window. The section "Bouncing Ball Windows" discusses these areas further.

Now we'll look at Bouncing Ball's two other types of simulation, both of which are based on the simulation described above.

Bifurcation Diagram

To create a *bifurcation diagram*, follow these steps (the first three steps simply restore initial values):

1. Use the **Table Parameters** command to restore the initial values, amplitude of 0.01 cm, frequency of 60 Hz, and damping coefficient of 0.5.

2. Use the **Initial Conditions** command to restore the original initial conditions, velocity of 10 cm/s, phase of 0.

3. Clear the current data by choosing **New** from the **File** menu.

4. Open the *Bifurcation Diagram* window by choosing **Bifurcation Diagram** from the **Windows** menu.

5. Choose **Bifurcation Diagram** from the **Special** menu and click OK.

The program will generate a diagram showing how the solutions to the Bouncing Ball system vary as the table's amplitude varies. Specifically, you'll see how the phase (vertical axis) changes with the table amplitude (horizontal axis). The bifurcation diagram stops automatically after it has moved through the specified amplitude range.

Basins of Attraction

To create a *basins of attraction* diagram, follow these steps (again, the first three steps simply restore the program to how it would be upon start-up):

1. Use the **Table Parameters** command to restore the initial values, amplitude of 0.01 cm, frequency of 60 Hz, and damping coefficient of 0.5.

2. Use the **Initial Conditions** command to restore the original initial conditions, velocity of 10 cm/s, phase of 0.

3. Clear the current data by choosing **New** from the **File** menu.

4. Open the *Basins of Attraction* window by choosing **Basins of Attraction** from the **Windows** menu.

5. Choose **Basins of Attraction** from the **Special** menu, input "8" in the Increment Size (or "step size") field, and click OK.

The program will generate a diagram showing which initial conditions eventually go to which types of solutions. Each different type of solution (sticking, period one, chaotic—really "other"—etc.) is represented by a different pattern or color in the basins of attraction chart, which is closely related to the impact map chart. A step size of 8 is used so that the diagram will be completed relatively quickly. The basins of attraction diagram stops automatically when it has discovered the eventual type of solution for all initial conditions in the charted range.

Sample Files

To help you quickly see some of its capabilities, Bouncing Ball is packaged with the following sample files. You can either double-click on these sample files or open them with the **Open** command in the **File** menu.

Period 2
A simulation that converges to a period two solution

Ball Sticks
A simulation in which, after several iterations, the ball becomes stuck to the table

Strange Attractor (lengthy)
An example of chaos (it contains 500 iterations, so it takes a while to open, especially on slower Macintoshes)

Sample Bifurcation Diagram
A bifurcation diagram in which the amplitude was incremented from 0.01 cm to 0.012 cm

Sample Basins of Attraction
A basins of attraction diagram, with velocity charted from –3.77 to 10 and a step size of 2

Complex Basins of Attraction
A basins of attraction diagram, with table amplitude 0.012, frequency 70, damping 0.5, velocity charted from –5.28 to 15, and a step size of 1. This diagram has basins for both a period one solution and a period three solution.

TYPES OF SIMULATION

Bouncing Ball can perform three types of simulation. The central simulation to Bouncing Ball is the straightforward one described in the beginning of the section "Getting Started," wherein you set values for the table parameters and the initial conditions, then watch the ball bouncing on the table in any combination of seven windows. This type of simulation will be called the "base" simulation.

The other two types of simulation, referred to as "special" simulations, are based on the base simulation. Each special simulation is displayed in its own window. The two types of special simulations are the bifurcation diagram and the basins of attraction diagram.

Base Simulation

The base simulation represents a computer version of the Bouncing Ball experiment. Using the **Initial Conditions** and **Table Parameters** menu commands, you set the experimental parameters of the experiment, then run the simulation.

As the simulation progresses, you can watch the data displayed in the *Trajectory*, *Animation*, *Impact Data* ("*Untitled*" at the beginning), *Impact Map*, *Phase Space*, *Impact Phases*, and *Time Series* windows. The section "Bouncing Ball Windows" describes how each of these windows displays the data.

Typical parameters for a base simulation are listed below. Note that in the Bouncing Ball program all measurements are in cgs (centimeters-grams-seconds) units.

Parameter	Typical experimental range
Table's amplitude	0–0.1 cm
Table's frequency	0–100 Hz
Damping coefficient	0–1
Initial velocity	0–100 cm/s
Initial phase	0–1

Bifurcation Diagram

In the *bifurcation diagram*, you select a *range* of values for one table parameter, either the frequency or the amplitude. The program then starts the Bouncing Ball system at the beginning of the table parameter range. It lets the base simulation run for a certain number of iterations (to eliminate the "transient" solution), then charts an impact value (phase or velocity) on the vertical axis versus the table parameter on the horizontal axis. It then increments the table parameter (by a value corresponding to one screen pixel) and repeats the process, until it reaches the end of the parameter range.

WARNING: The bifurcation diagram runs very slowly, especially on slow Macintoshes.

What one finds in the bifurcation diagram is that, as the table parameter is changed, the type of solution the Bouncing Ball system reaches can change. For instance, in the amplitude range from 0.01 to 0.012 cm, with frequency 60 Hz and a damping coefficient of 0.5, and initial conditions of velocity 10 cm/s and phase 0, the system will go through a period one solution, a period two solution, a period four solution, a period eight solution, and so on until it reaches chaos.

To create a bifurcation diagram, choose **Bifurcation Diagram** from the **Special** menu and complete the dialog box as described below.

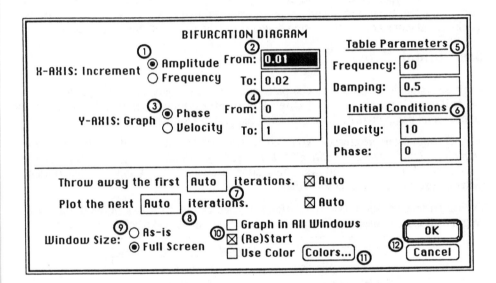

1.   Select the table parameter you want to increment: amplitude or frequency.

2.   Specify the range of values for the parameter chosen in step 1.

3.   Select the impact measurement you want to chart: phase or velocity.

4.   Specify the range of displayed values for the impact measurement selected in step 3.

5.   Specify the values for the remaining table parameters.

6.   Specify the initial conditions for the first iteration.

7.   Specify the number of iterations you want the simulation to run before charting the data, for each value of the table parameter. Checking the Auto box puts in the word "Auto" and gives a value based on the damping coefficient—the higher the damping coefficient, the longer it typically takes to work through the transient solution. In addition, with Auto checked, if after reaching the automatic value Bouncing Ball determines that it has not reached periodicity, but that it will probably reach periodicity if continued, it will increase this number of initial iterations. Checking Auto tends to give the most accurate results. *Note*: If Bouncing Ball determines that

periodicity has been reached before reaching the given number of iterations, it will automatically start charting the data.

8.   Specify the number of iterations you want the simulation to chart. Checking the Auto box puts in the word "Auto." If you choose the automatic option, Bouncing Ball charts several iterations, then charts an additional number of iterations based on the dispersion of those first charted iterations. If, because of a periodic solution, Bouncing Ball decides it does not need to display any more solutions, it stops. *Note*: In a chaotic system, in full-screen view on a small Macintosh, the automatic value described above can easily lead to 500 or more iterations, which can be quite time consuming. A value of 100 or so should lead to satisfactory results in most circumstances.

9.   Select the window size desired: if As-Is is selected, the size of the bifurcation diagram will be such that it will fit into the *Bifurcation Diagram* window as currently sized; if Full Screen is checked, the bifurcation diagram will take up almost the whole screen (it will just fit into a window that has been "zoomed out"). *Note*: The status of these radio buttons will not physically change the size of the *Bifurcation Diagram* window, but it will affect how the scroll bars of that window, if any, operate.

10.   Set these miscellaneous bifurcation diagram parameters:

      a.   Graph in All Windows—As remarked above, the bifurcation diagram actually creates a series of base simulations. Check this box if you wish to display the base simulations in all appropriate windows, as described in the "Base Simulation" part of this section. This allows you to see in detail what's happening at each value of the table parameter; however, it greatly slows down the creation of the bifurcation diagram.

      b.   (Re)Start—Check if you want to start the bifurcation diagram after clicking OK.

      c.   Use Color—Check if you want the bifurcation diagram to display in color. When displayed in color, each type of solution (refer to the "Types of Solutions" discussion of this section for a description of the different solution types) displays in a different color. If you do not have a color monitor, or if it is not set for 4- or more bit color (16 or more colors), this button will be disabled (grayed). *Note*: If your monitor is set for gray-scale, for 4 or more bits, you can implement "color"—it will simply use different shades of gray.

11.   If color is being used, pressing the Colors button will bring up the same dialog box as the **Colors** command from the **Special** menu.

12.   Press OK to complete the input, Cancel to cancel it.

Basins of Attraction

In the *basins of attraction* diagram you select a *range* of initial conditions to try. The program then runs a base simulation for the first initial condition in the range, looks to see what type of solution (attractor) that initial condition converges to, and graphically marks that initial condition with a pattern or color to depict it as belonging to the found attractor. The "Types of Solutions" discussion that follows describes the different types of solutions that Bouncing Ball recognizes. It then increments the initial conditions and repeats the process, until it has covered all initial conditions in the selected range. The basins of attraction chart is thus very similar to the impact map, in that both chart impact phases and velocities.

Until it finds the type of solution, Bouncing Ball plots and then erases data points as they are created. When it does find the solution type, Bouncing Ball then gives the initial condition point the appropriate pattern or color. If the "faster" option is on (see below), it will also give the appropriate pattern or color to all the data points it found along the way.

WARNING: Like the bifurcation diagram, the basins of attraction diagram runs very slowly, especially on slow Macintoshes.

One finds in the basins of attraction diagram exactly which initial conditions and regions of initial conditions converge to which types of solutions.

To create a basins of attraction diagram, choose **Basins of Attraction** from the **Special** menu and complete the dialog box by inputting the information listed below.

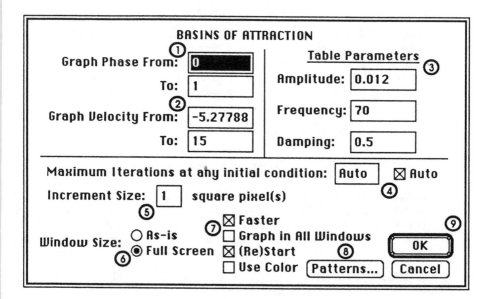

1.   Specify the range of values over which you want to increment the phase.

2.   Specify the range of values over which you want to increment the velocity.

3.   Specify the fixed table parameter values.

4.   Specify the maximum number of iterations you want the simulation to create at each initial condition. Checking the Auto box puts in the word "Auto" and yields a

value based on the damping coefficient; the higher the damping coefficient, the longer it typically takes to work through the transient solution. Also, with Auto checked, if after reaching that number of iterations Bouncing Ball determines that it has not reached periodicity, but that it will probably reach periodicity if continued, it will increase this number of iterations. Checking Auto tends to give the most accurate results. If after this number of iterations Bouncing Ball has found no periodic solution, it assumes that it is a chaotic (or "other") solution.

5. Specify the "step size," the size of the velocity and phase increments, in screen pixels. The larger the value (limited to the range 1-8), the faster the diagram will be completed, but the lower the resolution.

6. Select the window size desired: if As-Is is checked, the size of the basins of attraction will be such that it will fit into the *Basins of Attraction* window as currently sized; if Full Screen is checked, the basins of attraction will take up almost the whole screen (it will be such that it will just fit into a window that has been "zoomed out"). *Note*: The status of these buttons will not physically change the size of the *Basins of Attraction* window; it will affect how the scroll bars of that window, if any, operate.

7. Set these miscellaneous basins of attraction parameters:

    a. Faster—If this is checked, Bouncing Ball will remember the coordinates of each impact as it goes. Then, when the solution type is found, it will mark all of the impact values found along the way as initial conditions belonging to the basin of the solution type found. Because all these impacts are being collected, succeeding iterations can look to see if they are already in a basin; if several impacts in a row are all in the same basin, Bouncing Ball assumes that the current iteration is converging to the same solution as the initial conditions in that basin. Without this option, only the true initial condition is marked. Because of rounding, using this option gives less accurate results than the alternative, and one will receive results that are clearly off in places. However, this option speeds up the chart tremendously.

    b. Graph in All Windows—As remarked above, the basins of attraction diagram actually creates a series of base simulations. Check this box if you wish to display the base simulations in all appropriate windows, as described in the "Base Simulation" part of this section. This allows you to see in detail what's happening at each set of initial conditions; however, it greatly slows down the creation of the basins of attraction.

    c. (Re)Start—Check if you want to start the basins of attraction diagram after clicking OK.

    d. Use Color—Check if you want the basins of attraction to display in color. When displayed in color, each type of solution (sticking, period one, other (chaotic), etc.) displays in a different color; otherwise each type of solution is displayed in a different (gray) pattern. If you do not have a color monitor, or if it is not set for 4- or more bit color (16 or more colors), this button will be disabled (grayed). *Note*: If your monitor is set for gray-scale, for 4 or more bits, you can implement "color"—it will simply use different shades of gray. Also,

Bouncing Ball does allow you to go freely back and forth between color display and pattern display; a basins of attraction diagram created in black and white can later be displayed in color, without loss of resolution.

Note when patterns are being used: Because of the difficulties inherent in "coloring" different regions (using gray patterns) on a black and white monitor, all initial conditions corresponding to the same solution type may not be marked in the same way. For instance, with a "step size" of 2, each point is really a 2 × 2 rectangle of pixels. For points corresponding to a solution represented by the darkest gray pattern, some 2 × 2 pixel blocks will be all black, some partially white. However, the entire *region* corresponding to this solution type would accurately show the dark gray pattern. This implies that, when patterns are being used, it is not always possible to tell which specific pixel blocks correspond to which solutions.

8.   This button will be entitled "Colors..." or "Patterns...", depending on whether color is being used. Pressing it will bring up the same dialog box as the **Colors** command or the **Patterns** command, whichever is appropriate, from the **Special** menu.

9.   Press OK to complete the input, Cancel to cancel it.

Types of Solutions

In a basins of attraction or bifurcation diagram, Bouncing Ball recognizes the following types of solutions:

    Sticking

    Period One

    Period Two

    Period Three

    Period Four

    Period Eight

    Other

    Unphysical

"Sticking" solutions are solutions for which the ball eventually comes to rest with respect to the table, or "sticks" to it. "Period One" through "Period Eight" solutions are just what their names suggest; periodic solutions of the given periods. "Other" solutions, as you might guess, are those that do not fit into any other category. In practice, they are most often chaotic solutions, but can also be other 2^n-periodic solutions, such as period sixteen, or still other periodic solutions, such as period five. In addition, slow convergence and insufficient iterations can cause Bouncing Ball to think that, say, a period two solution is an "other" solution. "Unphysical" solutions, on the other hand, are those that represent physical impossibilities; in this case, these are initial conditions wherein the initial ball velocity is less than the initial table velocity, meaning the ball would move through the table (remember, the ball is initially on the table).

BOUNCING BALL WINDOWS

This section describes the nine windows Bouncing Ball uses to display data.

In all windows, any measurements shown are in cgs units.

Trajectory

The *Trajectory* window shows the position of the ball and the table as a function of time. It thus contains two traces. One trace represents the motion of the table and is sinusoidal. The other represents the motion of the ball and consists of a series of parabolas. The bottom of the window displays the initial conditions.

The scaling in this window depends on values entered with the **Trajectory Scaling** command from the **View** menu. You set the number of table periods that would be shown horizontally if the *Trajectory* window were at its largest size (zoomed out) and the number of table amplitudes that would be shown vertically under the same circumstances. Tick marks on the vertical axis are set at every two table amplitudes.

When the traces reach the right edge of the graph, by default the graph is cleared and the traces continue from the left edge. If you prefer, the traces can instead smoothly scroll to the left as data is added to the right. Use the **Scroll Trajectory** command in the **View** menu to toggle this scrolling. (Technically, scrolling can be made the default with the **Save Defaults** command.) Scrolling the traces takes considerably more time.

Impact Data ("Untitled")

This text window contains, for each impact, the sequential number of the impact, the outgoing ball velocity, the table phase, and the total time the simulation has run. Bouncing Ball provides no way to control the precision used to display the numbers in this window.

When a base simulation is running, text is being continuously created, so Bouncing Ball does not allow you to scroll through or select text. When the iteration is stopped, however, you may scroll through the text, make selections, and copy information to the Clipboard.

When a simulation is saved or opened, this window takes on the name of the corresponding file.

When approximately 32K of text is created (approximately 850 solutions), most of the text is eliminated from the window (but still stored and able to be saved), and an appropriate message appears at the beginning of the text. The preceding text may be viewed by playing back the solutions (**Play Back Solutions** from the **Control** menu) or by saving it and viewing the saved (text) file in a word processor or other application.

Impact Map

In the impact map, Bouncing Ball plots the outgoing ball velocity versus the table phase for each impact. The map displays the entire phase range on the horizontal axis, unless the plot is magnified. Vertically, the minimum velocity is equal to the minimum table velocity. The maximum velocity is set by default to either a "trapping region" value, or the initial ball velocity that would just barely take the ball's trace from the maximum table amplitude to the top of the *Trajectory* window in its largest position, whichever is smaller. This maximum velocity can be adjusted with the **Max. Velocity** command. Any of these values may be changed (and later restored) by magnifying the impact map, as described below and in the description of the **Impact Map Magnification** command.

Special Features: Clicking inside the impact map brings up the **Initial Conditions** dialog box with values corresponding to the point of the click. Dragging inside the impact map brings up the **Impact Map Magnification** dialog box, with values corresponding to the dragging endpoints. Hold the Shift key down to bypass the dialog box.

Animation

The ball and the table are seen in motion in this window. The animation is useful in visualizing the system, although it pauses slightly at each impact due to computational demands. The animation is somewhat smoother when playing back solutions than when generating them in the first place. It can also be made smoother by slowing down the simulation, with the **Simulation Speed** command.

The vertical scaling is the same as that used in the *Trajectory* window.

Phase Space

As the ball moves, this window displays its height versus its velocity. Note that the phase space is graphed only when this window is open.

The vertical scale is the same as that used in the *Trajectory* window. The horizontal scale is such that, when the window is at its largest size (zoomed out), it displays from the negative of the maximum velocity to the maximum velocity, with the maximum velocity determined as in the *Impact Map* window. For instance, if the maximum velocity has been set to 10, the horizontal axis will display from -10 to 10, when the window is at its largest size. Note in particular that, if the window is not at its largest horizontal size, it will display a smaller velocity range.

Impact Phases

This window contains an "embedded phase space," and looks and acts somewhat like the *Impact Map* window. It plots the phase of the current impact on the vertical axis versus the phase of a previous impact. Initially Bouncing Ball looks at the immediately preceding impact; the **Impact Phases Offset** command (**View** menu) allows you to specify which prior impact will be charted.

Unless magnified, the impact phases chart displays the entire phase range, as specified by the **Phase Range** command, on both axes.

Special Feature: Dragging inside the impact phases chart brings up the **Impact Phases Magnification** dialog box, with values corresponding to the dragging endpoints. Hold the Shift key down to bypass the dialog box.

Time Series

This window generates simulation data sampled at a user-specified number of times per table period. It generates the simulation time, the ball velocity, and the ball position above the lowest table position, as graphed in the *Trajectory* and *Phase Space* windows. You can set the sampling rate with the **Time Series** command from the **View** menu. Bouncing Ball provides no way to control the precision used to display the numbers in this window.

Generating time series data slows up the simulation far more than any other display of the data. For this reason, you must tell Bouncing Ball explicitly to begin generating the data, usually with the **Use Time Series** command from the **View** menu, and Bouncing Ball will guess when you wish to stop generating the data. There are thus several dialog boxes that arise in conjunction with this window.

This window acts the same as the *Impact Data* window in terms of scrolling, selection, and behavior when the amount of text approaches 32K.

Bifurcation Diagram

Bouncing Ball uses this window to create and display bifurcation diagrams, as discussed in the section "Types of Simulation." The bottom of the window displays the fixed bifurcation diagram parameters, as set in the **Bifurcation Diagram** dialog box, and the initial conditions and the value of the table parameter being used for the latest iteration.

When a bifurcation diagram is saved or opened, this window takes on the name of the corresponding file.

Special Features: Clicking inside the body of the bifurcation diagram brings up the **Bifurcation Diagram** dialog box. Dragging inside the bifurcation diagram brings up the same dialog box, with range values corresponding to the dragging endpoints (which effectively allows magnification, although, since it cannot magnify existing data, the data must be regenerated). Hold the Shift key down to bypass the dialog box and create the bifurcation diagram for the magnified area.

Basins of Attraction

Bouncing Ball uses this window to create and display basins of attraction diagrams, as discussed in the section "Types of Simulation." The bottom of the window displays the fixed basins of attraction parameters, as set in the **Basins of Attraction** dialog box and the latest initial conditions.

When a basins of attraction diagram is saved or opened, this window takes on the name of the corresponding file.

Special Features: Clicking inside the body of the basins of attraction diagram brings up the **Basins of Attraction** dialog box. Dragging inside the basins of attraction diagram brings up the same dialog box, with range values corresponding to the dragging endpoints (which effectively allows magnification, although, since it cannot magnify existing data, the data must be regenerated). Hold the Shift key down to bypass the dialog box and create the basins of attraction diagram for the magnified area.

Hot Spots

Hot spots are areas in Bouncing Ball windows that may be clicked to execute specific commands. Below is a summary of the hot spots in each window. (Clicking/dragging in the graphing regions of some windows has other effects; see the "Special Features" parts of the window-by-window discussion above.)

Window	Hot spot region	Command
Trajectory	Horizontal axis	**Initial Conditions**
	Vertical axis	**Trajectory Scaling**
Impact Data	Header area	**Table Parameters**
Impact Map	Horizontal axis	**Phase Range**
	Vertical axis	**Max. Velocity**
Animation[*]	Ball	**Initial Conditions**
	Table	**Table Parameters**
Phase Space	Horizontal axis	**Max. Velocity**
	Vertical axis	**Trajectory Scaling**
Impact Phases	Horizontal axis	**Phase Range**
	Vertical axis	**Impact Phases Offset**
Time Series	Header area	**Time Series**
Bifurcation Diagram	Horizontal axis	**Bifurcation Diagram**
	Vertical axis	**Bif. Diagram Phase Range** (if command is active) or **Bifurcation Diagram**
	Whole window (when blank)	**Bifurcation Diagram**
Basins of Attraction	Horizontal axis	**Basins of Attr. Phase Range** (if command is active) or **Basins of Attraction**
	Vertical axis	**Basins of Attraction**
	Whole window (when blank)	**Basins of Attraction**

---

[*] The *Animation* window hot spots only function "during" an impact; the hot spots are located where the ball and table collide.

BOUNCING BALL MENUS

In addition to the requisite **Apple** menu, Bouncing Ball has seven menus. The **File** menu contains the standard commands for file manipulation, printing, and quitting. The **Edit** menu contains the standard Macintosh editing commands (though Bouncing Ball only implements the **Copy** command; the rest are supplied for desk accessory support), and a few commands best described as preferences commands. The **Control** menu contains the commands that allow you to control Bouncing Ball's simulation: starting/stopping it, setting system parameters, and so on. The **View** menu allows you to control how you view the simulation; it contains commands to set scaling, hide old data, and so on. The **Special** menu contains the commands that control Bouncing Ball's special types of simulation, bifurcation diagrams and basins of attraction. The **Sound** menu allows you to control Bouncing Ball's sound. The **Windows** menu has commands to control window size and placement and to open windows.

The remainder of this section describes each of Bouncing Ball's menu commands.

File Menu

New

The **New** command stops the current simulation and clears the data from all windows. The **New** command closes the current file(s), if any.

Open

The **Open** command opens a simulation stored on the disk and displays it on the screen.

Since Bouncing Ball files (from base simulations) are stored as text files, the **Open** dialog box will display all text files. Opening a text file not in Bouncing Ball format will cause an error message and will clear all the windows. Similarly, Bouncing Ball cannot open its own time series files, so if you try to open a time series file, an error message will follow. See the **Save** command for more information on how Bouncing Ball saves simulations.

Check the Graph while opening (if appropriate) box to graph data in all applicable windows while opening. This option only works when a base simulation is being opened. If this box is not checked, data from a base simulation will be displayed in the *Impact Data*, *Impact Map*, and *Impact Phases* windows only. The **Save Defaults** command remembers the current setting of this command as the default setting.

Command-period will abort the opening of a base simulation file.

Opening Bouncing Ball files is somewhat different from standard Macintosh opening. Usually, when one opens a Macintosh file, it appears in *its own* window. Bouncing Ball files open into *pre-existing* windows. With the exception of size and position, these windows maintain their attributes when a new file is opened; think of these windows as tools to inspect whatever data is displayed, tools that are not reset for new data.

Save

The **Save** command saves the current simulation on the disk from which it was opened (or to which it was "saved as"). Bouncing Ball saves the simulation with the name used the last time you saved it, and replaces the old copy of the simulation on the disk. You can only use the **Save** command with simulations that you have previously saved with the **Save As** command; if the simulation has not previously been saved, this command acts like the **Save As** command.

Bouncing Ball decides what to save by looking at the active window. If the active window is the *Time Series* window, it saves the time series data. If the front window is the *Bifurcation Diagram* or *Basins of Attraction* window, and there is a current bifurcation diagram or basins of attraction diagram, respectively, it saves that data. In all other cases, it saves the base simulation data, the data that appears in the *Impact Data* window.

Bouncing Ball saves base simulations and time series data as text files, just as the simulation appears in the corresponding window (including, of course, any data that has scrolled out of the window). The data can thus be easily passed to other applications for further processing. Bouncing Ball also saves appropriate window sizes and positions, in such a way that these will not interfere with the transfer to other applications (this window information is stored in the resource fork of the file). Note that, except for window sizes and positions, only the values appearing in the text window (table parameters—amplitude, frequency, and damping—and solutions) are saved; other program parameters, such as scaling and sound parameters, are not saved.

When you are saving a base simulation or a time series for the first time or are using the **Save As** command, you can use the Delimiter pop-up menu to change the delimiter Bouncing Ball uses to save the file, that is, the character(s) that separate the pieces of data for a given solution or sampling (line) in the text file. Initially Bouncing Ball uses spaces, but it can also use tabs, commas, or carriage returns. The choice of delimiter depends on the application(s) you plan to export the data into. For instance, to export the data to a spreadsheet, tabs or commas are usually the best choice. If you do not plan to export the data to any other application, spaces are probably best. You can set the default delimiter with the **Save Defaults** command.

Bouncing Ball saves special simulations (bifurcation diagrams, basins of attraction) in non-text format. It saves all the parameters involved in the simulation, including the current state thereof, so that, if not completed, it can be continued during another session.

Save As

The **Save As** command saves new simulations or new versions of existing ones. After choosing **Save As**, the standard **Save As** dialog box will appear. If the simulation already has a name, Bouncing Ball proposes it in the text box. To accept the proposed name, click the Save button. To save the simulation under a different name, edit the proposed name or type a new one. Click the Drive and/or Eject buttons to save the simulation on another disk.

Refer to the **Save** command for a discussion of how Bouncing Ball saves files, including a discussion of the Delimiter pop-up menu.

Save Defaults

The **Save Defaults** command saves the current settings of various Bouncing Ball parameters in a default settings file. It also makes the current window configuration the default configuration. It creates a file entitled "Bouncing Ball Defaults" in the current system folder; when Bouncing Ball starts up, it looks in the system folder for this file and uses the configuration stored therein.

The default settings that Bouncing Ball sets are the following: the initial conditions and table parameters; the iteration parameters, the maximum number of table cycles in a base simulation and the number of decimal places of accuracy used; whether to implement the "graph while opening" option for opening saved Bouncing Ball base simulations; the delimiter used when saving base simulations; if running under MultiFinder, whether to run in the background, and, when running in the background, whether to run quickly or slowly, and whether to use sound; whether to suppress save warnings; whether to use the auto ⌘-key feature; the viewed phase range; whether to use the automatic maximum velocity in graphing, and, if not, what maximum velocity to use; how fast to run the base simulation; the scaling parameters for the *Trajectory* window; whether to scroll the *Trajectory* window; the index offset used in the *Impact Phases* window; whether to graph the *Phase Space* using the "faster" option; whether to generate time series data, and the time series sampling rate; for the special simulations, whether to graph in all windows and whether to use color; the pattern and color settings; and all the **Sound** menu settings.

If you'd like to have more than one "defaults file," you can copy or move the "Bouncing Ball Defaults" file out of the system folder (or even just rename it), create others similarly, then move whichever one you want to use into the system folder.

Note: to restore the original defaults, simply move the defaults file out of the system folder.

Page Setup

The **Page Setup** command allows you to select the paper size, paper orientation, and special printing effects for the document. The special printing effects available depend on which printer is currently selected with the Chooser desk accessory.

Print   The **Print** command prints the front window.

                                  Exactly what it prints depends on the window. If the
                                  window is a text window (*Impact Data* or *Time Series*), it
                                  prints all the text (but not text that has been stored
                                  elsewhere—see the discussion of the *Impact Data* window
                                  in the section "Bouncing Ball Windows"), printing on
                                  multiple pages if necessary.

                                  If the front window is a special window (*Bifurcation
                                  Diagram* or *Basins of Attraction*), Bouncing Ball prints
                                  all applicable information, even if the window is sized
                                  such that all information is not visible.

                                  For any other window, Bouncing Ball prints only what is
                                  seen in the window as currently sized.

Quit   The **Quit** command ends a Bouncing Ball session. It will
                                  prompt you to save any unsaved data; hold Option down
                                  when you execute the command for a forced quit—the
                                  program will not prompt you to save data.

Edit Menu

                                  With one exception, Bouncing Ball does not allow
                                  standard Macintosh editing commands; these non-
                                  functional commands will be disabled (grayed) during
                                  Bouncing Ball operation. One has no need to paste
                                  information into or cut information out of windows. The
                                  Copy command does work, however. Also, inside a
                                  dialog box, **Cut**, **Copy**, and **Paste** will work; they can
                                  be called with their Command-key equivalents. All
                                  commands are provided in the menu for compatibility
                                  with desk accessories.

Undo   Provided for desk accessory compatibility only. Disabled
                                  when the front window is a Bouncing Ball window.

Cut   Provided for desk accessory compatibility only. Disabled
                                  when the front window is a Bouncing Ball window.

Copy

For a text window (*Impact Data* or *Time Series*), the **Copy** command copies to the clipboard the current selection.

For a graphics window, the **Copy** command copies to the clipboard exactly what is seen in the window, as currently sized.

Paste

Provided for desk accessory compatibility only. Disabled when the front window is a Bouncing Ball window.

Clear

Provided for desk accessory compatibility only. Disabled when the front window is a Bouncing Ball window.

Suppress Save Warnings

By choosing the **Suppress Save Warnings** command, Bouncing Ball will not warn you when data is about to be lost due to quitting, erasing simulations, or creating new simulations. This command toggles "save-warning suppression"; choosing the command a second time will re-enable save warnings. The current setting of this option can be saved as the default setting with the **Save Defaults** command.

Auto ⌘-key

After choosing the **Auto ⌘-key** command, you can issue menu commands by typing their Command-key equivalents *without holding the Command key down*. For instance, you could issue the **Open** command by typing just "o"; you would not need to type "Command-o." This command toggles this feature; after choosing the command a second time, you will have to hold down the Command key to issue menu commands. The current setting of this option can be saved as the default setting with the **Save Defaults** command.

Background Operation

The **Background Operation** command allows you to tell Bouncing Ball whether to run in the background under MultiFinder. The command will only appear if you are running under MultiFinder. Initially Bouncing Ball does not run in the background, because some of its background processing interferes too much with the foreground application to be considered truly MultiFinder friendly. However, especially given the amount of time it takes to create the special diagrams, background processing can be a real asset.

The **Background Operation** dialog box allows you to specify whether you want Bouncing Ball to run in the background, and whether you want it to run slowly, minimizing its influence on the foreground application, or to run quickly, maximizing Bouncing Ball's processing speed. You also have the option of suppressing sound when Bouncing Ball is running in the background.

Running Bouncing Ball in the background is especially useful for the very time-consuming generation of basins of attraction and bifurcation diagrams.

The background options set here can be stored as the default settings with the **Save Defaults** command.

For more information on running Bouncing Ball in the background, refer to the "MultiFinder and Memory Considerations" part of the section "Tips and Techniques."

Control Menu

Begin Iteration

The **Begin Iteration** command starts a base simulation.

Stop Iteration
Continue Iteration

The **Stop Iteration** command stops a base simulation. After being stopped, the **Continue Iteration** will continue it. In some circumstances the simulation cannot be continued. This happens most frequently when the ball is stuck forever to the table and the simulation stops automatically.

Do One Impact

The **Do One Impact** command starts or continues a base simulation, whichever is appropriate, finds and graphs one impact, and stops the simulation.

Play Back Solutions

The **Play Back Solutions** command plays back the current simulation from the beginning. Because the collisions (roots of the equations of motion) do not have to be calculated again, playing back solutions is noticeably faster than generating them. It also tends to generate smoother graphics, especially in the *Animation* window.

Initial Conditions

The **Initial Conditions** command allows you to enter new initial conditions for the simulation. Check the (Re)Start Iteration box if you wish the iteration to begin after you click the OK button. Check the Clear Impacts box if you wish to clear the points plotted so far in the *Impact Map* and *Impact Phases* windows; doing this will in effect issue the **Clear Impacts** command from the **View** menu. The current initial conditions can be made the default ones with the **Save Defaults** command.

Table Parameters

The **Table Parameters** command allows you to change the table parameters—amplitude, frequency, and damping—of the simulation. Check the (Re)Start Iteration box if you wish the iteration to begin after clicking the OK button. Changing the table parameters will eliminate all data previously generated. The current table parameters can be made the default ones with the **Save Defaults** command.

Get Initial from Last

The **Get Initial from Last** command makes the outgoing ball velocity and phase of the last collision between the ball and the table the new initial conditions, bringing up the **Initial Conditions** dialog box.

Iteration

The **Iteration** command allows you to set the decimal places of accuracy used in the iteration, and the maximum number of cycles. The decimal places of accuracy controls the accuracy used to find the roots of the system, the points of collision—the fewer the decimal places, the less accurate but faster the iteration. It also controls how quickly the program decides that the ball is sticking to the table; the higher the accuracy, the longer it takes to decide that the ball is sticking. The maximum number of cycles determines the maximum number of table oscillations that may take place before the iteration ends. The current iteration parameters can be made the default ones with the **Save Defaults** command.

When creating bifurcation diagrams or basins of attraction diagrams, Bouncing Ball insists on at least ten decimal places of accuracy, overriding the current accuracy if necessary. The routines that determine when a periodic solution has been reached require such a level of accuracy.

Note that, because the Bouncing Ball system displays sensitive dependence on initial conditions, different accuracies can lead the same initial conditions to qualitatively different solutions.

View Menu

Phase Range

When you first start Bouncing Ball, it graphs phases (in the *Impact Map* and *Impact Phases* windows) from 0 to 1. While natural, this range is somewhat arbitrary. Often it is helpful to look at phases in another range, say, –0.5 to 0.5. The **Phase Range** command allows you to display a phase range of anywhere from –1 to 0, to 0 to 1, in increments of 0.01. Two buttons, entitled "–.5 – .5" and "0 – 1", allow instant selection of these common ranges. The **Save Defaults** command will save the current phase range as the default one.

Max. Velocity

The **Max. Velocity** command allows you to specify the maximum velocity graphed, in the *Impact Map* and *Phase Space* windows. Click Auto to use the default value, which is described in the discussion of the *Impact Map* window in the section "Bouncing Ball Windows." The setting set here can be made the default settings with the **Save Defaults** command.

Simulation Speed

The **Simulation Speed** command allows you to slow down the simulation so that the graphing in the *Trajectory*, *Animation*, and *Phase Space* windows can be watched more easily. The dialog box this command brings up gives you a scroll bar on which you can set the speed, from slow to fast. The slowest speed corresponds to approximately one table period every ten seconds. If the fastest speed is selected, the speed is limited only by the speed of the computer. The **Save Defaults** command will remember the last speed selected as the default speed.

Trajectory Scaling

The **Trajectory Scaling** command allows you to set the "number of table periods in full screen" and the "number of table amplitudes in full screen" in the *Trajectory* window. The number of table periods in full screen represents the number of periods that would be seen horizontally in the *Trajectory* window in its largest (zoomed out or "full screen") position. The number of table amplitudes in full screen represents the number of table amplitudes that would be seen in the *Trajectory* window in "full screen position." The latter scaling is also used in the *Animation* and *Phase Space* windows, as well as to determine a lower limit on the default maximum velocity shown in the *Impact Map* window. The **Save Defaults** command saves the current trajectory scaling parameters as the default ones.

Scroll Trajectory

Under the default settings, the graph in the *Trajectory* window clears itself and starts over when the traces reach the right edge. With the **Scroll Trajectory** command, you can tell Bouncing Ball to scroll the *Trajectory* window when it reaches the edge. This is a toggle command; successive executions of the command turn the scrolling off and on. While the scrolling has a nice effect, it tends to slow down graphing in that window by approximately a factor of four, even when the *Trajectory* window is closed, which can slow down the whole simulation by up to a factor of two, depending on which windows are open. The **Save Defaults** command makes the current value of this option the default value.

Clear Impacts

The **Clear Impacts** command clears the data in the *Impact Map* and *Impact Phases* windows, so that only subsequent impacts are displayed there. Use this command if the windows become too cluttered. To redisplay the impacts that were calculated prior to this command, choose **Show All Impacts** from this menu.

Show All Impacts

The **Show All Impacts** command undoes the effect of the **Clear Impacts** command above; all solutions that had been generated since the beginning of the simulation will be displayed in the *Impact Map* and *Impact Phases* windows.

Impact Phases Offset

The **Impact Phases Offset** command allows you to enter an integer that determines which previous impact phase the current impact phase is graphed against in the *Impact Phases* window; refer to the discussion of the window in the section "Bouncing Ball Windows." The current setting of this variable can be made the default setting with the **Save Defaults** command.

Impact Map
Magnification

The **Impact Map Magnification** command allows you to view, in the *Impact Map* window, only a portion of the entire impact map, gaining greater detail. You specify minimum and maximum values for both phase and velocity. Press any of the Auto boxes for a corresponding default value.

Click the NO MAGNIFICATION button to use the normal (unmagnified) view.

A convenient way to call up this command is by dragging around the area to be magnified in the *Impact Map* window. Doing this will bring up the **Impact Map Magnification** dialog box with appropriate endpoints.

Impact Phases
Magnification

The **Impact Phases Magnification** command allows you to view, in the *Impact Phases* window, only a portion of the entire impact phases chart, gaining greater detail. You specify minimum and maximum values for each phase axis. Press any of the Auto boxes for a corresponding default value.

Click the NO MAGNIFICATION button to use the normal (unmagnified) view.

A convenient way to call up this command is by dragging around the area to be magnified in the *Impact Phases* window. Doing this will bring up the **Impact Phases Magnification** dialog box with appropriate endpoints.

Faster Phase Space

The **Faster Phase Space** command noticeably speeds up graphing in the *Phase Space* window, at the expense of lessened accuracy. The **Save Defaults** command makes the current setting of this option the default one.

Clear Phase Space

The **Clear Phase Space** command clears the data in the *Phase Space* window. The cleared data cannot be restored (though it could be recreated).

Use Time Series

The **Use Time Series** command allows you to turn on and off the generation of time series data. If the *Time Series* window is not open and you ask to turn on the data, you will be asked if you want to open the window. The **Save Defaults** command makes the current setting of this option the default one.

Time Series

With the **Time Series** command, you can set the sampling rate of the time series data, in sampling times per period. Click the Generate Times Series Data box to start generating time series data. If you request that data be generated, and the *Time Series* window is closed, you will be asked if you want to open the window. The current sampling rate can be made the default one with the **Save Defaults** command.

Clear Time Series

The **Clear Time Series** command clears the data in the *Time Series* window. The data cannot be restored (though it could be recreated).

Special Menu

Start Bifurcation
  Diagram

The **Start Bifurcation Diagram** command begins the creation of the current bifurcation diagram. If the *Bifurcation Diagram* window is not open, you will be asked if you want to open the window.

Stop Bifurcation
  Diagram
Continue Bifurcation
  Diagram

While a bifurcation diagram is running, the **Stop Bifurcation Diagram** command stops it. **Continue Bifurcation Diagram** continues a bifurcation diagram that has been stopped before completion. If you issue the **Continue Bifurcation Diagram** command and the *Bifurcation Diagram* window is not open, you will be asked if you want to open the window.

Bifurcation Diagram

The **Bifurcation Diagram** command brings up the dialog box that allows you to set all the parameters of a bifurcation diagram. Please refer to the section "Types of Simulation" for a discussion of this dialog box. If you use this command to start a bifurcation diagram and the *Bifurcation Diagram* window is not open, you will be asked if you want to open the window.

Bif. Diagram Phase Range

The **Bif. Diagram Phase Range** command allows you to set the phase range displayed in an existing bifurcation diagram. This command is only active if there is a current bifurcation diagram, if phase is being graphed on the vertical axis (not velocity), and if the range of phases shown on the phase axis is the full range (whether from 0 to 1, –0.5 to 0.5, or any other range of extent 1.0).

The dialog box this command brings up works the same as the dialog box of the **Phase Range** command.

So that the graphs are drawn accurately, the phase range chosen is rounded so that it falls on an exact pixel boundary; thus a specified range of –0.31 to 0.69 might create an actual range of –0.3072884... to 0.6927116.... The latter is the range that would be shown in a subsequent call to the **Bifurcation Diagram** command.

Clear Bifurcation Diagram

The **Clear Bifurcation Diagram** command clears the *Bifurcation Diagram* window and resets all bifurcation diagram parameters. Because this command is destructive, you are given the option of saving the current bifurcation diagram first.

Graph in All Windows

The **Graph in All Windows** command instructs Bouncing Ball to toggle whether it displays the progress of the bifurcation diagram in all the windows. In constructing the bifurcation diagram, Bouncing Ball actually runs many base simulations; one can watch these simulations as the bifurcation diagram progresses. On the other hand, often one will not want to watch these simulations, as they significantly increase the time required to complete the bifurcation diagram. Use the **Save Defaults** command to make the current setting of this option the default setting.

Use Color

The **Use Color** command toggles whether the current bifurcation diagram is displayed in color. This menu item will be disabled (grayed) if the current monitor is not a color one with at least 4-bit color (16 colors) enabled. *Note*: If the monitor is set to display gray scale, at 4 bits or more, this option will be enabled, and, instead of colors, the bifurcation diagram will be displayed in shades of gray. Use the **Save Defaults** command to make the current setting of this option the default setting.

Start Basins of
   Attraction

The **Start Basins of Attraction** command begins the creation of the current basins of attraction diagram. If the *Basins of Attraction* window is not open, you will be asked if you want to open the window.

Stop Basins of
   Attraction
Continue Basins of
   Attraction

While a basins of attraction diagram is running, the **Stop Basins of Attraction** command stops it. **Continue Basins of Attraction** continues a basins of attraction diagram that has been stopped before completion. If you issue the **Continue Basins of Attraction** command and the *Basins of Attraction* window is not open, you will be asked if you want to open the window.

Basins of Attraction

The **Basins of Attraction** command brings up the dialog box that allows you to set all the parameters of a basins of attraction diagram. Please refer to the section "Types of Simulation" for a discussion of this dialog box. If you use this command to start a basins of attraction diagram and the *Basins of Attraction* window is not open, you will be asked if you want to open the window.

Basins of Attr. Phase
   Range

The **Basins of Attr. Phase Range** command allows you to set the phase range displayed in an existing basins of attraction diagram. This command is only active if there is a current basins of attraction and if the range of phases shown on the phase axis is the full range (whether from 0 to 1, –0.5 to 0.5, or any other range of extent 1.0).

The dialog box this command brings up works the same as the dialog box of the **Phase Range** command.

So that the graphs are drawn accurately, the phase range chosen is rounded so that it falls on an exact step size boundary; thus a specified range of –0.31 to 0.69 might create an actual range of –0.3072884... to 0.6927116.... The latter is the range that would be shown in a subsequent call to the **Basins of Attraction** command.

Clear Basins of Attraction

The **Clear Basins of Attraction** command clears the *Basins of Attraction* window and resets all basins of attraction parameters. Because this command is destructive, you are given the option of saving the current basins of attraction diagram first.

Graph in All Windows

The **Graph in All Windows** command instructs Bouncing Ball to toggle whether it displays the progress of the basins of attraction diagram in all the windows. In constructing the basins of attraction diagram, Bouncing Ball actually runs many base simulations; one can watch these simulations as the basins of attraction diagram progresses. On the other hand, often one will not want to watch these simulations, as they significantly increase the time required to complete the basins of attraction diagram. Use the **Save Defaults** command to make the current setting of this option the default setting.

Use Color

The **Use Color** command toggles whether the current basins of attraction diagram is displayed in color or with patterns. This menu item will be disabled (grayed) if the current monitor is not a color one with at least 4-bit color (16 colors) enabled. *Note*: If the monitor is set to display gray scale, at 4 bits or more, this option will be enabled, and, instead of colors, the basins of attraction will be displayed in shades of gray. Use the **Save Defaults** command to make the current setting of this option the default setting.

Patterns

The **Patterns** command allows you to specify the gray patterns used to represent the different types of solutions in the basins of attraction diagram, when color is off. Use the **Save Defaults** command to save the current pattern-solution relationship as the default one.

Colors

The **Colors** command allows you to specify the colors used to represent the different types of solutions in the basins of attraction diagram and the bifurcation diagram, when color is on. Use the **Save Defaults** command to save the current color-solution relationship as the default one.

Print Pattern List

The **Print Pattern List** command prints a legend of the patterns representing the different types of solutions.

Sound Menu

There are actually two **Sound** menus, one containing "simple options," the other "advanced options." The simple options only allow you to turn sounds off and on, while the advanced options allow you some control over the pitches of the sounds.

NOTE: To turn off all sound, eliminate the checks next to the **Phase Tones** command and the **Velocity Tones** command, if any, by choosing the appropriate command.

The **Save Defaults** command will remember all the current settings of this menu as the default settings.

Phase Tones
Velocity Tones

The **Phase Tones** and **Velocity Tones** commands toggle Bouncing Ball's sound. With "phase tones" on, Bouncing Ball creates, for each impact, a sound corresponding to the phase of that impact. With "velocity tones" on, Bouncing Ball creates a sound corresponding to the ball's velocity after impact, in relation to the minimum and maximum velocity values being used, as shown in the *Impact Map* window (unmagnified). See the other commands from this menu for descriptions of the sounds created. Initially, phase tones are on, velocity tones off.

Advanced Options

The **Advanced Options** command adds the following commands to the **Sound** menu.

One Octave

[Using advanced options] With the **One Octave** command, the frequency of the sound(s) Bouncing Ball creates (for each collision) ranges from 256 Hz to 512 Hz as the phase of the collision ranges from the low end of the phase range (as set by the **Phase Range** command) to the high end, or as the ball velocity after impact ranges from the minimum to the maximum velocity shown in the *Impact Map* window. See the **Phase Tones** and **Velocity Tones** commands above to turn sound on and off; see below for a description of how the sound frequencies are calculated.

Three Octaves

[Using advanced options] With the **Three Octaves** command, Bouncing Ball first calculates the frequency that would be generated using the **One Octave** command above. Then, for phase tones only, Bouncing Ball lowers that frequency an octave if the successive collisions are less than one period apart and raises it an octave if the successive collisions are more than two periods apart. The total range of tones is, therefore, three octaves.

Major
Minor
Chromatic
Continuous

[Using advanced options] These four commands determine the exact frequency of sound created by Bouncing Ball. The **Major** command divides the appropriate range (phase from 0 to 1, velocity from the minimum to the maximum velocity, as shown in the *Impact Map* window) into seven equal areas, corresponding to the seven tones of the major scale. The **Minor** command acts similarly, dividing the range into areas corresponding to the seven tones of the minor scale. The **Chromatic** command divides the range into twelve areas, corresponding to the twelve tones of the chromatic scale. With the **Continuous** command, Bouncing Ball creates a sound that ranges logarithmically from 256 to 512 Hz as the phase ranges from 0 to 1 or as the velocity ranges from the minimum to the maximum velocity.

Simple Options

The **Simple Options** command removes from this menu all commands except the first two, and adds the **Advanced Options** command, which can in turn be used to restore the removed commands.

Windows Menu

After the first three commands, which are described below, the **Windows** menu lists all of Bouncing Ball's windows, placing a diamond next to the windows that are open. Choosing one of these menu items brings the named window to the front, making it visible if necessary.

Close Window
Close All Windows

The **Close Window** command closes the front window. If the Option key is depressed, the command becomes **Close All Windows** and will close all open windows (holding down Option while clicking in the front window's close box will do the same thing). If there are no open windows, this command is grayed.

Arrange Windows

The **Arrange Windows** command arranges the currently-open windows in a standard format. If there are no open Bouncing Ball windows, this command is grayed.

Default Configuration

The **Default Configuration** command puts Bouncing Ball's windows in the configuration stored in the "Bouncing Ball Defaults" file in the current system folder (see **Save Defaults** command). If there is no such file, it sizes the *Trajectory*, *Impact Data*, *Impact Map*, and *Animation* windows equally, and puts one into each quarter of the screen.

TIPS AND TECHNIQUES

This section contains a series of tips and techniques to make your use of Bouncing Ball easier and more enjoyable.

Speeding Up the Simulation

Closing any or all of the following windows will speed up the simulation, with the windows listed first helping the most: *Time Series* (or just turning off the time series generation), *Phase Space*, *Animation*, and *Trajectory*. If you wish to keep the *Phase Space* window open, turning on the "Faster Phase Space" option (**Faster Phase Space** command, **View** menu) will help. Other techniques to speed up the simulation include turning the sound off and turning off trajectory scrolling (which initially is off—see the **Scroll Trajectory** command, **View** menu). Of course, if the simulation has been slowed with the **Simulation Speed** command (**View** menu), you can speed it up with that command. While opening base simulation files, leave the Graph while opening box unchecked; if it is checked, Bouncing Ball will maintain all windows as it opens, which can slow the operation down tremendously.

Bouncing Ball's special simulations, being composed as they are of many base simulations, tend to take a lot of time under any circumstances, especially with a slow Macintosh. Six hours is not at all uncommon for a bifurcation diagram, even more time for a basins of attraction diagram. To generate special simulations as fast as possible, do not select the appropriate "graph in all windows" command and turn the sound off. In addition, with basins of attraction diagrams, using a larger step size will speed it up tremendously, albeit with diminished resolution. Using the "Faster" option will also speed it up tremendously, especially if you include the actual attractors in the basins of attraction graph range. Finally, if you are using a color system, creating basins of attraction in color is a little slower than in black and white (patterns); you can create them in black and white and later display them in color.

MultiFinder and Memory Considerations

Bouncing Ball can run in the background under MultiFinder, but only if you explicitly instruct it to, with the **Background Operation** command from the **Edit** menu. See the description of that command for more information.

There are several considerations that should be given to memory usage with Bouncing Ball under MultiFinder. First of all, if you are not familiar with memory partitioning under MultiFinder, it is set with the Get Info command in the Finder; please refer to your MultiFinder manual for more information. Bouncing Ball's default memory partition is 900K. That should be sufficient to run all three types of simulations, and to create base simulations of a couple thousand iterations or more.

Of that 900K, about 250K is used to create basins of attraction diagrams. While generating these diagrams, Bouncing Ball maintains many off-screen bitmaps, which consume large amounts of memory. If you do not create basins of attraction diagrams, you can safely run Bouncing Ball with 600K; as little as 512K would, in fact, probably be sufficient.

If you are using a larger monitor or multiple monitors, you should increase the default partition size by the following amount: Multiply the number of pixels the larger monitor displays horizontally by the number of pixels it displays vertically to arrive at its resolution, its total number of pixels. In the case of multiple monitors, use the total numbers of pixels that would be displayed by a large monitor just barely encompassing all your monitors (as arranged with the Monitors cdev in the Control Panel). Subtract 200,000 from this resolution and multiply the resulting number by 0.002 (0.001 if you will not create any basins of attraction diagrams). The result of this calculation represents the additional memory (in "K") that should be added to Bouncing Ball's memory partition. The calculation is reproduced below:

Additional Memory = (Resolution – 200,000) * 0.002 (0.001 if no basins of attraction).

In addition, if you wish to create large base simulations (> 2000 iterations), you should increase the default partition size by 80K for each 1000 iterations.

Of course, depending on the memory available in your computer, the above calculations may imply certain restrictions on the operation of Bouncing Ball; for example, creating a basins of attraction diagram at full window size on a Macintosh with a large monitor but only 1 MB of memory, running without MultiFinder, should not be attempted.

One other consideration: If you open a bifurcation diagram created and sized in a window larger than your current monitor, Bouncing Ball may create "memory islands," which can cause it to be unable to fully use the available memory. Beware of this if you create special files on large monitors, then open them on small monitors. By the way, the enlarged memory values given above would partially apply to such situations. Thus, even though you may be using a small Macintosh monitor, you will require much of the memory of a monitor large enough to display the entire large-sized special file.

Bouncing Ball and Other Programs

Bouncing Ball can export data in several ways. Graphics and text can both be exported using the Copy command and the Clipboard. Files saved from Bouncing Ball's *Impact Data* or *Time Series* windows, being text files, can be easily shared with many programs. The Save As command allows you to specify the data delimiter, to ease data transfer. Bifurcation diagram and basins of attraction files, on the other hand, use a proprietary format. The graphics in these files are stored as PICT resources; an ambitious user could use a resource editor to access these graphics and transfer them to another application, perhaps to gain greater color control, or to make color separations.

For the most part, it is not possible to import data into Bouncing Ball. Technically, Bouncing Ball can try to open any text file, and if that file contains data in Bouncing Ball format, it will open it. If the data is not actual Bouncing Ball data, however, this is likely to result in erratic performance.

Bouncing Ball appears to coexist peacefully with all popular INITs, cdevs, and desk accessories. If you begin a simulation and leave the computer alone, and if you use a screen saver, the screen saver will blank the screen while Bouncing Ball is creating the simulation. Bouncing Ball continues to create data after the screen saver is activated. When the screen saver is deactivated, Bouncing Ball will properly display all the data.

Miscellaneous

Warning or confirmation dialog boxes are generated by certain commands. For example, several time series commands and special commands trigger dialog boxes. For many of these commands, holding the Shift key down when executing the command will bypass the appropriate dialog box, automatically making the default response. For some of these, holding the Option key down when executing the command will effectively answer "No" to the dialog box. For instance, if you hold Shift down while executing the **Quit** command, the program will automatically save all current data before quitting, while if you hold Option down, the program will quit while neither prompting you to save data nor saving data. Shift has priority over Option; if both are held down, the Option key will be ignored. Either Shift or both Shift and Option work in this fashion with the following commands: **New**, **Open**, **Quit**, **Table Parameters**, **Get Initial from Last**, **Clear Bifurcation Diagram**, **Clear Basins of Attraction**, both **Graph in All Windows** commands, commands that start or continue basins of attraction or bifurcation diagrams, and several time series-related commands. Shift also works when magnifying in the *Impact Map*, *Impact Phases*, *Bifurcation Diagram*, and *Basins of Attraction* windows (to bypass the corresponding dialog box), and when clicking initial conditions in the *Impact Map* window.

In any dialog box, pressing Command-letter will have the effect of clicking the "first" control (button, radio button, check box) whose title begins with that letter. The order of these controls is determined internally; in general, the OK (default) and Cancel (or analogous) buttons come first, followed by the others in order from top to bottom, left to right. In a dialog box with no text input (edit text) fields, pressing a letter without Command will click the first radio button or check box whose title begins with that letter. Command-period or Escape will cancel dialog boxes. Tab takes you forward through the text input fields; Shift-Tab takes you backward through these fields.

Command-period (⌘-.) will stop the current simulation, whether a base simulation, a bifurcation diagram, or a basins of attraction diagram. It will also stop the file-opening of a base simulation.

In the **Phase Range** dialog boxes, pressing the Left and Right Arrow keys decrements/increments the phase range by 0.01; pressing the Up and Down keys decrements/increments by 0.1. Similarly, in the **Simulation Speed** dialog box, pressing the Left and Right Arrow keys moves the scroll bar's "thumb" by a small amount in the corresponding direction; pressing the Up and Down keys moves the thumb by a larger amount.

For Advanced Users: If you wish to change the patterns used to display the ball, the table, and the table's frame in the *Animation* window (default: black and 50% gray, respectively), you can use a resource editor to modify the first three patterns, respectively, in 'PAT#' ID 132 (entitled "Ball, Table").

Program Problems

Perhaps Bouncing Ball's biggest problem is that it only handles mouse clicks and other input at impact time. For instance, while a single trajectory is being shown in the *Animation* window, mouse clicks will not be acted upon. They are stored up and are handled at the time the impact is made. This problem is accentuated when the ball bounces high off the table.

Bifurcation diagrams and basins of attraction diagrams may not print properly on an Apple LaserWriter SC™ with some versions of the printer driver. If you use the latest version of the printer driver, they should print fine.

While using color, in particular on a basins of attraction diagram, the Macintosh's cursor may disappear for a few seconds. Don't be alarmed—it will come back.

If you find any bugs or other program problems, please notify the publisher.

SELECTED ERROR MESSAGES

Most of Bouncing Ball's error messages should be self-explanatory. Below are a few error messages that may require additional explanations and/or suggestions.

Could not load solution patterns. Will use smaller set.

Bouncing Ball was unable to read the patterns it uses in the basins of attraction diagram from its own resource fork ('PAT#' ID 131, "Eight Grays"). It will substitute the standard 100% (black), 75%, 50%, 25%, and 0% (white) grays the Macintosh automatically provides (note that this procedure yields only five different grays, less than the usual eight; however, in general this will be enough to give each type of solution found a different pattern). If you are so inclined, you could use a resource editor to inspect that resource; there should be eight progressively lighter patterns in it.

Initial velocity less than table velocity

When starting an iteration, the ball must start out going as fast as or faster than the table; otherwise, the ball would sink *through* the table. You may get this error message at times when, analytically, the ball and the table should be traveling at exactly the same speed. If this happens, just increase the initial ball velocity by a small amount.

Not a Bouncing Ball file

The text file Bouncing Ball is trying to open does not have a valid Bouncing Ball header (which contains the amplitude, frequency, and damping coefficient). This could be because the file has nothing to do with Bouncing Ball, since Bouncing Ball's **Open** dialog includes all text files—refer to the discussion of the **Open** command. If the file should be a proper Bouncing Ball file, you could use a word processor to look at the file, compare it to a file that Bouncing Ball can open, make any required changes, save it as a text file, then open it inside Bouncing Ball again.

Root finding error; Bouncing Ball will assume sticking solution
Root finding error: no zero on interval

Either of these messages means that Bouncing Ball was unable to find a solution for the current iteration. Ideally these messages will never come up; if they do, it is likely due to a problem elsewhere in the program. The first message appears if you are running a bifurcation diagram or a basins of attraction diagram; in this case, the program assumes it has found a sticking solution (this error is most likely to occur in a sticking region, where the bounces are small) and proceeds just as if it had found a normal sticking solution. The second message appears while running a base iteration; in this situation the iteration will stop and may not be continued (though it could be restarted).

Sorry—not enough memory to run Bouncing Ball

If you are running under MultiFinder, this command means that Bouncing Ball's memory partition is too small. If not, the Macintosh being used does not have sufficient memory to run Bouncing Ball. Refer to the discussion of memory usage in the section "Tips and Techniques."

Unable to open entire file

One of the solutions in the file Bouncing Ball is opening is not in valid format. Each solution should consist of the hit number (an integer) and three floating point numbers, the phase, the velocity, and the time. To correct this problem, inspect the file with a word processor, looking where Bouncing Ball stopped loading the data, and try to find an invalid number, or an incomplete line, or something of that nature. If you are able to find and correct the problem, save it as a text file and re-open it in Bouncing Ball.

Bouncing Ball Index

Quadratic Map User's Guide

TABLE OF CONTENTS

INTRODUCTION

Quadratic Map is a program written for the Apple® Macintosh™ computer that iterates the quadratic map

$$x_{n+1} = \lambda x_n (1 - x_n).$$

Quadratic Map 1.0 requires the use of a Macintosh with at least 512 kilobytes (512K) of memory, at least the 128K version of ROM (all Macintoshes except the original 128K Macintosh and the original Macintosh 512 have such a version), and the 6.0 System set (System 6.0 and Finder 6.1) or later.

This manual explains how to use Quadratic Map. Before you start, you should already know how to perform basic operations with your Macintosh. You should know how to use the Finder™ and the mouse, manipulate windows, scroll, pull down menus, and choose commands. If you need information about any of these topics, consult your owner's guide.

This manual has six main sections:

- Getting Started
- Types of Iteration
- Quadratic Map Windows
- Quadratic Map Menus
- Tips and Techniques
- Selected Error Messages

Getting Started shows you how to create iterations and describes the sample files that accompany Quadratic Map.

Types of Iteration describes the two types of iterations Quadratic Map can run.

Quadratic Map Windows describes the windows Quadratic Map uses to display data from the iterations.

Quadratic Map Menus contains a directory of all Quadratic Map menu commands, arranged in the order they appear on the pull-down menus.

Tips and Techniques describes various ways to tailor your use of Quadratic Map.

Selected Error Messages lists the error messages that Quadratic Map generates and describes what to do when they appear.

GETTING STARTED

1. Start the Macintosh by turning it on. If you do not have a hard disk, insert a Macintosh system disk.

2. Insert the Quadratic Map disk and open it, if it is not already open.

3. Open the Quadratic Map folder.

4. Open the Quadratic Map application.

5. After the program starts up, choose **Initial Condition** from the **Control** menu. Change the initial condition to some other value, or leave it the same, then click the OK button.

The iteration should now be running. Note that there are four windows currently displayed. The upper left window, *Phase Space*, contains a plot of the $n+1$st value versus the nth value. The upper right window, *Untitled* (when you open an iteration or save one, this window takes on the name of the file; it will often be called the *Iteration Data* window), lists the system parameter—lambda—and, for each iteration, the sequential number of the iteration and the iteration value. The lower left window, *Embedded Phase Space*, initially contains a plot of the $n+2$nd value versus the nth value. The lower right window, *Orbit*, plots, for each iteration, the iteration value versus the sequential number of the iteration.

In addition to the initial value, you can control the system parameter lambda by choosing **System Parameter** from the **Control** menu.

For instance, to illustrate the period-doubling route to chaos in the quadratic map system, we increase lambda in successive iterations. First reset the initial condition to the original value, 0.5. The initial value for lambda, which leads to a period one orbit, is 3. To generate a period two orbit, increase lambda to 3.2 by choosing **System Parameter** from the **Control** menu and changing lambda from 3 to 3.2. Now click OK to run the iteration with this new parameter value. Follow the same procedure to increase lambda to 3.9 to produce a chaotic trajectory.

An alternate way to choose a new initial condition is to click in the body of the phase space; the **Initial Condition** dialog box will come up with an initial value corresponding to the point of the mouse-click (as long as the click doesn't hit the 45° line $x_{n+1} = x_n$ or the parabola that forms the basis for the quadratic map). In addition, clicking in certain areas of the various windows calls forth appropriate dialog boxes. These active areas are known as "hot spots." Such hot spots are located along the two axes of the *Phase Space* window. The section "Quadratic Map Windows" discusses these areas further.

Now we'll look at Quadratic Map's other type of iteration, which is based on the iteration described above.

Bifurcation Diagram

To create a *bifurcation diagram*, follow these steps:

1.   Use the **Initial Condition** command to restore the original initial value, 0.5.

2.   Use the **System Parameter** command to change the value of the system parameter lambda to 2.

3.   Clear the current data by choosing **New** from the **File** menu.

4.   Open the *Bifurcation Diagram* window by choosing **Bifurcation Diagram** from the **Windows** menu.

5.   Choose **Bifurcation Diagram** from the **Special** menu and click OK.

The program will generate a diagram showing how the solutions to the Quadratic Map system vary as the system parameter lambda varies. The bifurcation diagram stops automatically after it has moved through the specified lambda range.

Sample Files

To help you quickly see some of its capabilities, Quadratic Map is packaged with the following sample files. You can either double-click on these sample files or open them with the **Open** command in the **File** menu.

Period 2

An iteration that converges to a period two solution; $\lambda = 3.2$

Period 16 (lengthy)

An iteration that converges to a period sixteen solution; $\lambda = 3.568$ (it contains 700 iterations, so it takes a while to open, especially on slower Macintoshes)

Strange Attractor (lengthy)

An example of chaos; $\lambda = 4$ (it contains 1000 iterations, so it takes a while to open, especially on slower Macintoshes)

Bifurcation Diagram 0–4

A bifurcation diagram in which lambda is incremented from 0 to 4

Bifurcation Diagram 2.9–4

A bifurcation diagram that essentially magnifies the "period-doubling route to chaos" shown in the first bifurcation diagram file

Types of Iteration

Quadratic Map can perform two types of iteration. The central iteration to Quadratic Map is the straightforward one described in the beginning of the section "Getting Started," wherein you set the system parameter and the initial condition, then watch the iteration in any combination of four windows. This type of iteration will be called the "base" iteration.

The other type of iteration, sometimes referred to as a "special" iteration, is the bifurcation diagram. It is based on the base iteration and is displayed in its own window.

Base Iteration

The base iteration is a computer simulation of a logistic map iteration. Using the **Initial Condition** and **System Parameter** menu commands, you set the parameters, then let the computer iterate the solutions.

As the iteration progresses, you can watch the data displayed in the *Phase Space*, *Iteration Data* ("*Untitled*" at the beginning), *Embedded Phase Space*, and *Orbit* windows. The section "Quadratic Map Windows" describes how each of these windows displays the data.

Bifurcation Diagram

In the *Bifurcation Diagram*, you select a *range* of values for the system parameter lambda. The program then starts the Quadratic Map system at the beginning of this range. It lets the base iteration run for a certain number of iterations (to eliminate the "transient" solution), then charts the values of the remaining iterations on the vertical axis versus lambda on the horizontal axis. It then increments lambda (by a value corresponding to one screen pixel) and repeats the process, until it reaches the end of the parameter range.

WARNING: The bifurcation diagram runs fairly slowly, especially on slow Macintoshes.

What one finds in the bifurcation diagram is that, as the parameter is changed, the type of solution the Quadratic Map system reaches can change. For instance, in the lambda range of 2–4, with an initial value of 0.5, the system will go through a period one solution, a period two solution, a period four solution, a period eight solution, and so on until it reaches chaos.

To create a bifurcation diagram, choose **Bifurcation Diagram** from the **Special** menu and complete the dialog box as described below.

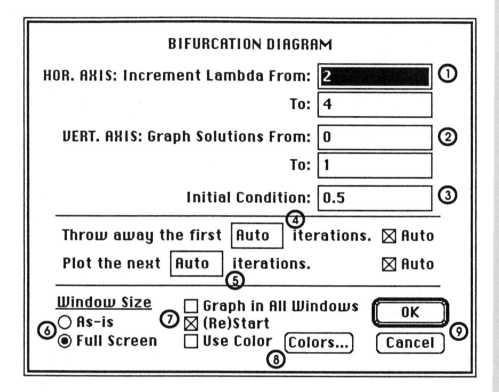

1.   Specify the range of values for the system parameter lambda.

2.   Specify the range of iteration values to be displayed.

3.   Specify the initial value for the first iteration.

4.   Specify the number of iterations you want the iteration to run before charting the data, for each value of lambda. Checking the Auto box puts in the word "Auto" and gives a value based on lambda—higher lambdas give higher numbers of iterations. In addition, with Auto checked, if after reaching the automatic value Quadratic Map determines that it has not reached periodicity, but that it will probably reach periodicity if continued, it will increase this number of initial iterations. Checking Auto tends to give the most accurate results. *Note*: If Quadratic Map determines that periodicity has been reached before reaching the given number of iterations, it will automatically start charting the data.

5.   Specify the number of iterations you want Quadratic Map to chart at each value of lambda. Checking the Auto box puts in the word "Auto." If you choose the

automatic option, Quadratic Map charts several iterations, then charts an additional number of iterations based on the dispersion of those first charted iterations. If, because of a periodic solution, Quadratic Map decides it does not need to display any more solutions, it stops. *Note*: In a chaotic system, in full-screen view on a small Macintosh, the automatic value described above can easily lead to 500 or more iterations, which can be quite time consuming. A value of 100 or so should lead to satisfactory results in most circumstances.

6. Select the window size desired: if As-Is is selected, the size of the bifurcation diagram will be such that it will fit into the *Bifurcation Diagram* window as currently sized; if Full Screen is checked, the bifurcation diagram will take up almost the whole screen (it will just fit into a window that has been "zoomed out"). *Note*: The status of these radio buttons will not physically change the size of the *Bifurcation Diagram* window, but it will affect how the scroll bars of that window, if any, operate.

7. Set these miscellaneous bifurcation diagram parameters:

    a. Graph in All Windows—As remarked above, the bifurcation diagram actually creates a series of base iterations. Check this box if you wish to display the base iterations in all appropriate windows, as described in the "Base Iteration" part of this section. This allows you to see in detail what's happening at each value of the system parameter; however, it greatly slows down the creation of the bifurcation diagram.

    b. (Re)Start—Check if you want to start the bifurcation diagram after clicking OK.

    c. Use Color—Check if you want the bifurcation diagram to display in color. When displayed in color, each type of solution (refer to the "Types of Solutions" discussion of this section for a description of the different solution types) displays in a different color. If you do not have a color monitor, or if it is not set for 4- or more bit color (16 or more colors), this button will be disabled (grayed). *Note*: If your monitor is set for gray-scale, for 4 or more bits, you can implement "color"—it will simply use different shades of gray.

8. If color is being used, pressing the Colors button will bring up the same dialog box as the **Colors** command from the **Special** menu.

9. Press OK to complete the input, Cancel to cancel it.

Types of Solutions

In a bifurcation diagram, Quadratic Map recognizes the following types of solutions:

Unbounded

Period One

Period Two

Period Three

Period Four

Period Eight

Period Sixteen

Other

"Unbounded" solutions are solutions for which the iteration goes to negative infinity. For positive lambda and an initial condition between 0 and 1, lambda must be greater than 4 for this to occur. "Period One" through "Period Sixteen" solutions are just what their names suggest; periodic solutions of the given periods. "Other" solutions, as you might guess, are those that do not fit into any other category. In practice, they are most often chaotic solutions, but can also be other 2^n-periodic solutions, such as period 32, or still other periodic solutions, such as period five. In addition, slow convergence and insufficient iterations can cause Quadratic Map to think that, say, a period two solution is an "other" solution.

QUADRATIC MAP WINDOWS

This section describes the five windows Quadratic Map uses to display data.

Phase Space

For each iteration, Quadratic Map plots in the phase space the value of the current iteration against the value of the previous one. By default the map displays the entire graph range on both axes, as set by the **Graph Range** command. Either range may be changed (and later restored) by magnifying the phase space as described below and in the description of the **Phase Space Magnification** command.

In addition to the points plotted, the *Phase Space* window can also show the parabola the quadratic map is based on (and on which all solutions in this window lie), the 45° line $x_{n+1} = x_n$, the outline of the rectangle that goes from $x_{n+1} = 0$ to $x_{n+1} = 1$ and from $x_n = 0$ to $x_n = 1$, and vertical and horizontal lines that show the path from one solution to the next, alternatively with arrows showing the direction of iteration. Use the **Phase Space** command (**View** menu) to implement these options.

Finally, the window can also display a number line which represents in one dimension the solutions as they are created. Again use the **Phase Space** command to display and set the width of this number line.

Special Features: Clicking inside the phase space will do one of three things. If the click falls close to the parabola defining the quadratic map, it will toggle the display of the parabola. Otherwise, if the click falls close to the 45° line mentioned above, it will toggle the display of that line. If the click hits neither of the above, it will bring up the **Initial Condition** dialog box with a value corresponding to the point of the click. Dragging inside the phase space brings up the **Phase Space Magnification** dialog box, with values corresponding to the dragging endpoints. Hold the Shift key down to bypass the dialog box.

Iteration Data ("Untitled")

This text window contains, for each iteration, the sequential number and the value of the iteration. Quadratic Map provides no way to control the precision used to display the numbers in this window.

When a base iteration is running, text is being continuously created, so Quadratic Map does not allow you to scroll through or select text. When the iteration is stopped, however, you may scroll through the text, make selections, and copy information to the Clipboard.

When an iteration is saved or opened, this window takes on the name of the corresponding file.

When approximately 32K of text is created (approximately 1450 solutions), most of the text is eliminated from the window (but still stored and able to be saved), and an appropriate message appears at the beginning of the text. The preceding text may be viewed by playing back the solutions (**Play Back Solutions** from the **Control** menu) or by saving it and viewing the saved (text) file in a word processor or other application.

Embedded Phase Space

For each iteration, Quadratic Map plots in the embedded phase space the value of the current iteration against the value of a previous one. You can tell Quadratic Map which previous iteration to use (how many iterations back—the number is called the *embedding*) with the **Embedded Phase Space** command in the **View** menu; the first time you start Quadratic Map, it is the second preceding iteration (2 iterations back). Note that the latter value gives it the same display as the *Phase Space* window. By default the map displays the entire graph range on both axes, as set by the **Graph Range** command. Either range may be changed (and later restored) by magnifying the embedded phase space, as described below and in the description of the **Embed. Phase Space Magnification** command.

In addition to the points plotted, the *Embedded Phase Space* window can also show the 2^n-degree polynomial that results from composing the quadratic map a number of times equal to the embedding (all points will fall on this polynomial), the 45° line $x_{n+k} = x_n$, where k equals the embedding, and the outline of the rectangle that goes from $x_{n+k} = 0$ to $x_{n+k} = 1$ and from $x_n = 0$ to $x_n = 1$. Use the **Embedded Phase Space** command to implement these options.

Special Features: Clicking inside the embedded phase space will do one of two things. If the click falls close to the 2^n-degree polynomial described above, it will toggle the display of the polynomial. Otherwise, if the click falls close to the 45° line mentioned above, it will toggle the display of that line. Dragging inside the embedded phase space brings up the **Embed. Phase Space Magnification** dialog box, with values corresponding to the dragging endpoints. Hold the Shift key down to bypass the dialog box.

Orbit

The *Orbit* window shows the value of the iteration as a function of the iteration index n. It contains a series of dots, moving from left to right, that represent the values of the iteration. The bottom of the window displays the initial condition.

The scaling in this window depends on the values set with the **Graph Range** command from the **View** menu. With that command, you set the numerical range shown by the vertical axis of this window; it automatically scales to fit this range into the window. In addition, the dots may be made larger, and space can be added horizontally between the dots, with the **Orbit** command.

When the dots reach the right edge of the graph, by default the graph is cleared and the traces continue from the left edge. If you prefer, the dots can instead smoothly scroll to the left as data is added to the right. Use the **Scroll Orbit** command in the **View** menu to toggle this scrolling. (Technically, you can make scrolling the default with the **Save Defaults** command.) Scrolling takes considerably more time.

Special Feature: Clicking inside the graph area brings up the **Orbit** dialog box. NOTE: This is not technically considered a hot spot, since it does not invert color when clicked, but it functions like one, and is listed parenthetically in the Hot Spots part of this section.

Bifurcation Diagram

Quadratic Map uses this window to create and display bifurcation diagrams, as discussed in the section "Types of Iteration." The bottom of the window displays the fixed bifurcation diagram parameters, as set in the **Bifurcation Diagram** dialog box, and the initial condition and the value of lambda for the latest iteration.

When a bifurcation diagram is saved or opened, this window takes on the name of the corresponding file.

Special Features: Clicking inside the body of the bifurcation diagram brings up the **Bifurcation Diagram** dialog box. Dragging inside the bifurcation diagram brings up the same dialog box, with range values corresponding to the dragging endpoints (which effectively allows magnification; however, since it cannot magnify existing data, the data must be regenerated). Hold the Shift key down to bypass the dialog box and create the bifurcation diagram for the magnified area.

Hot Spots

Hot spots are areas in Quadratic Map windows that may be clicked in order to execute specific commands. Below is a summary of the hot spots in each window. (Clicking/dragging in the graphing regions of some windows has other effects; see the "Special Features" parts of the window-by-window discussion above.) As used below, "margin" means that region between the graph area and the corresponding edge of the window (not including scroll bars or scroll bar lines).

Window	Hot spot region	Command
Phase Space	Bottom margin	**Phase Space**
	Left margin (while not magnifying)	**Graph Range**
	Left margin (while magnifying)	**Phase Space Magnification**
Iteration Data	Header area	**System Parameter**
Embedded Phase Space	Bottom margin	**Embedded Phase Space**
	Left margin (while not magnifying)	**Graph Range**
	Left margin (while magnifying)	**Embed. Phase Space Magnification**
Orbit	Bottom margin	**Initial Condition**
	Left margin	**Graph Range**
	(Graph area)	**Orbit**
Bifurcation Diagram	Bottom margin	**Bifurcation Diagram**
	Left margin	**Bif. Diagram Phase Range** (if command is active) or **Bifurcation Diagram**
	Whole window (when blank)	**Bifurcation Diagram**

QUADRATIC MAP MENUS

In addition to the requisite **Apple** menu, Quadratic Map has seven menus. The **File** menu contains the standard commands for file manipulation, printing, and quitting. The **Edit** menu contains the standard Macintosh editing commands (though Quadratic Map only implements the **Copy** command; the rest are supplied for desk accessory support), and a few commands best described as preferences commands. The **Control** menu contains the commands that allow you to control Quadratic Map's iteration: starting/stopping it, setting parameters, and so on. The **View** menu allows you to control how you view the iteration; it contains commands to set scaling, hide old data, and so on. The **Special** menu contains the commands that control Quadratic Map's special type of iteration, bifurcation diagrams. The **Sound** menu allows you to control Quadratic Map's sound. The **Windows** menu has commands to control window size and placement and to open windows.

The remainder of this section describes each of Quadratic Map's menu commands.

File Menu

New

> The **New** command stops the current iteration and clears the data from all windows. The **New** command closes the current file(s), if any.

Open

> The **Open** command opens an iteration stored on the disk and displays it on the screen.
>
> Since Quadratic Map files (from base iterations) are stored as text files, the **Open** dialog box will display all text files. Opening a text file not in Quadratic Map format will cause an error message and will clear all the windows. See the **Save** command for more information on how Quadratic Map saves iterations.
>
> Command-period will abort the opening of a base iteration file.
>
> Opening Quadratic Map files is somewhat different from standard Macintosh opening. Usually, when one opens a Macintosh file, it appears in *its own* window. Quadratic Map files open into *pre-existing* windows. With the exception of size and position, these windows maintain their attributes when a new file is opened; think of these windows as tools to inspect whatever data is displayed, tools that are not reset for new data.

Save

The **Save** command saves the current iteration on the disk from which it was opened (or to which it was "saved as"). Quadratic Map saves the iteration with the name used the last time you saved it, and replaces the old copy of the iteration on the disk. You can only use the **Save** command with iterations that you have previously saved with the **Save As** command; if the iteration has not previously been saved, this command acts like the **Save As** command.

Quadratic Map decides what to save by looking at the active window. If the active window is the *Bifurcation Diagram* window, and there is a current bifurcation diagram, it saves that data. In all other cases, it saves the base iteration data, the data that appears in the *Iteration Data* window.

Quadratic Map saves base iterations as text files, just as the iteration appears in the *Iteration Data* window. The data can thus be easily passed to other applications for further processing. Quadratic Map also saves appropriate window sizes and positions, in such a way that these will not interfere with the transfer to other applications (this window information is stored in the resource fork of the file). Note that, except for window sizes and positions, only the values appearing in the text window (system parameter lambda and solutions) are saved; other program parameters, such as scaling and sound parameters, are not saved.

When you are saving a base iteration for the first time or are using the **Save As** command, you can use the Delimiter pop-up menu to change the delimiter Quadratic Map uses to save the file, that is, the character(s) that separate the pieces of data for a given solution (line) in the text file. When you first start Quadratic Map, it uses spaces, but it can also use tabs, commas, or carriage returns. The choice of delimiter depends on the application(s) you plan to export the data into. For instance, to export the data to a spreadsheet, tabs or commas are usually the best choice. If you do not plan to export the data to any other application, spaces are probably best. You can set the default delimiter with the **Save Defaults** command.

Quadratic Map saves bifurcation diagram files in non-text format. It saves all the parameters involved in the iteration, including the current state thereof, so that, if not completed, it can be continued during another session.

Save As

The **Save As** command saves new iterations or new versions of existing ones. After choosing **Save As**, the standard **Save As** dialog box will appear. If the iteration already has a name, Quadratic Map proposes it in the text box. To accept the proposed name, click the Save button. To save the iteration under a different name, edit the proposed name or type a new one. Click the Drive and/or Eject buttons to save the iteration on another disk.

Refer to the **Save** command for a discussion of how Quadratic Map saves files, including a discussion of the Delimiter pop-up menu.

Save Defaults

The **Save Defaults** command saves the current settings of various Quadratic Map parameters in a default settings file. It also makes the current window configuration the default configuration. It creates a file entitled "Quadratic Map Defaults" in the current system folder; when Quadratic Map starts up, it looks in the system folder for this file and uses the configuration stored therein.

The default settings that Quadratic Map sets are the following: the initial condition and system parameter; the maximum number of iterations in a base iteration; the delimiter used when saving base iterations; if running under MultiFinder, whether to run in the background, and, when running in the background, whether to run quickly or slowly, and whether to use sound; whether to suppress save warnings; whether to use the auto ⌘-key feature; whether to use the automatic graph range, and, if not, what graph range to use; how fast to run the base iteration; all the parameters set in the **Phase Space**, **Embedded Phase Space**, and **Orbit** commands (**View** menu); whether to scroll the graph in the *Orbit* window; for bifurcation diagrams, whether to graph in all windows and whether to use color; the color settings; and all the **Sound** menu settings.

If you'd like to have more than one "defaults file," you can copy or move the "Quadratic Map Defaults" file out of the system folder (or even just rename it), create others similarly, then move whichever one you want to use into the system folder.

Note: to restore the original defaults, simply move the defaults file out of the system folder.

Page Setup

The **Page Setup** command allows you to select the paper size, paper orientation, and special printing effects for the document. The special printing effects available depend on which printer is currently selected with the Chooser desk accessory.

Print

The **Print** command prints the front window.

Exactly what it prints depends on the window. If the window is the text window (*Iteration Data*), it prints all the text (but not text that has been stored elsewhere—see the discussion of the *Iteration Data* window in the section "Quadratic Map Windows"), printing on multiple pages if necessary.

If the front window is the *Bifurcation Diagram* window, Quadratic Map prints all applicable information, even if the window is sized such that all information is not visible.

For any other window, Quadratic Map prints only what is seen in the window as currently sized.

Quit

The **Quit** command ends a Quadratic Map session. It will prompt you to save any unsaved data; hold Option down when you execute the command for a forced quit—the program will not prompt you to save data.

Edit Menu

With one exception, Quadratic Map does not allow standard Macintosh editing commands; these non-functional commands will be disabled (grayed) during Quadratic Map operation. One has no need to paste information into or cut information out of windows. The **Copy** command does work, however. Also, inside a dialog box, **Cut**, **Copy**, and **Paste** will work; they can be called with their Command-key equivalents. All commands are provided in the menu for compatibility with desk accessories.

Undo   Provided for desk accessory compatibility only. Disabled when the front window is a Quadratic Map window.

Cut   Provided for desk accessory compatibility only. Disabled when the front window is a Quadratic Map window.

Copy   For the *Iteration Data* window, the **Copy** command copies to the clipboard the current selection.

For any other window, the **Copy** command copies to the clipboard exactly what is seen in the window, as currently sized.

Paste   Provided for desk accessory compatibility only. Disabled when the front window is a Quadratic Map window.

Clear   Provided for desk accessory compatibility only. Disabled when the front window is a Quadratic Map window.

Suppress Save Warnings   By choosing the **Suppress Save Warnings** command, Quadratic Map will not warn you when data is about to be lost due to quitting, erasing iterations, or creating new iterations. This command toggles "save-warning suppression"; choosing the command a second time will re-enable save warnings. The current setting of this option can be saved as the default setting with the **Save Defaults** command.

Auto ⌘-key

After choosing the **Auto ⌘-key** command, you can issue menu commands by typing their Command-key equivalents *without holding the Command key down*. For instance, you could issue the **Open** command by typing just "o"; you would not need to type "Command-o." This command toggles this feature; after choosing the command a second time, you will have to hold down the Command key to issue menu commands. The current setting of this option can be saved as the default setting with the **Save Defaults** command.

Background Operation

The **Background Operation** command allows you to tell Quadratic Map whether to run in the background under MultiFinder. The command will only appear if you are running under MultiFinder.

The **Background Operation** dialog box allows you to specify whether you want Quadratic Map to run in the background, and whether you want it to run slowly, minimizing its influence on the foreground application, or to run quickly, maximizing Quadratic Map's processing speed. You also have the option of suppressing sound when Quadratic Map is running in the background.

Running Quadratic Map in the background is especially useful for the very time-consuming generation of bifurcation diagrams.

The background options set here can be stored as the default settings with the **Save Defaults** command.

For more information on running Quadratic Map in the background, refer to the "MultiFinder and Memory Considerations" part of the section "Tips and Techniques."

Control Menu

Begin Iteration

The **Begin Iteration** command starts a base iteration.

Stop Iteration
Continue Iteration

The **Stop Iteration** command stops a base iteration. After being stopped, the **Continue Iteration** will continue it. When a solution is unbounded and goes to negative infinity, it stops automatically and cannot be continued.

Do One Iteration

The **Do One Iteration** command starts or continues a base iteration, whichever is appropriate, finds and graphs the next iteration, then stops the iteration.

Play Back Solutions

The **Play Back Solutions** command plays back the current iteration from the beginning.

Initial Condition

The **Initial Condition** command allows you to enter a new initial condition for the iteration. Check the (Re)Start Iteration box if you wish the iteration to begin after you click OK. Check the Clear Phase Spaces box if you wish to clear the points plotted so far in the *Phase Space* and *Embedded Phase Space* windows; doing this will in effect issue the **Clear Iterations** command from the **View** menu. The current initial condition can be made the default one with the **Save Defaults** command.

System Parameter

The **System Parameter** command allows you to change the system parameter lambda of the iteration. Check the (Re)Start Iteration box if you wish the iteration to begin after clicking OK. Changing the system parameter will eliminate all data previously generated. The current system parameter can be made the default one with the **Save Defaults** command.

Get Initial from Last

The **Get Initial from Last** command makes the last value of the iteration the new initial condition, bringing up the **Initial Condition** dialog box.

Maximum Iterations

The **Maximum Iterations** command allows you to set the maximum number of solutions a base iteration will generate. The current maximum iterations can be made the default one with the **Save Defaults** command.

View Menu

Graph Range

When you first start it up, Quadratic Map graphs data (in the *Phase Space*, *Embedded Phase Space*, and *Orbit* windows) in the following manner. For $0 \le \lambda \le 4$, Quadratic Map graphs from 0 to 1. For $\lambda > 4$, Quadratic Map graphs from 0 to $\lambda/4$. For $0 > \lambda \ge -4$, Quadratic Map graphs from -1 to 1. For $\lambda < -4$, Quadratic Map graphs from $\lambda/4$ to $|\lambda/4|$. The **Graph Range** command allows you to change the range of values graphed. For the *Phase Space* and *Embedded Phase Space* windows, the graph range given here can be overridden by magnification. The graph range set here can be made the default graph range with the **Save Defaults** command.

Iteration Speed

The **Iteration Speed** command allows you to slow down the iteration so that the graphing can be watched more easily. The dialog box this command brings up gives you a scroll bar on which you can set the speed, from slow to fast. The slowest speed corresponds to approximately one iteration every five seconds. If the fastest speed is selected, the speed is limited only by the speed of the computer. The **Save Defaults** command will remember the last speed selected as the default speed.

Phase Space

The **Phase Space** command gives you control over the display in the *Phase Space* window; refer to the discussion of that window in the section "Quadratic Map Windows" for further description of these options.

This command allows you to display a number line to graph the data, and set the number line's width (in pixels). It also allows you to turn on and off the display of the parabola the quadratic map is based on, the 45° line $x_{n+1} = x_n$, the outline of the rectangle that goes from $x_{n+1} = 0$ to $x_{n+1} = 1$ and from $x_n = 0$ to $x_n = 1$, and vertical and horizontal lines that show the path from one solution to the next, alternatively with arrows showing the direction of iteration. If you are turning on display of these latter lines, click Show from n=0 to display all the lines from the beginning of the iteration; otherwise, only lines for solutions yet to be generated will be displayed. Finally, click Same Size Axes to use symmetric axes.

The **Save Defaults** command makes the current values of the parameters set with this command the default values.

Embedded Phase Space

The **Embedded Phase Space** command gives you control over the display in the *Embedded Phase Space* window; refer to the discussion of that window in the section "Quadratic Map Windows" for further description of these options.

Most importantly, this command allows you to set the embedding of the embedded phase space, the number that determines to which past iteration you are comparing the current iteration. It also allows you to turn on and off the display of the 2^n-degree polynomial that results from composing the quadratic map a number of times equal to the embedding (all points will fall on this polynomial), the 45° line $x_{n+k} = x_n$, where k equals the embedding, and the outline of the rectangle that goes from $x_{n+k} = 0$ to $x_{n+k} = 1$ and from $x_n = 0$ to $x_n = 1$. Finally, click Same Size Axes to use symmetric axes.

The **Save Defaults** command makes the current values of the parameters set with this command the default values.

Orbit

The **Orbit** command gives you control over the display in the *Orbit* window. You can set the size (in square pixels) of the points it plots in this window, and the horizontal space (in pixels) between these points.

The **Save Defaults** command makes the current values of the parameters set with this command the default values.

Scroll Orbit

Under the default settings, the plot in the *Orbit* window clears itself and starts over when it reaches the right edge. With the **Scroll Orbit** command, you can tell Quadratic Map to scroll the *Orbit* window when it reaches the edge. This is a toggle command; successive executions of the command turn the scrolling off and on. While the scrolling has a nice effect, it tends to slow down down the whole iteration considerably, even when the *Orbit* window is closed. The **Save Defaults** command makes the current value of this option the default value.

Clear Iterations

The **Clear Iterations** command clears the data in the *Phase Space* and *Embedded Phase Space* windows, so that only subsequent iterations are displayed there. Use this command if the windows become too cluttered. To redisplay the iterations that were calculated prior to this command, choose **Show All Iterations** from this menu.

Show All Iterations

The **Show All Iterations** command undoes the effect of the **Clear Iterations** command above; all solutions that had been generated since the beginning of the iteration will be displayed in the *Phase Space* and *Embedded Phase Space* windows.

Clear Iteration Lines

When the *Phase Space* window is displaying "iteration lines," the lines that show graphically the construction of the quadratic map, the display can get quite cluttered quite quickly. The **Clear Iteration Lines** command clears these lines, so that only those from subsequent iterations are displayed. To redisplay the lines that were drawn prior to this command, choose **Show All Iteration Lines** from this menu. This command is dimmed (grayed) when iteration lines are not being displayed.

Show All Iteration Lines

The **Show All Iteration Lines** command undoes the effect of the **Clear Iterations** command above; the lines corresponding to all solutions that had been generated since the beginning of the iteration will be displayed.

Phase Space Magnification

The **Phase Space Magnification** command allows you to view, in the *Phase Space* window, any part of the entire phase space; this allows you to gain greater detail. You specify minimum and maximum values for both axes. Press any of the Auto boxes for a corresponding default value.

Click the NO MAGNIFICATION button to use the normal (unmagnified) view.

A convenient way to call up this command is by dragging around the area to be magnified in the *Phase Space* window. Doing this will bring up the **Phase Space Magnification** dialog box with appropriate endpoints.

Embed. Phase Space Magnification

The **Embed. Phase Space Magnification** command allows you to view, in the *Embedded Phase Space* window, any part of the entire embedded phase space; this allows you to gain greater detail. You specify minimum and maximum values for both axes. Press any of the Auto boxes for a corresponding default value.

Click the NO MAGNIFICATION button to use the normal (unmagnified) view.

A convenient way to call up this command is by dragging around the area to be magnified in the *Embedded Phase Space* window. Doing this will bring up the **Embed. Phase Space Magnification** dialog box with appropriate endpoints.

Special Menu

Start Bifurcation Diagram

The **Start Bifurcation Diagram** command begins the creation of the current bifurcation diagram. If the *Bifurcation Diagram* window is not open, you will be asked if you want to open the window.

Stop Bifurcation Diagram
Continue Bifurcation Diagram

While a bifurcation diagram is running, the **Stop Bifurcation Diagram** command stops it. **Continue Bifurcation Diagram** continues a bifurcation diagram that has been stopped before completion. If you issue the **Continue Bifurcation Diagram** command and the *Bifurcation Diagram* window is not open, you will be asked if you want to open the window.

Bifurcation Diagram

The **Bifurcation Diagram** command brings up the dialog box that allows you to set all the parameters of a bifurcation diagram. Please refer to the section "Types of Iteration" for a discussion of this dialog box. If you use this command to start a bifurcation diagram and the *Bifurcation Diagram* window is not open, you will be asked if you want to open the window.

Clear Bifurcation Diagram

The **Clear Bifurcation Diagram** command clears the *Bifurcation Diagram* window and resets all bifurcation diagram parameters. Because this command is destructive, you are given the option of saving the current bifurcation diagram first.

Graph in All Windows

The **Graph in All Windows** command instructs Quadratic Map to toggle whether it displays the progress of the bifurcation diagram in all the windows. In constructing the bifurcation diagram, Quadratic Map actually runs many base iterations; one can watch these iterations as the bifurcation diagram progresses. On the other hand, often one will not want to watch these iterations, as they significantly increase the time required to complete the bifurcation diagram. Use the **Save Defaults** command to make the current setting of this option the default setting.

Use Color

The **Use Color** command toggles whether the current bifurcation diagram is displayed in color. This menu item will be disabled (grayed) if the current monitor is not a color one with at least 4-bit color (16 colors) enabled. *Note*: If the monitor is set to display gray scale, at 4 bits or more, this option will be enabled, and, instead of colors, the bifurcation diagram will be displayed in shades of gray. Use the **Save Defaults** command to make the current setting of this option the default setting.

Colors

The **Colors** command allows you to specify the colors used to represent the different types of solutions in the bifurcation diagram, when color is on. Use the **Save Defaults** command to save the current color-solution relationship as the default one.

Sound Menu

NOTE: To turn off all sound, eliminate the checks next to the **x_n Tones** command and the **x_n–k Tones** command, if any, by choosing the appropriate command.

x_n Tones
x_n–k Tones

The **x_n Tones** and **x_n–k Tones** commands toggle Quadratic Map's sound. With "x_n tones" on, Quadratic Map creates, for each iteration, a sound corresponding to where the iteration falls in the graph range, as defined by the **Graph Range** command. With "x_n–k tones" on, Quadratic Map creates a sound in the same way that corresponds to the $n–k$th iteration, where k equals the embedding of the *Embedded Phase Space* window, as set by the **Embedded Phase Space** command. When you first start Quadratic Map, x_n tones are on, x_n–k tones off. You can save the current settings as the default settings with the **Save Defaults** command.

The frequency of the sound(s) Quadratic Map creates (for each iteration) ranges from 256 Hz to 512 Hz as the iteration ranges from the low end of the graph range to the high end. Values out of the graph range will generate frequencies out of the 256 to 512 Hz range. See the other commands from this menu for descriptions of the sounds created.

Major
Minor
Chromatic
Continuous

These four commands determine the exact frequency of sound created by Quadratic Map. The **Major** command divides the graph range, as defined by the **Graph Range** command, into seven equal areas, corresponding to the seven tones of the major scale. To find the tone for a particular value, Quadratic Map looks at which area the value falls in, and chooses the corresponding tone of the major scale. The **Minor** command acts similarly, dividing the range into areas corresponding to the seven tones of the minor scale. The **Chromatic** command divides the range into twelve areas, corresponding to the twelve tones of the chromatic scale. With the **Continuous** command, Quadratic Map creates a sound that ranges logarithmically from 256 to 512 Hz as the iteration value ranges from the beginning of the graph range to the end.

The **Save Defaults** command will make the current "frequency-finding method" the default one.

Windows Menu

After the first three commands, which are described below, the **Windows** menu lists all of Quadratic Map's windows, placing a diamond next to the windows that are open. Choosing one of these menu items brings the named window to the front, making it visible if necessary.

Close Window
Close All Windows

The **Close Window** command closes the front window. If the Option key is depressed, the command becomes **Close All Windows** and will close all open windows (holding down Option while clicking in the front window's close box will do the same thing). If there are no open windows, this command is grayed.

Arrange Windows

The **Arrange Windows** command arranges the currently-open windows in a standard format. If there are no open Quadratic Map windows, this command is grayed.

Default Configuration

The **Default Configuration** command puts Quadratic Map's windows in the configuration stored in the "Quadratic Map Defaults" file in the current system folder (see **Save Defaults** command). If there is no such file, it sizes the *Phase Space*, *Iteration Data*, *Embedded Phase Space*, and *Orbit* windows equally, and puts one into each quarter of the screen.

Tips and Techniques

This section contains a series of tips and techniques to make your use of Quadratic Map easier and more enjoyable.

Speeding Up the Iteration

To speed up Quadratic Map iterations, there are a few steps you can take. Turning the sound off yields a considerable increase in speed. Turning off orbit scrolling (which initially is off—see the **Scroll Orbit** command, **View** menu) also helps. Of course, if the iteration has been slowed with the **Iteration Speed** command (**View** menu), you can speed it up with that command.

Bifurcation diagrams tend to take a lot of time under any circumstances, especially with a slow Macintosh. To generate them as fast as possible, do not select the **Graph in All Windows** command and turn the sound off.

MultiFinder and Memory Considerations

Quadratic Map can run in the background under MultiFinder. The way it runs in the background is set with the **Background Operation** command from the **Edit** menu. See the description of that command for more information.

There are several considerations that should be given to memory usage with Quadratic Map under MultiFinder. First of all, if you are not familiar with memory partitioning under MultiFinder, it is set with the Get Info command in the Finder; please refer to your MultiFinder manual for more information. Quadratic Map's default memory partition is 512K. That should be sufficient to run both types of iterations, and to create base iterations of a couple thousand iterations or more.

If you are using a larger monitor or multiple monitors, you should increase the default partition size by the following amount: Multiply the number of pixels the larger monitor displays horizontally by the number of pixels it displays vertically to arrive at its resolution, its total number of pixels. In the case of multiple monitors, use the total numbers of pixels that would be displayed by a large monitor just barely encompassing all your monitors (as arranged with the Monitors cdev in the Control Panel). Subtract 200,000 from this resolution and multiply the resulting number by 0.0007. The result of this calculation represents the additional memory (in "K") that should be added to Quadratic Map's memory partition. The calculation is reproduced below:

$$\text{Additional Memory} = (\text{Resolution} - 200{,}000) * 0.0007.$$

In addition, if you wish to create large base iterations (> 2000 iterations), you should increase the default partition size by 40K for each 1000 iterations.

Of course, depending on the memory available in your computer, the above calculations may imply certain restrictions on the operation of Quadratic Map.

One other consideration: If you open a bifurcation diagram created and sized in a window larger than your current monitor, Quadratic Map may create "memory islands," which can cause it to be unable to fully use the available memory. Beware of this if you create bifurcation diagram files on large monitors, then open them on small monitors. By the way, the enlarged memory values given above would partially apply to such situations. Thus, even though you may be using a small Macintosh monitor, you will require much of the memory of a monitor large enough to display the entire large-sized bifurcation diagram file.

Quadratic Map and Other Programs

Quadratic Map can export data in several ways. Graphics and text can both be exported using the **Copy** command and the Clipboard. Files saved from Quadratic Map's *Iteration Data* window, being text files, can be easily shared with many programs. The **Save As** command allows you to specify the data delimiter, to ease data transfer. Bifurcation diagram files, on the other hand, use a proprietary format. The graphics in these files are stored as PICT resources; an ambitious user could use a resource editor to access these graphics and transfer them to another application, perhaps to gain greater color control, or to make color separations.

For the most part, it is not possible to import data into Quadratic Map. Technically, Quadratic Map can try to open any text file, and if that file contains data in Quadratic Map format, it will open it. If the data is not actual Quadratic Map data, however, this is likely to result in erratic performance.

Quadratic Map appears to coexist peacefully with all popular INITs, cdevs, and desk accessories. If you begin a simulation and leave the computer alone, and if you use a screen saver, the screen saver will blank the screen while Quadratic Map is creating the simulation. Quadratic Map continues to create data after the screen saver is activated. When the screen saver is deactivated, Quadratic Map will properly display all the data.

Miscellaneous

Warning or confirmation dialog boxes are generated by certain commands. For many of these commands, holding the Shift key down when executing the command will bypass the appropriate dialog box, automatically making the default response. For some of these, holding the Option key down when executing the command will effectively answer "No" to the dialog box. For instance, if you hold Shift down while executing the **Quit** command, the program will automatically save all current data before quitting, while if you hold Option down, the program will quit while neither prompting you to save data nor saving data. Shift has priority over Option; if both are held down, the Option key will be ignored. Either Shift or both Shift and Option work in this fashion with the following commands: **New, Open, Quit, System Parameter, Get Initial from Last, Clear Bifurcation Diagram, Graph in All Windows**, and commands that start or continue bifurcation diagrams. Shift also works when magnifying in the *Phase Space*, *Embedded Phase Space*, and *Bifurcation Diagram* windows (to bypass the corresponding dialog box), and when clicking an initial condition in the *Phase Space* or *Embedded Phase Space* windows.

In any dialog box, pressing Command-letter will have the effect of clicking the "first" control (button, radio button, check box) whose title begins with that letter. The order of these controls is determined internally; in general, the OK (default) and Cancel (or analogous) buttons come first, followed by the others in order from top to bottom, left to right. In a dialog box with no text input (edit text) fields, pressing a letter without Command will click the first radio button or check box whose title begins with that letter. Command-period or Escape will cancel dialog boxes. Tab takes you forward through the text input fields; Shift-Tab takes you backward through these fields.

Command-period (⌘-.) will stop the current iteration, whether a base iteration or a bifurcation diagram. It will also stop the file-opening of a base iteration.

In the **Iteration Speed** dialog box, pressing the Left and Right Arrow keys moves the scroll bar's "thumb" by a small amount in the corresponding direction; pressing the Up and Down keys moves the thumb by a larger amount.

Program Problems

Bifurcation diagrams may not print properly on an Apple LaserWriter SC™ with some versions of the printer driver. If you use the latest version of the printer driver, they should print fine.

If you find any bugs or other program problems, please notify the publisher.

SELECTED ERROR MESSAGES

Most of Quadratic Map's error messages should be self-explanatory. Below are a few error messages that may require additional explanations and/or suggestions.

Not a Quadratic Map file

The text file Quadratic Map is trying to open does not have a valid Quadratic Map header (which contains the system parameter lambda). This could be because the file has nothing to do with Quadratic Map, since Quadratic Map's **Open** dialog includes all text files—refer to the discussion of the **Open** command. If the file should be a proper Quadratic Map file, you could use a word processor to look at the file, compare it to a file that Quadratic Map can open, make any required changes, save it as a text file, then open it inside Quadratic Map again.

Sorry—not enough memory to run Quadratic Map

If you are running under MultiFinder, this command means that Quadratic Map's memory partition is too small. If not, the Macintosh being used does not have sufficient memory to run Quadratic Map. Refer to the discussion of memory usage in the section "Tips and Techniques."

Unable to open entire file

One of the solutions in the file Quadratic Map is opening is not in valid format. Each solution should consist of the iteration number (an integer) and the iteration value, a floating point number. To correct this problem, inspect the file with a word processor, looking where Quadratic Map stopped loading the data, and try to find an invalid number, or an incomplete line, or something of that nature. If you are able to find and correct the problem, save it as a text file and re-open it in Quadratic Map.

QUADRATIC MAP INDEX